Lecture Notes in Computer Science 12303

More information about this series at http://www.springer.com/series/7407

Christina Boucher · Sharma V. Thankachan (Eds.)

String Processing and Information Retrieval

27th International Symposium, SPIRE 2020
Orlando, FL, USA, October 13–15, 2020
Proceedings

 Springer

Editors
Christina Boucher ⓘ
CISE Department
University of Florida
Gainesville, FL, USA

Sharma V. Thankachan ⓘ
Department of Computer Science
University of Central Florida
Orlando, FL, USA

ISSN 0302-9743 ISSN 1611-3349 (electronic)
Lecture Notes in Computer Science
ISBN 978-3-030-59211-0 ISBN 978-3-030-59212-7 (eBook)
https://doi.org/10.1007/978-3-030-59212-7

LNCS Sublibrary: SL1 – Theoretical Computer Science and General Issues

This Springer imprint is published by the registered company Springer Nature Switzerland AG
The registered company address is: Gewerbestrasse 11, 6330 Cham, Switzerland

Preface

The 27th International Symposium on String Processing and Information Retrieval (SPIRE 2020), held October 13–15, 2020, was hosted online in a virtual way due to COVID-19. SPIRE started in 1993 as the South American Workshop on String Processing, therefore it was held in Latin America until 2000 when SPIRE traveled to Europe. From then on, SPIRE meetings have been held in Australia, Japan, the UK, Spain, Italy, Finland, Portugal, Israel, Brazil, Chile, Colombia, Mexico, Argentina, Bolivia, and Peru. In this edition, we continued the long and well-established tradition of encouraging high-quality research at the broad nexus of algorithms and data structures for sequences and graphs, data compression, databases, data mining, information retrieval, and computational biology. As per usual, SPIRE 2020 continues to provide an opportunity to bring together specialists and young researchers working in these areas.

This volume contains 21 papers, out of a total of 32 submissions accepted to be presented at SPIRE 2020. Each submission received three or four reviews. We thank all authors who submitted their work for consideration to SPIRE 2020. We also thank the Program Committee and the external reviewers, whose many thorough reviews helped us select the papers presented. The success of the scientific program is due to their hard work. In addition to the accepted papers, the scientific program included three invited lectures, given by Laxmi Parida (IBM, USA), Laura Dietz (University of New Hampshire, USA), and Michael A. Bender (Stony Brook University, USA). We thank the invited speakers for accepting our invitation and for their excellent presentations at the conference.

To complete the event, this year for the fifth year running, SPIRE 2020 had a Best Paper Award, sponsored by Springer, that was announced at the conference. Alongside Springer, we thank the Web4Good board for their financial support.

October 2020

Christina Boucher
Sharma V. Thankachan

Organization

Steering Committee

Alistair Moffat	The University of Melbourne, Australia
Berthier Ribeiro-Neto	Federal University of Minas Gerais, Brazil
Gabriele Fici	Università di Palermo, Italy
Gonzalo Navarro	University of Chile, Chile
Marinella Sciortino	Università di Palermo, Italy
Nieves R. Brisaboa	University of A Coruña, Spain
Nivio Ziviani	Federal University of Minas Gerais, Brazil
Ricardo Baeza-Yates	NTENT and Universitat Pompeu Fabra, Spain
Rossano Venturini	Università di Pisa, Italy
Simon J. Puglisi	University of Helsinki, Finland
Travis Gagie	Dalhousie University, Canada

General Co-chairs

Christina Boucher	University of Florida, USA
Sharma V. Thankachan	University of Central Florida, USA

Program Committee

Amihood Amir	Bar-Ilan University, Israel
Lorraine Ayad	King's College London, UK
Golnaz Badkobeh	Goldsmiths University of London, UK
Hideo Bannai	Kyushu University, Japan
Djamal Belazzougui	CERIST, Algeria
Philip Bille	Technical University of Denmark, Denmark
Sankardeep Chakraborty	RIKEN, Japan
Rayan Chikhi	CNRS, France
Charles Clarke	University of Waterloo, Canada
Simone Faro	Università di Catania, Italy
Gabriele Fici	Università di Palermo, Italy
Travis Gagie	Dalhousie University, Canada
Arnab Ganguly	University of Wisconsin-Whitewater, USA
Pawel Gawrychowski	University of Wroclaw, Poland
Simon Gog	Karlsruhe Institute of Technology, Germany
Wing-Kai Hon	National Tsing Hua University, Taiwan
Tomohiro I	Kyushu Institute of Technology, Japan
Shunsuke Inenaga	Kyushu University, Japan
Giuseppe F. Italiano	LUISS Guido Carli, Italy
Dominik Kempa	University of California, Berkeley, USA

Tomasz Kociumaka	Bar-Ilan University, Israel
Tsvi Kopelowitz	Bar-Ilan University, Israel
Dominik Köppl	Kyushu University and JSPS, Japan
M. Oguzhan Kulekci	Istanbul Technical University, Turkey
Susana Ladra	University of A Coruña, Spain
Thierry Lecroq	University of Rouen, France
Inbok Lee	Korea Aerospace University, South Korea
Moshe Lewenstein	Bar-Ilan University, Israel
Zsuzsanna Lipták	University of Verona, Italy
Veli Mäkinen	University of Helsinki, Finland
Giovanni Manzini	University of Eastern Piedmont, Italy
Camille Marchet	CRIStAL, France
Juan Mendivelso	Universidad Nacional de Colombia, Colombia
Laurent Mouchard	University of Rouen, France
Gonzalo Navarro	University of Chile, Chile
Yakov Nekrich	Michigan Technological University, USA
Kunsoo Park	Seoul National University, South Korea
Nadia Pisanti	Università di Pisa, Italy
Solon P. Pissis	CWI, The Netherlands
Nicola Prezza	LUISS Guido Carli, Italy
Simon J. Puglisi	University of Helsinki, Finland
Jakub Radoszewski	University of Warsaw, Poland
Leena Salmela	University of Helsinki, Finland
Srinivasa Rao Satti	Seoul National University, South Korea
Marinella Sciortino	Università di Palermo, Italy
Rahul Shah	Louisiana State University, USA
Jouni Sirén	University of California, Santa Cruz, USA
Jens Stoye	Bielefeld University, Germany
Yasuo Tabei	RIKEN, Japan
Rossano Venturini	Università di Pisa, Italy
Bojian Xu	Eastern Washington University, USA
Binhai Zhu	Montana State University, USA

Additional Reviewers

Juliusz Straszyński	Yuto Nakashima
Michelle Sweering	Itai Boneh
Takaaki Nishimoto	Mikhail Rubinchik
Daniil Galaktionov	Bastien Cazaux
Wiktor Zuba	Seungbum Jo
Gwenaël Richomme	Taku Onodera
Pascal Ochem	Frantisek Franek
Giuseppe Romana	Takuya Mieno
Shunsuke Kanda	Manuela Montangero

Local Arrangements

Sumit Kumar Jha	University of Central Florida, USA
Daniel Gibney	University of Central Florida, USA
Sahar Hooshmand	University of Central Florida, USA

Publicity Chair

Massimiliano Rossi	University of Florida, USA

Contents

Data Structures

Contextual Pattern Matching

Gonzalo Navarro(✉) (ID)

CeBiB—Center for Biotechnology and Bioengineering,
Department of Computer Science, University of Chile,
Beauchef 851, Santiago, Chile
gnavarro@dcc.uchile.cl

Abstract. The research on indexing repetitive string collections has
focused on the same search problems used for regular string collections,
though they can make little sense in this scenario. For example, the basic
pattern matching query "list all the positions where pattern P appears"
can produce huge outputs when P appears in an area shared by many
documents. All those occurrences are essentially the same.

In this paper we propose a new query that can be more appropriate
in these collections, which we call *contextual pattern matching*. The basic
query of this type gives, in addition to P, a context length ℓ, and asks to
report the occurrences of all *distinct* strings XPY, with $|X| = |Y| = \ell$.
While this query is easily solved in optimal time and linear space, we
focus on using space related to the repetitiveness of the text collection
and present the first solution of this kind. Letting $\bar{r}$ be the maximum of
the number of runs in the BWT of the text $T[1..n]$ and of its reverse, our
structure uses $O(\bar{r} \log(n/\bar{r}))$ space and finds the c contextual occurrences
XPY of (P, ℓ) in time $O(|P| \log \log n + c \log n)$. We give other space/time
tradeoffs as well, for compressed and uncompressed indexes.

1 Introduction

About a decade ago, it was realized that many of the fastest-growing text collections of the "data deluge" were highly repetitive [17]. Since then, a number of research results have focused on developing indexes whose size is related to some good measure of compressibility for highly repetitive string collections [21]. Today one can find indexes built on measures like the size of the Lempel-Ziv parse [4,9,11,16], of a grammar generating only the text [8,25], of a string attractor [7,23], the number of runs in the Burrows-Wheeler Transform (BWT) [6] of the text [12,17], or the size of an automaton [5] recognizing text substrings [1,2].

All these indexes are devoted to the basic *pattern matching* query: given a short pattern string $P[1..m]$, output all the occ positions where it occurs in the text $T[1..n]$. Some indexes have managed to solve this problem in optimal time, $O(m + occ)$, using space bounded by some function of the above measures [1,12], whereas others have low polylogarithmic factors multiplying m or occ.

Supported in part by Fondecyt grant 1-200038 and Basal Funds FB0001, Chile.

C. Boucher and S. V. Thankachan (Eds.): SPIRE 2020, LNCS 12303, pp. 3–10, 2020.
https://doi.org/10.1007/978-3-030-59212-7_1

While very reasonable in general, this query can be pretty useless in a highly repetitive text collection. A pattern P that appears inside a highly repeated text area will be reported myriad times, wasting a lot of effort to produce and to handle the result. We are not aware of many efforts to propose queries that are better adapted to a scenario of high repetitiveness.

In this paper we make a first step in this direction. We propose a query called *contextual pattern matching* which, in addition to P, gives a context length ℓ. We then want one element of output per distinct context where P appears, that is, all the positions where P appears preceded by the same string X of length ℓ and followed by the same string Y of length ℓ shall be reported only once.

Definition 1. *The* contextual pattern matching *problem on a text $T[1..n]$ is, given a pair $(P[1..m], \ell)$, return a position in T for each of the c distinct strings XPY occurring in T, for all X, Y such that $|X| = |Y| = \ell$. For the occurrences near the extremes of T, assume T is preceded and followed by ℓ copies of the special symbol $\$$, which cannot appear in P.*

It is not hard to solve this query in optimal time $O(m + c)$ if we use linear space, $O(n)$, by using suffix trees [26] and other linear-space auxiliary structures. We are interested, however, in using space related to a relevant repetitiveness measure. We show that, if we call $\bar{r}$ the maximum of the number of equal-letter runs in the BWT of T or its reverse, then a data structure using $O(\bar{r} \log(n/\bar{r}))$ space can solve contextual pattern matching in time $O(m \log \log n + c \log n)$. We also show how any compressed text index can be extended with $O(n)$ *bits* and efficiently solve this query; this can be interesting for mildly repetitive texts.

2   Preliminaries

We index a text $T[0..n]$ over alphabet $[1..\sigma]$, where $T[0] = T[n] = \$$ is a special terminator smaller than all the other alphabet symbols. The *suffix array* [18] $SA[1..n]$ of T lists all the suffixes $T[i..n]$ for $i \geq 1$ in lexicographic order, and the *LCP array*, $LCP[1..n]$, gives the length of the longest common prefix between consecutive suffix array entries, $LCP[i] = lcp(T[SA[i]..n], T[SA[i-1]..n])$.

One relevant measure of repetitiveness is called r, the number of equal-letter runs in the Burrows-Wheeler Transform (BWT) of $T[1..n]$. The BWT [6] is a reordering of the symbols of T obtained by collecting the symbol preceding the lexicographically sorted suffixes of T. That is, if $SA[1..n]$ is the suffix array of T, then $BWT[i] = T[SA[i] - 1]$. For example, it is known that $r = O(\gamma \log^2 n)$ [14], where γ is the smallest attractor of T [15].

Gagie et al. [12] introduce data structures of size $O(r)$ that can find the suffix array range of any pattern $P[1..m]$ in time $O(m \log \log(\sigma + n/r)) \subseteq O(m \log \log n)$, and of size $O(r \log(n/r))$ that can compute any entry $SA[i]$, $SA^{-1}[i]$, and $LCP[i]$, in time $O(\log(n/r))$. The $O(\log(n/r))$-space data structures are binary context-free grammars of height $O(\log(n/r))$ built on the differential versions of the arrays, for example, $DSA[i] = SA[i] - SA[i-1]$ in the case of the suffix array. The grammars exploit the fact that these differential sequences inherit the repetitiveness of the text.

3   Our Solution

We present a suffix-array-oriented solution that solves a stronger variant of the problem: we give the c suffix array ranges of all the distinct contexts XPY where P occurs in T. We can then report one text position for each, but also determine how many times each context occurs, and report its occurrences one by one.

We store the r-bounded data structures of Gagie et al. [12] for both $T[0..n]$ and its reverse $T^{rev}[0..n]$. We call $\bar{r}$ the maximum of the number of equal-letter runs in the BWT of T and of T^{rev}, therefore the structures we use take space $O(\bar{r}\log(n/\bar{r}))$. The general strategy to solve a query $(P[1..m], \ell)$ is as follows:

1. We first find, in $O(m \log \log n)$ time, the suffix array range $[rs..re]$ of P^{rev} (i.e., P read backwards) in the suffix array SA' of T^{rev}.
2. We then partition $[rs..re]$ into $k \le c$ maximal consecutive intervals $[rs_i, re_i]$ where the suffixes in each interval share their first $m + \ell$ symbols, that is, $T^{rev}[SA'[p]..SA'[p] + m + \ell - 1] = P^{rev}X_i^{rev}$ for all $rs_i \le p \le re_i$.
3. We map each interval $SA'[rs_i, re_i]$ to the interval $SA[ds_i..de_i]$ corresponding to the suffixes that start with X_iP.
4. We partition each interval $SA[ds_i..de_i]$ into k_i maximal consecutive subintervals $SA[ds_i^j..de_i^j]$ where the suffixes in each subinterval share their first $m + 2\ell$ symbols, $T[SA[p]..SA[p] + m + 2\ell - 1] = X_iPY_j$ for all $ds_i^j \le p \le de_i^j$.
5. We report the $c = \sum_{i=1}^{k} k_i$ resulting subintervals $SA[ds_i^j..de_i^j]$ and, if desired, a text position $SA[p]$ with $ds_i^j \le p \le de_i^j$ for each.

We now solve the two nonobvious subproblems of our general strategy. The first, in points 2 and 4, is to partition a suffix array interval into subintervals of suffixes sharing their first t symbols. The second, in point 3, is how to map an interval of the suffix array of T^{rev} into the corresponding interval in the suffix array of T. The solutions we find have a complexity of $O(\log n)$ per item output, which leads to our promised result.

Theorem 1. *Let T be a text of length n, and let $\bar{r}$ be the maximum of the number of equal letter runs of its BWT and the BWT of its reverse. Then there is a data structure of size $O(\bar{r}\log(n/\bar{r}))$ that finds the c contextual occurrences of $(P[1..m], \ell)$ in time $O(m \log \log n + c \log n)$.*

The data structures [12] can be built in $O(n)$ time and space, or in $O(n \log n)$ time and $O(\bar{r}\log(n/\bar{r}))$ space, the same as the final space of the structures. The extra data we add next do not change the space nor construction complexities.

Example. Figure 1 shows an example on the text $T[0..17] = \$alabaralalabarda\$$, where we search for $P = $ a with context length $\ell = 1$. Step 1 finds the interval $SA'[rs..re] = SA'[2..9]$ of all the occurrences of $P^{rev} = $ a on T^{rev}. Step 2 finds the places where $LCP'[p] < m + \ell = 2$ (see Sect. 3.1), for $p \in [2..9]$, namely $2, 3, 5, 6, 9$. These are the starting positions of the intervals $[rs_i, re_i] = [2, 2], [3, 4], [5, 5], [6, 8], [9, 9]$, and correspond to the contexts

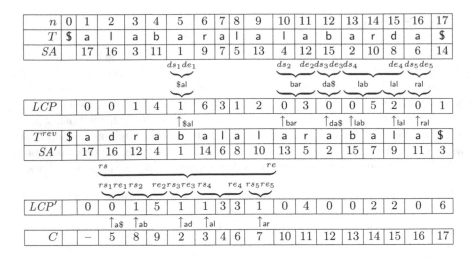

Fig. 1. Example trace.

$Prev\,X_i^{rev} = $ a$, ab, ad, al, ar. Step 3 maps those intervals to SA (see Sect. 3.2), $[ds_i, de_i] = [5,5], [10,11], [12,12], [13,15], [16,16]$; they retain the same order of SA' only because $\ell = 1$. Step 4 splits each interval at subintervals starting wherever $LCP[p] < m + 2\ell = 3$, namely positions $5, 10, 12, 13, 15, 16$. Therefore, the resulting subintervals (i.e., the output) are $[5,5], [10,11], [12,12], [13,14], [15,15], [16,16]$, corresponding to the contexts al, bar, da, lab, lal, ral.

We also show the array C used in Sect. 3.3; note that each ds_i corresponds to mapping the minimum position of C in $[rs_i, re_i]$.

3.1 Partitioning a Suffix Array Interval

Given a range $[s..e]$ of the suffix array of a string S, and a length t, we must partition it into maximal subranges $[s_1..e_1], \ldots, [s_k..e_k]$ where the suffixes starting in each subrange share their first t symbols. Note that the positions $s_2, \ldots, s_k$ are the values in $[s..e]$ where $LCP[i] < t$, where LCP is the LCP array of S.

We store a binary grammar of height $h = O(\log n)$ on $DLCP[i] = LCP[i] - LCP[i-1]$, with $DLCP[1] = 0$ [12]. For each grammar nonterminal A, expanding to a sequence $exp(A)$ of positive and negative integers of $DLCP$ (and for terminals A, assuming $exp(A) = A$), we store

- $w(A) = |exp(A)|$, the number of consecutive $DLCP$ cells A expands to;
- $s(A)$, the sum of the differential values in $exp(A)$, $s(A) = \sum_{j=1}^{w(A)} exp(A)[j]$.
- $m(A)$, the minimum cumulative value reached inside $exp(A)$, that is, $m(A) = \min_{1 \le i \le w(A)} \sum_{j=1}^{i} exp(A)[j]$.

With $w(A)$ and $s(A)$, a standard procedure descends in time $O(h)$ to the sth and eth leaves in the parse tree of $DLCP$, finding both (1) the value of $LCP[s]$ and (2) the $O(h)$ maximal nodes (regarding ancesorship) that cover $DLCP[s..e]$

in the parse tree. To reach the xth leaf, we descend from the root node and move to the left child A if $w(A) \geq x$, otherwise we move to the right child and decrease x by $w(A)$. To find (1) we add up the values $s(A)$ of the left children A every time we descend to the right child in the path to the sth leaf. To find (2) we do the paths to the sth and eth leaves and, once they diverge at a node v, we collect the right children when we go left in our path from v to the sth leaf, and the left children when we go right in our path from v to the eth leaf.

Let A_i be the $O(h)$ maximal parse tree nodes that cover $DLCP[s..e]$. They start in $[s..e]$ at positions $p_1 = s$ and $p_{i+1} = p_i + w(A_i)$. The LCP values at the positions p_i are $l_1 = LCP[s]$ and $l_{i+1} = l_i + s(A_i)$. Note then that each A_i where $l_i + m(A_i) < t$ contains at least one position s_j where $LCP[s_j] < t$; the others can be discarded.

For each A_i where $l_i + m(A_i) < t$, we consider its rule $A_i \rightarrow BC$. Note that B and C start at $p = p_i$ and $p' = p_i + w(B)$ and their first LCP values are $l = l_i$ and $l' = l_i + s(B)$, respectively. We recursively continue with B if $l + m(B) < t$ and with C if $l' + m(C) < t$ (we can continue by both). When we arrive at a terminal grammar symbol, we can report its value p as a new position s_j.

We then report the positions $s_2, \ldots, s_k$ in left-to-right order by considering $A_1, A_2, \ldots$ in turn and considering B before C when $A_i \rightarrow BC$. Since every time we consider a node we know that it contains an answer, the total time is $O(h)$ plus $O(h)$ for each of the $k - 1$ starting positions $s_2, \ldots, s_k$. The total time is then $O(k\,h) \subseteq O(k \log n)$, that is, $O(\log n)$ per range we output.

3.2 Mapping Suffix Array Intervals

Given the suffix array interval $SA'[s'..e']$ of T^{rev}, consisting of all the suffixes that start with a string of length t, we want to find the corresponding suffix array interval $SA[s..e]$ of T. With the suffix array SA' of T^{rev} and the inverse suffix array SA^{-1} of T, we can translate any such suffix, say $p = SA^{-1}[n - SA'[s'] - (m + \ell - 1)]$ (or $p = SA^{-1}[1]$ if $n - SA'[s'] - (m + \ell - 1) \leq 0$). Our index stores the structures to compute those in time $O(\log n)$ [12].

We know that $s \leq p \leq e$, so the task is to extend p in both directions: $s \leq p$ is the largest position where $LCP[s] < t$ and $e \geq p$ is the smallest position where $LCP[e + 1] < t$. We show how to find e; the case of s is analogous.

Just as in Sect. 3.1, we compute $LCP[p]$ and find the $O(h)$ maximal nodes $A_1, \ldots$ that cover the area $DLCP[p..n]$. We then compute the values p_i and l_i, and scan $A_1, \ldots$ for the first A_i such that $l_i + m(A_i) < t$. Then, if $A_i \rightarrow BC$, we continue by B if $l_i + m(B) < t$; otherwise we continue by C with values $p = p_i + w(B)$ and $l = l_i + s(B)$. In $O(h)$ time we reach a terminal symbol, whose position p is, precisely, $e + 1$. The total time is then $O(h) = O(\log n)$.

3.3 Running on Other Indexes

If we are willing to store uncompressed data structures of $O(n)$ space, we can find the interval of point (1) in RAM-optimal time $O(m/\log_\sigma n)$ using an enhanced suffix tree [22] on T^{rev}. The k intervals $[rs_i, re_i]$ of point (2) can be found in $O(k)$

time using range minimum queries on the LCP array of T^{rev}, LCP': $rmq(i,j) = \min_{i \le p \le j} LCP'[p]$. We use the standard procedure for 3-sided queries: compute $p = rmq(rs, re)$ and, if $LCP'[p] < t$, recurse on $[rs, p-1]$, report p, and recurse on $[p+1, re]$. Queries rmq take constant time even using $2n + o(n)$ bits of space [10]. Each such interval $SA'[rs_i, re_i]$ can then be mapped (point 3) to $SA[ds_i, de_i]$ by storing an array $C[1..n]$ with $C[i] = SA^{-1}[n - SA'[i]]$ and building an rmq data structure on C, so that $ds_i = SA^{-1}[n - SA'[rmq_C(rs_i, re_i)]] - (m + \ell - 1)$ and $de_i = ds_i + (re_i - rs_i)$. (Note that we build C on the values $SA^{-1}[n - SA'[i]]$, not $SA^{-1}[n - SA'[i] - (m - \ell + 1)]$, because the latter depend on ℓ and all the suffixes in this range share their first $m + \ell$ symbols anyway, so the lexicographic comparison is the same.) Finally, point (4) on each $SA[ds_i, ds_i]$ is solved as for point (2), now on the LCP array of T. The total time is then the optimal $O(m / \log_\sigma n + c)$.

Theorem 2. *Let T be a text of length n over an alphabet of size σ. Then there is a data of size $O(n)$ that finds the c contextual occurrences of $(P[1..m], \ell)$ in time $O(m / \log_\sigma n + c)$.*

More generally, if we have an index that finds the suffix array range $[rs..re]$ for P in T^{rev}, and can extract any cell of SA, SA^{-1}, and SA', we can use it for contextual reporting using our general solution. We need $O(n)$ extra bits for the various rmq data structures. Note we do not need to store C explicitly because we can simulate it using SA' and SA^{-1}. Further, the arrays LCP' and LCP are simulated with other $2n + o(n)$ bits if we have access to SA' and SA [24]. We then have the following result.

Theorem 3. *Let T be a text of length n and an index on T^{rev} using S bits of space that finds the suffix array range of $P[1..m]$ in time $t_s(m)$, and computes any cell of SA, SA', or SA^{-1} in time t_{SA}, where SA and SA' are the suffix arrays of T and T^{rev}, respectively. Then there is a data structure using $S + O(n)$ bits of space that finds the c contextual occurrences of $(P[1..m], \ell)$ in time $O(t_s(m) + c t_{SA})$.*

Building on an index [3] that uses $nH_k(T^{rev}) + o(n \log \sigma) + O(n)$ bits of space for any $k < \alpha \log_\sigma n$ and constant $0 < \alpha < 1$, where $H_k(S) < \log \sigma$ is the kth order empirical entropy of string S [19], we have $t_s(m) = O(m)$ and $t_{SA} = O(\log n)$. The index provides access to SA' and $(SA')^{-1}$ by storing their values at regular intervals of T^{rev}, of length $s = \Theta(\log n)$ in our case, and marking the sampled positions of SA' in a bitvector. It provides a way to move in constant time from i such that $SA'[i] = j$ to $i' = LF(i)$ such that $SA'[i'] = j - 1$. Thus, if $SA'[i]$ is not sampled, it can move $s' < s$ times until finding a sampled cell $SA'[LF^{s'}(i)] = j'$, and then $SA'[i] = j' + s'$. The same LF function is used $j' - j <$ s times, for $j' = \lceil j/s \rceil \cdot s$, to find $(SA')^{-1}[j]$, by starting from the sampled value $(SA')^{-1}[j']$ and tracing it back to $(SA')^{-1}[j] = LF^{j'-j}((SA')^{-1}[j'])$. Enhancing it to computing values of SA and SA^{-1} (which correspond to T) requires to store their sampled values as well, because $T^{rev}[j] = T[n - j]$. Finally, because $H_k(T) = H_k(T^{rev})$ [20, Sec. 11.3.2], we have the following result.

Theorem 4. *Let T be a text of length n over an alphabet of size σ, with kth order empirical entropy $H_k(T)$, for any $k < \alpha \log_\sigma n$ and constant $0 < \alpha < 1$.*

Then there is a data structure of $nH_k(T) + o(n \log \sigma) + O(n)$ bits that finds the c contextual occurrences of $(P[1..m], \ell)$ in time $O(m + c \log n)$.

We can speed up this index by using *compact* space, $O(n \log \sigma)$ bits (i.e., proportional to a plain representation of T). In this case, any cell of SA or SA^{-1} (and of SA' by building the structures on T^{rev} as well) can be computed in time $O(\log_\sigma^\epsilon n)$ for any constant $\epsilon > 0$ [13]. Further, this index finds the suffix array interval of P in almost RAM-optimal time, $O(m/\log_\sigma n + \log_\sigma^\epsilon n)$.

Theorem 5. *Let T be a text of length n over an alphabet of size σ. Then there is a data structure using $O(n \log \sigma)$ bits that finds the c contextual occurrences of $(P[1..m], \ell)$ in time $O(m/\log_\sigma n + (c+1)\log_\sigma^\epsilon n)$, for any constant $\epsilon > 0$.*

4 Conclusions

We have proposed a query that should be more meaningful than standard pattern locating in the case of highly repetitive text collections. Instead of simply locating all the positions of $T[1..n]$ where $P[1..m]$ appears, we give a context length ℓ and ask for the occurrences of all the c distinct strings XPY in the text, for any X, Y where $|X| = |Y| = \ell$. If P occurs inside a highly repeated substring, many essentially identical occurrences will be reported one by one with the standard locating, whereas we will report only a single suffix array range comprising all the occurrences of the same context XPY.

While the query can be solved in $O(n)$ space and RAM-optimal $O(m/\log_\sigma n + c)$ time, we focus on using space proportional to the repetitiveness of T. We use one such measure, the number $r(S)$ of equal-letter runs of the Burrows-Wheeler Transform of the string S. Within space $O(\bar{r} \log(n/\bar{r}))$, where $\bar{r} = \max(r(T), r(T^{rev}))$, we solve the problem in time $O(m \log \log n + occ \log n)$. We also show how to adapt our general strategy to any compressed text index.

This is a first step towards studying queries that make more sense on highly repetitive text collections, possibly deviating from the classical ones used for regular collections. Some relevant remaining questions are: Can the obtained space/time tradeoffs be improved? Are there other relevant and challenging queries that are better suited to highly repetitive text collections?

References

1. Belazzougui, D., Cunial, F.: Representing the suffix tree with the CDAWG. In: Proceedings of the 28th CPM, pp. 7:1–7:13 (2017)
2. Belazzougui, D., Cunial, F., Gagie, T., Prezza, N., Raffinot, M.: Composite repetition-aware data structures. In: Proceedings of the 26th CPM, pp. 26–39 (2015)
3. Belazzougui, D., Navarro, G.: Alphabet-independent compressed text indexing. ACM Trans. Algorithms **10**(4), Article no. 23 (2014)
4. Bille, P., Ettienne, M.B., Gørtz, I.L., Vildhøj, H.W.: Time-space trade-offs for Lempel-Ziv compressed indexing. Theor. Comput. Sci. **713**, 66–77 (2018)

5. Blumer, A., Blumer, J., Haussler, D., McConnell, R.M., Ehrenfeucht, A.: Complete inverted files for efficient text retrieval and analysis. J. ACM **34**(3), 578–595 (1987)
6. Burrows, M., Wheeler, D.: A block sorting lossless data compression algorithm. Technical report 124, Digital Equipment Corporation (1994)
7. Christiansen, A.R., Ettienne, M.B., Kociumaka, T., Navarro, G., Prezza, N.: Optimal-time dictionary-compressed indexes. CoRR 1811.12779 (2019)
8. Claude, F., Navarro, G.: Improved grammar-based compressed indexes. In: Calderón-Benavides, L., González-Caro, C., Chávez, E., Ziviani, N. (eds.) SPIRE 2012. LNCS, vol. 7608, pp. 180–192. Springer, Heidelberg (2012). https://doi.org/10.1007/978-3-642-34109-0_19
9. Ferrada, H., Kempa, D., Puglisi, S.J.: Hybrid indexing revisited. In: Proceedings of the 20th ALENEX, pp. 1–8 (2018)
10. Fischer, J., Heun, V.: Space-efficient preprocessing schemes for range minimum queries on static arrays. SIAM J. Comput. **40**(2), 465–492 (2011)
11. Gagie, T., Gawrychowski, P., Kärkkäinen, J., Nekrich, Y., Puglisi, S.J.: LZ77-based self-indexing with faster pattern matching. In: Pardo, A., Viola, A. (eds.) LATIN 2014. LNCS, vol. 8392, pp. 731–742. Springer, Heidelberg (2014). https://doi.org/10.1007/978-3-642-54423-1_63
12. Gagie, T., Navarro, G., Prezza, N.: Fully-functional suffix trees and optimal text searching in BWT-runs bounded space. J. ACM **67**(1), Article no. 2 (2020)
13. Grossi, R., Vitter, J.S.: Compressed suffix arrays and suffix trees with applications to text indexing and string matching. SIAM J. Comput. **35**(2), 378–407 (2006)
14. Kempa, D., Kociumaka, T.: Resolution of the Burrows-Wheeler Transform conjecture. CoRR 1910.10631 (2019)
15. Kempa, D., Prezza, N.: At the roots of dictionary compression: string attractors. In: Proceedings of the 50th STOC, pp. 827–840 (2018)
16. Kreft, S., Navarro, G.: On compressing and indexing repetitive sequences. Theor. Comput. Sci. **483**, 115–133 (2013)
17. Mäkinen, V., Navarro, G., Sirén, J., Välimäki, N.: Storage and retrieval of highly repetitive sequence collections. J. Comput. Biol. **17**(3), 281–308 (2010)
18. Manber, U., Myers, G.: Suffix arrays: a new method for on-line string searches. SIAM J. Comput. **22**(5), 935–948 (1993)
19. Manzini, G.: An analysis of the Burrows-Wheeler transform. J. ACM **48**(3), 407–430 (2001)
20. Navarro, G.: Compact Data Structures - A Practical Approach. Cambridge University Press, Cambridge (2016)
21. Navarro, G.: Indexing highly repetitive string collections. CoRR abs/2004.02781 (2020)
22. Navarro, G., Nekrich, Y.: Time-optimal top-k document retrieval. SIAM J. Comput. **46**(1), 89–113 (2017)
23. Navarro, G., Prezza, N.: Universal compressed text indexing. Theor. Comput. Sci. **762**, 41–50 (2019)
24. Sadakane, K.: Compressed suffix trees with full functionality. Theory Comput. Syst. **41**(4), 589–607 (2007)
25. Takabatake, Y., Tabei, Y., Sakamoto, H.: Improved ESP-index: a practical self-index for highly repetitive texts. In: Gudmundsson, J., Katajainen, J. (eds.) SEA 2014. LNCS, vol. 8504, pp. 338–350. Springer, Cham (2014). https://doi.org/10.1007/978-3-319-07959-2_29
26. Weiner, P.: Linear pattern matching algorithms. In: Proceedings of the 14th FOCS, pp. 1–11 (1973)

Navigating Forest Straight-Line Programs in Constant Time

Carl Philipp Reh$^{(\boxtimes)}$ and Kurt Sieber

Universität Siegen, Siegen, Germany
{reh,sieber}@informatik.uni-siegen.de

Abstract. We present a data structure of linear size that allows to perform navigation steps and subtree equality checks in grammar-compressed forests in constant time. Navigation steps include going to the parent, to the left/right neighbor or to the first/last child.

Keywords: Grammar-compressed forests · Forest straight-line programs · Algorithms for compressed forests

1 Introduction

SLPs (straight-line programs) are context-free grammars that produce a single string. They can use nonterminals to identify repetition and therefore be much smaller than the string that they represent. For example, the SLP with $A_0 \to ab$ an d $A_{i+1} \to A_i A_i$ for $1 \leq i < 10$ is a succinct representation of the string $(ab)^{1024}$. *Compression algorithms* try to find a succinct representation for a given input string. Some of them produce SLPs directly, while the output of others can be efficiently translated into SLPs [9]. Since SLPs can compress exponentially, i.e. an SLP of size n might produce a string of size $\Theta(2^n)$, answering a query about the represented string by uncompressing the SLP has exponential runtime. It is however often possible to implement algorithms that do better. A simple example is to print the character at a given position, which can be easily done in time $\mathcal{O}(h)$ where h is the height of the syntax tree of the SLP. Another example is navigating on the string of an SLP: We can start at the first position, go to the next position, go to the previous position or print the character at the current position. Each of these operations takes constant time if we allow linear time preprocessing. This was shown in [12], which extended a result from [8]. The authors use the Word RAM model, i.e. operations on integers whose number of bits is logarithmic in the length of the input (the SLP) require constant time, which is also the model we assume throughout the paper.

SLPs have been extended to TSLPs (tree straight-line programs) which allow to compress trees (see [11] for a survey). Instead of strings, a TSLP can store trees and *tree contexts*, i.e. trees with a single hole x. For example, a very tall tree $\underbrace{a(\cdots a(\,b\,)\cdots)}_{2^n \text{ times}}$ can be compressed by first applying the tree context $a(x)$ 2^n times

© Springer Nature Switzerland AG 2020
C. Boucher and S. V. Thankachan (Eds.): SPIRE 2020, LNCS 12303, pp. 11–26, 2020.
https://doi.org/10.1007/978-3-030-59212-7_2

to itself, which requires n productions, and then replacing x with b. Navigation has been extended to TSLPs in [12], where a navigation step can go to a specific child, to the parent node, or print the character at the current node. Each of these operations requires constant time after linear time preprocessing. Another operation has also been introduced in [12]: After navigating to two (different) nodes in the tree, we can ask if the subtrees at these nodes are equal. This problem appears in several contexts, see for example [5]. Adding the ability to do subtree equality checks however comes at a cost: First, the preprocessing time is now polynomial instead of linear, but the data structure that is precomputed still has linear size. Second, navigation steps now need to keep track of how deep into the tree they went, which requires integers whose number of bits is logarithmic in the tree size. We still refer to this as being constant time, which was also done in [12] and [2]. Grammar-compressed multigraphs are another data structure for which constant time traversal was implemented [13].

A shortcoming of TSLPs is that they can only compress vertically but not horizontally. This is a limiting factor because a lot of tree-like documents (i.e. Wikipedia articles) tend to be very wide but not very tall. A workaround is to transform a tree into its *fcns* (first-child next-sibling) encoding. This is basically a head-tail representation of a tree, e.g. a tree $a(bcd)$ is represented as $a(b(\bot, c(\bot, d(\bot, \bot))), \bot)$, where $\bot$ means that there is no first child or next sibling. This way, TSLP compression can work better because horizontal repetition is turned into vertical repetition. However, it can be tricky to design algorithms that try to answer queries *on the original tree* using only the TSLP for the tree's fcns encoding. For example, it is mentioned in [12] that using the fcns we can support in constant time the navigation operations of going to the first child or to the next sibling. But how we can go to the parent or to the last child (if it is at all possible) in constant time is unclear. A better way to deal with arbitrarily wide trees is to compress them more directly. An FSLP (forest straight-line program), introduced in [7], is basically an extension of a TSLP: It can compress vertically using tree contexts like a TSLP can, but it can also compress horizontally. For example, the tree $a(c^{1024} \cdots a(c^{1024}(b)) \cdots)$, which is a slight variation of the previous example, has a very direct succinct representation: We can compress c^{1024} and thus also the contexts $a(c^{1024}x)$.

In this work, we first introduce FSLP navigation without subtree equality checks, which only requires linear preprocessing time. The allowed operations are: go to the first/last child, the next/previous sibling or to the parent node, or print the current character. Then, we add an extension that allows subtree equality checks but requires polynomial preprocessing time, while still producing a structure of linear size. In both cases, we achieve the same results as the ones previously shown for TSLPs in [12]. While implementing our data structures and algorithms, we use SLP navigation as a black box instead of extending its structure like it was done in [12].

Another formalism for forest compression that is very similar to FSLPs are Top Dags [1]. However, the most basic trees that can be represented with Top Dags are of the form $a(b)$ and the most basic contexts are $a(b(x))$. This leads to cases where the smallest Top Dag is by a factor of the alphabet size larger than the smallest FSLP [7]. Since it is also possible to transform Top Dags into equivalent FSLPs in linear time [7], and we allow linear time preprocessing, all the results of this paper basically extend to Top Dags.

2   Preliminaries

For a string $w = a_1 \ldots a_n \in \Theta^n$ with $n \geq 0$ over some alphabet Θ we write $w[i] = a_i$ for $1 \leq i \leq n$, $w[i : j] = a_i \ldots a_j$ for $1 \leq i \leq j \leq n$, $w[: i] = w[1 : i]$ and $w[i :] = w[i : n]$. In [7] the authors unified the view of several SLP-like structures, which we are also going to use in this work. What was formerly called SLP is now called SSLP (string straight-line program) which frees up the name SLP to be used for something else: An SLP can compress *any* expression by giving names to common subexpressions. This unifies the view of SSLPs, TSLPs, FSLPs and Top Dags.

2.1   Algebras and Straight-Line Programs

An *algebra* $\mathcal{A} = (\mathcal{U}, \mathcal{I})$ consists of a universe (the carrier set) $\mathcal{U}$ and a set of operations $\mathcal{I}$, where each $f \in \mathcal{I}$ is a partial function $f \colon \mathcal{U}^k \to \mathcal{U}$ for some $k \in \mathbb{N}$. Instead of allowing partial functions we could also have defined multi-sorted algebras (see for example [4]), but we feel that at least in this paper this would have unnecessarily complicated the notation. The *expressions* $\mathcal{E}$ over $\mathcal{A}$ consist of terms over the elements from $\mathcal{I}$, i.e. if $f \in \mathcal{I}$ and f is k-ary then $f(e_1, \ldots, e_k) \in \mathcal{E}$ for all $e_1, \ldots, e_k \in \mathcal{E}$. Evaluating an expression $e \in \mathcal{E}$ is written as $[\![e]\!]_{\mathcal{A}}$ and is defined in the usual way. For example, consider $\mathcal{U} = \mathbb{N}$ and $\mathcal{I} = \{+, 1\}$, where $+ \colon \mathbb{N}^2 \to \mathbb{N}$ and $1 \in \mathbb{N}$, with the usual meaning. The expressions are $\mathcal{E} = \{1, +(1, 1), +(+(1, 1), 1), +(1, +(1, 1)), \ldots\}$, and, for example, $[\![+(+(1, 1), 1)]\!]_{\mathcal{A}} = (1 + 1) + 1 = 3$.

For a set of variables V we write $\mathcal{E}_V$ for the set of *expressions with variables* over $\mathcal{A}$. A variable can occur anywhere where a nullary function symbol can occur. For example, if $X, Y \in V$ then $+(+(1, X), Y) \in \mathcal{E}_V$.

A *straight-line program* (SLP for short) over $\mathcal{A}$ is a tuple $G = (V, \mathrm{rhs})$, where V is a finite set of variables and $\mathrm{rhs} \colon V \to \mathcal{E}_V$ is the right-hand side mapping. The relation $\{(A, B) \in V^2 \mid B \text{ appears in } \mathrm{rhs}(A)\}$ must be acyclic. This way, we can transform every expression from $e \in \mathcal{E}_V$ into one from $\mathcal{E}$ by recursively applying rhs. The result can then be evaluated in the algebra $\mathcal{A}$, which we denote with $[\![e]\!]_G \in \mathcal{U}$. For example, if $V = \{X, Y\}$ with $\mathrm{rhs}(X) = +(1, Y)$ and $\mathrm{rhs}(Y) = +(1, 1)$, then $[\![+(X, Y)]\!]_G = [\![+(+(1, +(1, 1)), +(1, 1))]\!]_{\mathcal{A}} = 5$. The *size of an expression* is $|f(e_1, \ldots, e_n)| = 1 + |e_1| + \cdots + |e_n|$, where $|A| = 1$ for a variable A, and the *size of* G is $|G| = \sum_{A \in V} |\mathrm{rhs}(A)|$. The size of G in our example is $|G| = |+(1, Y)| + |+(1, 1)| = 3 + 3 = 6$.

2.2   String Straight-Line Programs

A *string straight-line program* (SSLP for short) over some alphabet Θ is an SLP over the algebra $(\Theta^*, \{\varepsilon, \circ\} \cup \Theta)$, where ε evaluates to the empty string, $\circ$ is string concatenation, and every symbol from Θ is a constant that evaluates to itself. Instead of $v \circ w$ we often simply write vw. Consider the SSLP $G = (\{A, B, C\}, \text{rhs})$ over $\{a, b\}$ with $\text{rhs}(A) = BB$, $\text{rhs}(B) = CCb$ and $\text{rhs}(C) = aa$, then $[\![A]\!]_G = aaaabaaaab$.

2.3   Forest Straight-Line Programs

We fix an alphabet Σ for the rest of the paper that is used to label nodes in forests. A *forest* is a list of trees $t_1 \ldots t_n$ $(n \geq 0)$, while a *tree* is of the form $a(f)$ where $a \in \Sigma$ and f is a forest. Here, a is the root character and its children are the roots of the forest f. The set of forests is denoted by $\mathcal{F}$ and the set of trees by $\mathcal{T}$. *Forests with a hole* are defined as follows: x is a forest with a hole, and $t_1 \ldots t_n$ $(n \geq 1)$ is a forest with a hole if $t_1, \ldots, t_{i-1}, t_{i+1}, \ldots, t_n$ are trees and t_i is a tree with a hole for some $1 \leq i \leq n$. Here, x can be thought of as a placeholder that appears exactly once in a forest with a hole. Trees with a hole are of the form $a(f)$, where $a \in \Sigma$ and f is a forest with a hole. We write $\mathcal{F}_x$ for the set of forests with a hole and $\mathcal{T}_x$ for the set of trees with a hole. The *forest algebra* $\mathcal{A}_{\mathcal{F}} = (\mathcal{F} \cup \mathcal{F}_x, \mathcal{I}_{\mathcal{F}})$ has the following operations $\mathcal{I}_{\mathcal{F}}$:

- $[\![\varepsilon]\!]_{\mathcal{F}} = \varepsilon$—the empty forest.
- $[\![x]\!]_{\mathcal{F}} = x$—a single hole.
- $[\![fg]\!]_{\mathcal{F}} = [\![f]\!]_{\mathcal{F}} [\![g]\!]_{\mathcal{F}}$—horizontal concatenation, only defined if $[\![f]\!]_{\mathcal{F}} \notin \mathcal{F}_x$ or $[\![g]\!]_{\mathcal{F}} \notin \mathcal{F}_x$.
- $[\![a]\!]_{\mathcal{F}} = a(x)$—a single node with a hole.
- $[\![f\langle g\rangle]\!]_{\mathcal{F}} = [\![f]\!]_{\mathcal{F}}[[\![g]\!]_{\mathcal{F}}]$—substitution, only defined if $[\![f]\!]_{\mathcal{F}} \in \mathcal{F}_x$.

The notation $f[g]$ means that x in f is replaced with g. The idea of forest algebras goes back to [3].

An SLP $F = (V, \text{rhs})$ over $\mathcal{A}_{\mathcal{F}}$ is called a *forest straight-line program* or FSLP for short. As a short-hand notation we write $V_0 = \{A \in V \mid [\![A]\!]_F \in \mathcal{F}\}$ for the variables that produce forests without a hole and $V_1 = \{A \in V \mid [\![A]\!]_F \in \mathcal{F}_x\}$ for the variables that produce forests with a hole. *Normal form FSLPs* were introduced in [7], that restrict the rhs-forms that may be used. We are also going to make use of this normal form, since it simplifies the upcoming constructions. First, let $V_0^{\perp} \subseteq V_0$ be defined as

$$V_0^{\perp} = \{A \in V_0 \mid \text{rhs}(A) = a\langle C\rangle, \, a \in \Sigma, \, C \in V_0\}$$
$$\cup \{A \in V_0 \mid \text{rhs}(A) = B\langle C\rangle, \, B \in V_1, \, C \in V_0\}$$

For a normal form FSLP we require the following:

- rhs(A) for every $A \in V_1$ is of the form $a\langle LxR \rangle$, where $a \in \Sigma$, $L, R \in V_0$, or of the form $B\langle C \rangle$, where $B, C \in V_1$.
- rhs(A) for every $A \in V_0$ is of the form ε, or of the form BC, where $B, C \in V_0$, or of the form $a\langle C \rangle$, where $a \in \Sigma$ and $C \in V_0$, or of the form $B\langle C \rangle$, where $B \in V_1$ and $C \in V_0^{\perp}$.

Note that $[\![A]\!]_F \in \mathcal{T}$ for all $A \in V_0^{\perp}$ and $[\![B]\!]_F \in \mathcal{T}_x$ for all $B \in V_1$, i.e., only tree contexts appear in a normal form FSLP instead of arbitrary forest contexts.

Lemma 1. *An FSLP* $F = (V, \text{rhs})$ *can be transformed in linear time into an FSLP* F' *that is in normal form such that* $[\![A]\!]_F = [\![A]\!]_{F'}$ *for every* $A \in V_0$.

This was shown in [7]. Since we allow linear time preprocessing, we assume from now on that every FSLP is in normal form.

Example 1. Suppose that $a, b, c \in \Sigma$ and let $n \in \mathbb{N}$. Let $F_n = (V, \text{rhs})$ with

$$V = \{E, A^{\ell}, A^r\} \cup \{A_i, B_i \mid 0 \le i \le n\} \cup \{C_i^k, D_i^k, G_i^k, H_i^k \mid k \in \{\ell, r\}, 0 \le i \le n\},$$

and rhs(E) = ε, rhs(A^{ℓ}) = $a\langle E \rangle$, rhs(A^r) = $b\langle E \rangle$, rhs(A_0) = $A^{\ell}A^r$, rhs(B_0) = $b\langle ExE \rangle$,

$$\text{rhs}(A_{i+1}) = A_i A_i \text{ for } 0 \le i < n,$$
$$\text{rhs}(B_{i+1}) = B_i \langle B_i \rangle \text{ for } 0 \le i < n,$$
$$\text{rhs}(C_i^{\ell}) = c\langle A_i x A^r \rangle \text{ for } 0 \le i \le n,$$
$$\text{rhs}(C_i^r) = c\langle A^{\ell} x A_i \rangle \text{ for } 0 \le i \le n,$$
$$\text{rhs}(D_i^k) = B_i \langle C_i^k \rangle \text{ for } k \in \{\ell, r\}, 0 \le i \le n,$$
$$\text{rhs}(G_i^k) = C_i^k \langle D_i^k \rangle \text{ for } k \in \{\ell, r\}, 0 \le i \le n,$$
$$\text{rhs}(H_i^k) = G_i^k \langle A^k \rangle \text{ for } k \in \{\ell, r\}, 0 \le i \le n.$$

F_n is in normal form with $V_0^{\perp} = \{A^{\ell}, A^r\} \cup \{H_i^k \mid 0 \le i \le n, k \in \{\ell, r\}\}$. For $0 \le i \le n$ we have $[\![A_i]\!]_{F_n} = (ab)^{2^i}$ and $[\![B_i]\!]_{F_n} = \underbrace{b(\cdots b(}_{2^i} x) \cdots)$,

$$[\![H_i^{\ell}]\!]_{F_n} = [\![G_i^{\ell}\langle A^{\ell} \rangle]\!]_{F_n} = [\![C_i^{\ell}\langle D_i^{\ell}\langle A^{\ell} \rangle\rangle]\!]_{F_n} = [\![C_i^{\ell}\langle B_i^{\ell}\langle C_i^{\ell}\langle A^{\ell} \rangle\rangle\rangle]\!]_{F_n}$$
$$= c((ab)^{2^i} \underbrace{b(\cdots b(}_{2^i} c((ab)^{2^i} ab)\underbrace{) \cdots)}_{2^i} b) \text{ and}$$

$$[\![H_i^r]\!]_{F_n} = c(a\underbrace{b(\cdots b(}_{2^i} c(ab(ab)^{2^i}))\underbrace{) \cdots)}_{2^i}(ab)^{2^i}).$$

Note that the trees $[\![H_n^k]\!]_{F_n}$, where $k \in \{\ell, r\}$, have both exponential width and height in n. See Fig. 1 for an illustration.

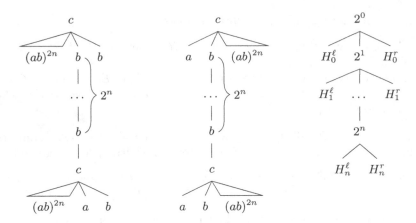

Fig. 1. The trees $[\![H_n^\ell]\!]_{F_n}$ on the left and $[\![H_n^r]\!]_{F_n}$ in the middle from Example 1, and the tree t_M from Example 4 on the right.

3   Navigation

The goal of this section is to prove the following theorem:

Theorem 1. *Let $F = (V, \mathrm{rhs})$ be an FSLP in normal form. We can in linear time precompute some data structure of linear size in $|F|$, such that the following operations work in constant time, where $\mathcal{N}(F)$ is the set of node representations, that we will define later:*

- $r_\lhd \colon V_0 \to \mathcal{N}(F) \cup \{\bot\}$: *Return the root of the first tree in the forest represented by an input variable.*
- $\nearrow \colon \mathcal{N}(F) \to \mathcal{N}(F) \cup \{\bot\}$: *Return the first child of the current node.*
- $\to \colon \mathcal{N}(F) \to \mathcal{N}(F) \cup \{\bot\}$: *Return the right sibling of the current node.*
- $r_\rhd \colon V_0 \to \mathcal{N}(F) \cup \{\bot\}$: *Return the root of the last tree in the forest represented by an input variable.*
- $\searrow \colon \mathcal{N}(F) \to \mathcal{N}(F) \cup \{\bot\}$: *Return the last child of the current node.*
- $\leftarrow \colon \mathcal{N}(F) \to \mathcal{N}(F) \cup \{\bot\}$: *Return the left sibling of the current node.*
- $\uparrow \colon \mathcal{N}(F) \to \mathcal{N}(F) \cup \{\bot\}$: *Return the parent of the current node.*
- $z \colon \mathcal{N}(F) \to \Sigma$: *Get the symbol at the current node.*

The special value $\bot$ is used to indicate that an operation may fail. For example, going to the first child of a leaf node returns $\bot$.

    The implementation of these operations will make use of SSLP traversals, that can already be done in constant time, which was proven in [12].

Lemma 2. *Let $G = (V, \mathrm{rhs})$ be an SSLP over some alphabet Θ. We can precompute some data structure in linear time in $|G|$, such that the following operations work in constant time, where $\mathcal{N}(G)$ is the set of positions:*

- $\triangleleft\colon V \to \mathcal{N}(G) \cup \{\bot\}$: *Go to the first position in the string derived by a given variable.*
- $\triangleright\colon V \to \mathcal{N}(G) \cup \{\bot\}$: *Go to the last position in the string derived by a given variable.*
- $z\colon \mathcal{N}(G) \to \Theta$: *Get the symbol at the current position.*
- $\to\colon \mathcal{N}(G) \to \mathcal{N}(G) \cup \{\bot\}$: *Go to the next position.*
- $\leftarrow\colon \mathcal{N}(G) \to \mathcal{N}(G) \cup \{\bot\}$: *Go to the previous position.*

Each element $\gamma \in \mathcal{N}(G)$ represents a position in the string $[\![A]\!]_G$ for a certain variable $A \in V$. We denote this variable A by S_γ.

Like in [12], we first define the *spine SSLP* $F_\square = (V_\square, \mathrm{rhs}_\square)$ over $\Sigma_\square$ by

$$\Sigma_\square = \{a\langle LxR\rangle \mid \mathrm{rhs}(A) = a\langle LxR\rangle, \ A \in V\}$$
$$\cup \{a\langle C\rangle \mid \mathrm{rhs}(A) = a\langle C\rangle, \ A \in V\},$$
$$V_\square = V_1 \cup V_0^\perp,$$
$$\mathrm{rhs}_\square(A) = \begin{cases} BC & \text{if } \mathrm{rhs}(A) = B\langle C\rangle, \ A \in V_1, \\ a\langle LxR\rangle & \text{if } \mathrm{rhs}(A) = a\langle LxR\rangle, \\ a\langle C\rangle & \text{if } \mathrm{rhs}(A) = a\langle C\rangle, \\ \mathrm{rhs}(B) & \text{if } \mathrm{rhs}(A) = B\langle C\rangle, \ A \in V_0^\perp. \end{cases}$$

The idea of the spine SSLP is that its symbols $\Sigma_\square$ are exactly the rhs-expressions of F where symbols from Σ occur. Navigating to one of the symbols in $[\![A]\!]_{F_\square}$ for $A \in V_0^\perp$ is essentially the same as navigating to a specific node in $[\![A]\!]_F$. If $\mathrm{rhs}(A) = a\langle C\rangle$ this string only consists of a single symbol $a\langle C\rangle$. If instead $\mathrm{rhs}(A) = B\langle C\rangle$, the word $[\![A]\!]_{F_\square}$ is of the form $a_1\langle L_1 x R_1\rangle \ldots a_m\langle L_m x R_m\rangle$. Navigating such a word left or right corresponds to navigating up or down in the tree $[\![A]\!]_F$, while following the x position. Standing on the symbol $a_i\langle L_i x R_i\rangle$ means that the current node is labelled with a_i. When we walk past the last element, $a_m\langle L_m x R_m\rangle$, we have to continue to navigate in $[\![C]\!]_F$. Let us thus define the vertical navigation structure as $\mathcal{V}(F) = \mathcal{N}(F_\square)^+$, which enables us to chain multiple navigations in $F_\square$ together. We can then define the following navigation steps, which we will later use to implement the actual navigation.

- $\triangle\colon V_0^\perp \to \mathcal{V}(F)$: Go to the root node,
- $\uparrow\colon \mathcal{V}(F) \to \mathcal{V}(F) \cup \{\bot\}$: Go to the parent node.
- $\downarrow\colon \mathcal{V}(F) \to \mathcal{V}(F) \cup \{\bot\}$: Go to the child node at the x position.
- $z\colon \mathcal{V}(F) \to \Sigma_\square$: Get the current symbol.

Implementing these is straight-forward: We set $\triangle(A) = \triangleleft(A)$ for $A \in V_0^\perp$. Let $v \in \mathcal{N}(F_\square)^*$ and $\gamma \in \mathcal{N}(F_\square)$. We set $z(v\gamma) = z(\gamma)$. The other two operations are defined as follows:

$$\uparrow(v\gamma) = \begin{cases} v \leftarrow(\gamma) & \text{if } \leftarrow(\gamma) \neq \bot, \\ v & \text{if } \leftarrow(\gamma) = \bot \text{ and } v \neq \varepsilon, \\ \bot & \text{if } \leftarrow(\gamma) = \bot \text{ and } v = \varepsilon, \end{cases}$$

$$\downarrow(v\gamma) = \begin{cases} v \to (\gamma) & \text{if } \to(\gamma) \neq \bot, \\ v\gamma \lhd (C) & \text{if } \to(\gamma) = \bot \text{ and } \mathrm{rhs}(S_\gamma) = B\langle C\rangle, \\ \bot & \text{if } \to(\gamma) = \bot \text{ and } \mathrm{rhs}(S_\gamma) = a\langle C\rangle. \end{cases}$$

Defining the operations on $\mathcal{V}(F)$ in isolation not only makes the definition of the actual navigation easier, but it also provides the benefit that we only have to change these when we add the ability to do subtree equality checks, while the definition of the actual navigation will stay the same.

What is left to do is to add the ability to do horizontal navigations as well. Horizontal navigations can happen on any of the L and R in $a\langle LxR\rangle$ as well as C in $a\langle C\rangle$. For this, we define another auxiliary SSLP, called the *rib SSLP* $F_\boxminus = (V_\boxminus, \mathrm{rhs}_\boxminus)$ over $\Sigma_\boxminus$ by $\Sigma_\boxminus = \{\underline{A} \mid A \in V_0^\bot\}$, $V_\boxminus = V_0$ and

$$\mathrm{rhs}_\boxminus(A) = \begin{cases} \varepsilon & \text{if } \mathrm{rhs}(A) = \varepsilon, \\ BC & \text{if } \mathrm{rhs}(A) = BC, \\ \underline{A} & \text{if } A \in V_0^\bot. \end{cases}$$

We make a copy of all $A \in V_0$ which we simply call $\underline{A}$. This is because we actually want an SSLP navigation on $\Sigma_\boxminus$ to end in a symbol from $V_0^\bot$. Without making the copy, we would not be able to assign an rhs value to such a symbol. Note that for each $A \in V_0$ we have $[\![A]\!]_{F_\boxminus} = A_1 \ldots A_n$, where $A_i \in V_0^\bot$. Thus $[\![A_i]\!]_F$ is the ith tree in the forest $[\![A]\!]_F$, i.e. $[\![A]\!]_F = [\![\overline{A_1}]\!]_F \cdots [\![\overline{A_n}]\!]_F \in \mathcal{T}^n$.

Example 2. Using F_n from Example 1, we have

$$\Sigma_\boxdot = \{b\langle ExE\rangle, a\langle E\rangle, b\langle E\rangle\} \cup \{c\langle A_i x A^r\rangle, c\langle A^\ell x A_i\rangle \mid 0 \leq i \leq n\},$$

$$\Sigma_\boxminus = \{\underline{A}^\ell, \underline{A}^r\} \cup \{\underline{H}_i^k \mid 0 \leq i \leq n, k \in \{\ell, r\}\} \text{ and for example}$$

$$[\![H_i^\ell]\!]_{F_\boxdot} = c\langle A_i x A^r\rangle (b\langle ExE\rangle)^{2^i} c\langle A_i x A^r\rangle \text{ for } 0 \leq i \leq n \text{ and}$$

$$[\![A_i]\!]_{F_\boxminus} = (\underline{A}^\ell \underline{A}^r)^{2^i} \text{ for } 0 \leq i \leq n.$$

For horizontal navigations, we have to remember if we started in an L or R in $a\langle LxR\rangle$ or in C in $a\langle C\rangle$, which we record as ℓ, r and m, respectively. Therefore, the horizontal navigation structure is $\mathcal{H}(F) = \{\ell, m, r\} \times \mathcal{N}(F_\boxminus)$. The idea for the whole navigation is then to interleave navigations on $F_\boxminus$ with navigations on $F_\boxdot$, so we define $\mathcal{N}(F) = (\mathcal{H}(F) \times \mathcal{V}(F))^+$.

We first introduce a short-hand notation to create a navigation on $F_\boxminus$ followed by a navigation on $F_\boxdot$. Let $\lhd_d \colon V_0 \to (\mathcal{H}(F) \times \mathcal{V}(F)) \cup \{\bot\}$ for every $d \in \{\ell, m, r\}$ be defined by

$$\lhd_d(A) = \begin{cases} ((d, \lhd(A)), \triangle(D)) & \text{if } \lhd(A) \neq \bot \text{ and } \mathrm{z}(\lhd(A)) = \underline{D}, \\ \bot & \text{if } \lhd(A) = \bot. \end{cases}$$

We are now going to implement the operations from Theorem 1. Suppose that the current state is $w((d, h), v) \in \mathcal{N}(F)$, where $w \in (\mathcal{H}(F) \times \mathcal{V}(F))^*$, $(d, h) \in \mathcal{H}(F)$, so $d \in \{\ell, m, r\}$ and $h \in \mathcal{N}(F_\boxminus)$, and $v \in \mathcal{V}(F)$. To query the current symbol, we define $\mathrm{z}(w((d, h), v)) = a$ if $\mathrm{z}(v) = a\langle LxR \rangle$ or $\mathrm{z}(v) = a\langle C \rangle$. To implement $\nearrow$, we have to consider the following cases: If we are on a symbol of the form $\mathrm{z}(v) = a\langle LxR \rangle$ we can either have $[\![L]\!]_F \neq \varepsilon$ or $[\![L]\!]_F = \varepsilon$. In the first case we go to the first symbol of $[\![L]\!]_{F_\boxminus}$ and record ℓ. This symbol is of the form $\underline{A}$, where $A \in V_0^\perp$, so we have to go to the root node of $[\![A]\!]_F$. We therefore append $\triangleleft_\ell(L) = ((\ell, \triangleleft(L)), \triangle(A))$. In case $[\![L]\!]_F = \varepsilon$ we reach the x of $a\langle LxR \rangle$, which means that we have to move the current navigation on $\mathcal{V}(F)$ down one position, i.e. we replace v with $\downarrow(v)$. If we are on a symbol of the form $a\langle C \rangle$ we can again have that $[\![C]\!]_F = \varepsilon$, in which case there is nowhere to go. If $[\![C]\!]_F \neq \varepsilon$ we go to the first symbol $\underline{A}$ of $[\![C]\!]_{F_\boxminus}$, record m and again start a navigation to the root of $[\![A]\!]_{F_\square}$, so we append $\triangleleft_m(C) = ((m, \triangleleft(C)), \triangle(A))$. Altogether, the function is

$$
\nearrow(w((d, h), v)) = \begin{cases} w((d, h), v) \triangleleft_\ell(L) & \text{if } \mathrm{z}(v) = a\langle LxR \rangle \text{ and } \triangleleft_\ell(L) \neq \perp, \\ w((d, h), \downarrow(v)) & \text{if } \mathrm{z}(v) = a\langle LxR \rangle \text{ and } \triangleleft_\ell(L) = \perp, \\ w((d, h), v) \triangleleft_m(C) & \text{if } \mathrm{z}(v) = a\langle C \rangle \text{ and } \triangleleft_m(C) \neq \perp, \\ \perp & \text{if } \mathrm{z}(v) = a\langle C \rangle \text{ and } \triangleleft_m(C) = \perp, \end{cases}
$$

The function $\searrow$ is symmetric and is therefore omitted. Going to the left or right neighbor is more involved. Since going to the right neighbor is basically a mirrored version of going to the left neighbor, we will focus on the former. We start by trying to move the current navigation on $\mathcal{V}(F)$ one position up, i.e. replace v with $\uparrow(v)$, because we need to know what is to the right of the current node. If moving up succeeds, the current symbol is of the form $a\langle LxR \rangle$, which means that we were standing on the x position and thus the next tree we have to go to is the first tree of $[\![R]\!]_F$ if $[\![R]\!]_F \neq \varepsilon$. We then go to the first symbol $\underline{A}$ of $[\![R]\!]_{F_\boxminus}$, record r, and go to the root node of $[\![A]\!]_F$, which means we append $\triangleleft_r(R) = ((r, \triangleleft(R)), \triangle(A))$. If $[\![R]\!]_F = \varepsilon$, there is nowhere to go. If moving up does not succeed, it means that v points to the root node. We are therefore on an $\underline{A_i}$ of the previous horizontal navigation h, which is of the form $\underline{A_1} \dots \underline{A_n}$. If $i < n$, then we move this navigation one to the right and go to the root node of $[\![A_{i+1}]\!]_F$, so we replace $((d, h), v)$ with $((d, \rightarrow(h)), \triangle(A_{i+1}))$. If $i = n$, then we have to look at the current symbol of the last navigation from w. In case $w = \varepsilon$, there is nowhere to go. Now suppose w ends in $v' \in \mathcal{V}(F)$. Suppose that the current symbol of v' is of the form $\mathrm{z}(v') = a\langle LxR \rangle$ and $d = \ell$. This means that we left the navigation on L to the right and end up on x, so we have to move the vertical navigation one position down, i.e. we replace v' with $\downarrow(v')$. In case $d = r$, we left the navigation on R to the right, so there is nowhere to go. If the current symbol is instead of the form $\mathrm{z}(v') = a\langle C \rangle$ there is also nowhere to go, since we were on the last tree of $[\![C]\!]_F$. Altogether, we have the function

$$\rightarrow(w((d,h),v)) = \begin{cases} w((d,h),u) \lhd_r(R) & \text{if } \uparrow(v) = u,\ z(u) = a\langle LxR\rangle \\ & \text{and } \lhd_r(R) \neq \bot, \\ \bot & \text{if } \uparrow(v) = u,\ z(u) = a\langle LxR\rangle \\ & \text{and } \lhd_r(R) = \bot, \\ w((d,h'),\triangle(D)) & \text{if } \uparrow(v) = \bot,\ \rightarrow(h) = h' \neq \bot \\ & \text{and } z(h') = \underline{D}, \\ w'((d',h'),\downarrow(v')) & \text{if } \uparrow(v) = \bot,\ \rightarrow(h) = \bot, \\ & w = w'((d',h'),v'), \\ & w' \in (\mathcal{H}(F) \times \mathcal{V}(F))^*,\ (d',h') \in \mathcal{H}(F), \\ & v' \in \mathcal{V}(F) \text{ and } d = \ell, \\ \bot & \text{if } \uparrow(v) = \bot,\ \rightarrow(h) = \bot,\ w = \varepsilon \\ & \text{or } d \neq \ell. \end{cases}$$

Going to the parent node is straight-forward. We try to replace v with $\uparrow(v)$. If this is not possible, we remove (d, h) and v, if $w \neq \varepsilon$. If $w = \varepsilon$, then there is nowhere to go. Formally, this is

$$\uparrow(w(((d,h),v)) = \begin{cases} w((d,h),\uparrow(v)) & \text{if } \uparrow(v) \neq \bot, \\ w & \text{if } \uparrow(v) = \bot \text{ and } w \neq \varepsilon, \\ \bot & \text{if } \uparrow(v) = \bot \text{ and } w = \varepsilon, \end{cases}$$

For starting a navigation in $S \in V_0$, we go to the first tree of $[\![S]\!]_F$ and record m in case $[\![S]\!]_F \neq \varepsilon$. This navigation ends on a symbol $\underline{A}$, and we go to the root node of $[\![A]\!]_F$. Thus we start with $r_\lhd(S) = \lhd_m(S) = ((m, \lhd(S)), \triangle(A))$. If $[\![S]\!]_F = \varepsilon$ there is nowhere to go, so $r_\lhd(S) = \bot$. Going to the last tree is symmetric.

Example 3. Consider F_n from Example 1. We have

$$r_\lhd(H_n^\ell) = ((m, \lhd(H_n^\ell)), \triangle(H_n^\ell))$$

and since $z(\triangle(H_n^\ell)) = c\langle A_n x A^r\rangle$, we obtain $z(r_\lhd(H_n^\ell)) = c$. Now consider applying $\diagup$ to this node: Since $z(\lhd(A_n)) = \underline{A}^\ell$, we have

$$\diagup(r_\lhd(H_n^\ell)) = ((m, \lhd(H_n^\ell)), \triangle(H_n^\ell))((\ell, \lhd(A_n)), \triangle(A^\ell)).$$

Using z on this structure yields a, because $z(\triangle(A^\ell)) = a\langle E\rangle$.

4   Navigation with Equality Checks

In this section we change our navigation structure, which we again call $\mathcal{N}(F)$, to include subtree equality checks.

Theorem 2. *Using polynomial time preprocessing, we can precompute some data structure of linear size in $|F|$ such that in addition to the operations from Theorem 1 the following operation, which checks if two subtrees rooted at the given input nodes are equal:*

$$\mathrm{eq} \colon \mathcal{N}(F) \times \mathcal{N}(F) \to \{0,1\}.$$

We ensure that our FSLP is *reduced* which means that there are no $A \neq A' \in V$ such that $[\![A]\!]_F = [\![A']\!]_F$, which can be tested using a result from [7]. We give a similar characterization of equal subtrees as the one found in [12]. Let us write $A_{\boxdot}$ instead of $[\![A]\!]_{F_{\boxdot}}$. We define $V_0^{\boxdot} = \{A \in V_0^{\perp} \mid \mathrm{rhs}(A) = B\langle C\rangle\}$. For $A \in V_0^{\boxdot}$ with $\mathrm{rhs}(A) = B\langle C\rangle$ let the *i'th subtree* be defined as $A_{\triangle}(i) = [\![B_{\boxdot}[i]\langle \cdots B_{\boxdot}[\ell(A)]\langle C\rangle \cdots \rangle]\!]_F$, where $1 \leq i \leq \ell(A) + 1$ and $\ell(A) = |B_{\boxdot}|$. Note that $A_{\triangle}(1) = [\![A]\!]_F$ and $A_{\triangle}(\ell(A) + 1) = [\![C]\!]_F$. Now let $i \geq 2$ be the smallest number such that there is a $D \in V_0^{\perp}$ with $A_{\triangle}(i) = [\![D]\!]_F$. We call i the *split index* of A, written as $\mathrm{si}(A)$, and D the *split variable* of A, written as $\mathrm{sv}(A)$. Since $A_{\triangle}(\ell(A)+1) = [\![C]\!]_F$ and $C \in V_0^{\perp}$, the split index and split variable always exist. The idea to implement the navigation that also supports subtree equality checks is to always stay below the split index. When we reach the split index, we simply continue to navigate in the split variable, which preserves subtree equality. We now only have to characterize the equal subtrees below split indices.

Lemma 3. *Let $t, t' \in \mathcal{T}$, $a, a' \in \Sigma$ and $L, L', R, R' \in V_0$, with*

1. $a\langle LxR\rangle \neq a'\langle L'xR'\rangle$, *and*
2. $[\![a\langle LxR\rangle]\!]_F[t] = [\![a'\langle L'xR'\rangle]\!]_F[t']$.

Then there are $D, D' \in V_0^{\perp}$ with $[\![D]\!]_F = t'$ and $[\![D']\!]_F = t$.

Proof. Since F is in normal form, there are variables

$$\{L_1, \ldots, L_n, R_1, \ldots, R_m, L'_1, \ldots, L'_{n'}, R'_1, \ldots, R'_{m'}\} \subseteq V_0^{\perp}$$

with $[\![L]\!]_F = [\![L_1 \ldots L_n]\!]_F$, $[\![R]\!]_F = [\![R_1 \ldots R_m]\!]_F$, $[\![L']\!]_F = [\![L'_1 \ldots L'_{n'}]\!]_F$ and $[\![R']\!]_F = [\![R'_1 \ldots R'_{m'}]\!]_F$. From Point 2 we obtain $a = a'$ and $[\![L]\!]_F t [\![R]\!]_F = [\![L']\!]_F t' [\![R']\!]_F$. From $[\![L]\!]_F = [\![L']\!]_F$ we would obtain $t = t'$ and $[\![R]\!]_F = [\![R']\!]_F$ which is in contradiction to Point 1 because F is reduced. Hence we must have $[\![L]\!]_F \neq [\![L']\!]_F$ which implies that $[\![L_1 \ldots L_n]\!]_F \in \mathcal{T}^n$ is a proper prefix of $[\![L'_1 \ldots L'_{n'}]\!]_F \in \mathcal{T}^{n'}$ or vice versa. In the first case we have $t = [\![L'_{n+1}]\!]_F$ and $t' = [\![R_{n'-n}]\!]_F$, in the second case $t' = [\![L_{n'+1}]\!]_F$ and $t = [\![R'_{n-n'}]\!]_F$.

Lemma 4. *Let $A, A' \in V_0^{\boxdot}$, $1 \leq i < \mathrm{si}(A)$ and $1 \leq i' < \mathrm{si}(A')$. We have $A_{\triangle}(i) = A'_{\triangle}(i')$ if and only if*

1. $A_{\boxdot}[i : \mathrm{si}(A) - 2] = A'_{\boxdot}[i' : \mathrm{si}(A') - 2]$, *and*
2. $[\![A_{\boxdot}[\mathrm{si}(A) - 1]\langle \mathrm{sv}(A)\rangle]\!]_F = [\![A'_{\boxdot}[\mathrm{si}(A') - 1]\langle \mathrm{sv}(A')\rangle]\!]_F$.

Proof. It is easy to see that Points 1 and 2 imply $A_\triangle(i) = A'_\triangle(i')$. To prove the opposite direction, we use induction on $m = \min\{\mathrm{si}(A) - i - 1,\ \mathrm{si}(A') - i' - 1\}$. Assume that $A_\triangle(i) = A'_\triangle(i')$. Let $m = 0$, which means that either $i = \mathrm{si}(A) - 1$ or $i' = \mathrm{si}(A') - 1$. We assume that $i = \mathrm{si}(A) - 1$ and show that $i' = \mathrm{si}(A') - 1$. By definition of si and sv we have $[\![A_\square[\mathrm{si}(A) - 1]\langle \mathrm{sv}(A)\rangle]\!]_F = A_\triangle(i)$. Since $A'_\triangle(i') = [\![A'_\square[i']]\!]_F[A'_\triangle(i' + 1)]$, we obtain

$$[\![A_\square[\mathrm{si}(A) - 1]\langle \mathrm{sv}(A)\rangle]\!]_F = [\![A'_\square[i']]\!]_F[A'_\triangle(i' + 1)].$$

If $A_\square[\mathrm{si}(A) - 1] = A'_\square[i']$ we have $[\![\mathrm{sv}(A)]\!]_F = A'_\triangle(i'+1)$. If $A_\square[\mathrm{si}(A) - 1] \neq A'_\square[i']$, then we obtain from Lemma 3 that there is a $D \in V_0^\perp$ with $[\![D]\!]_F = A'_\triangle(i'+1)$. In both cases, there is a variable that evaluates to $A'_\triangle(i' + 1)$, and since $i' < \mathrm{si}(A')$, we must have that $i' + 1 = \mathrm{si}(A')$ and $A'_\triangle(i' + 1) = [\![\mathrm{sv}(A')]\!]_F$. Therefore, we obtain that $A_\square[i : \mathrm{si}(A) - 2] = \varepsilon$, $A'_\square[i' : \mathrm{si}(A') - 2] = \varepsilon$ and

$$[\![A_\square[\mathrm{si}(A) - 1]\langle \mathrm{sv}(A)\rangle]\!]_F = [\![A'_\square[\mathrm{si}(A') - 1]\langle \mathrm{sv}(A')\rangle]\!]_F.$$

The symmetric case, in which $i' = \mathrm{si}(A') - 1$, uses the same arguments. Now let $m > 0$, so $i < \mathrm{si}(A) - 1$ and $i' < \mathrm{si}(A') - 1$. Since $A_\triangle(i) = A'_\triangle(i')$ we obtain that $[\![A_\square[i]]\!]_F[A_\triangle(i + 1)] = [\![A'_\square[i']]\!]_F[A'_\triangle(i' + 1)]$. If $A_\square[i] \neq A'_\square[i']$ we would obtain from Lemma 3 that there are $D, D' \in V_0^\perp$ with $[\![D]\!]_F = A'_\triangle(i' + 1)$ and $[\![D']\!]_F = A_\triangle(i + 1)$ which contradicts $i < \mathrm{si}(A) - 1$ as well as $i' < \mathrm{si}(A') - 1$. Therefore, we must have $A_\square[i] = A'_\square[i']$. From $A_\triangle(i) = A'_\triangle(i')$ and this fact we can conclude that $A_\triangle(i + 1) = A'_\triangle(i' + 1)$. Therefore by induction we have $A_\square[i + 1 : \mathrm{si}(A) - 2] = A'_\square[i' + 1 : \mathrm{si}(A') - 2]$, as well as Point 2. Together with $A_\square[i] = A'_\square[i']$ we also obtain Point 1.

To use Lemma 4 for equality checks, we still have to argue that we can implement some data structure of linear size that allows to perform these checks in constant time. To check Point 2 of Lemma 4 we do the following: Let

$$\sim = \{(A, A') \in V_0^\square \times V_0^\square \mid [\![A_\square[\mathrm{si}(A) - 1]\langle \mathrm{sv}(A)\rangle]\!]_F = [\![A'_\square[\mathrm{si}(A') - 1]\langle \mathrm{sv}(A')\rangle]\!]_F\}.$$

This relation is an equivalence relation. We assign each equivalence class a natural number and precompute a mapping that takes an element to its equivalence class, represented as this number. This mapping requires linear space and we can test if two elements belong to the same equivalence class in constant time. Let $A, A' \in V_0^\square$, $1 \leq i < \mathrm{si}(A)$ and $1 \leq i' < \mathrm{si}(A')$. We only have to check Point 1 of Lemma 4 if Point 2 is true, so suppose that $(A, A') \in \sim$. We now have to test if $A_\square[i : \mathrm{si}(A) - 2] = A'_\square[i' : \mathrm{si}(A') - 2]$. This can only be true if $k := \mathrm{si}(A) - 2 - i = \mathrm{si}(A') - 2 - i'$. Let $\mathrm{suff}(A, A')$ be the length of the longest common suffix of $A_\square[: \mathrm{si}(A) - 2]$ and $A'_\square[: \mathrm{si}(A') - 2]$, which can be computed in polynomial time using a result from [14]. We then have $A_\square[i : \mathrm{si}(A) - 2] = A'_\square[i' : \mathrm{si}(A') - 2]$ if and only if $k \leq \mathrm{suff}(A, A')$. Storing suff explicitly for all elements belonging to the same equivalence class $M \in V_0^\square/\!\sim$ requires quadratic space. Instead, we compute for each M a tree t_M that has linear size in $|M|$, which we can use to query suff in constant time. In this tree, the elements from M are the leaves and the *lowest common ancestor* of two leaves $A \neq A' \in M$ is labelled with $\mathrm{suff}(A, A')$. Lowest common ancestor queries can be performed in constant time after linear time preprocessing, using the result from [15].

The trees t_M can be constructed as follows: We start with any $A \neq A' \in M$ and make A and A' children of a node labelled with $\mathrm{suff}(A, A')$. Now suppose we have constructed a tree for some elements of M. To add a new $A \in M$ to the tree, we take a leaf A' where $\mathrm{suff}(A, A')$ is maximal. We then find the closest ancestor node a of A' whose parent p is labelled with $m \leq \mathrm{suff}(A, A')$, or in case this does not exist then a is the root node. If $m = \mathrm{suff}(A, A')$ then A becomes a new child of a. If $m < \mathrm{suff}(A, A')$ then we add a new node between a and p, label it with $\mathrm{suff}(A, A')$ and add A as its second child. If a is the root node then we add a new parent to a, label it with $\mathrm{suff}(A, A')$ and add A as its second child.

It remains to argue why we can precompute si and sv in polynomial time, which we can do as follows: For every $A \in V_0^{\square}$ and $A' \neq A \in V_0^{\perp}$ we test if there is an $1 < i \leq \ell(A)$ with $A_{\triangle}(i) = [\![A']\!]_F$, which is done as follows: Since $|A_{\triangle}(1)| > \cdots > |A_{\triangle}(\ell(A))|$ we can use binary search to test if there is an $1 < i \leq \ell(A)$ such that $|A_{\triangle}(i)| = |[\![A']\!]_F|$. For a given i computing $|A_{\triangle}(i)|$ can be done in polynomial time because we can compute an FSLP G with variable X such that $[\![X]\!]_G = A_{\triangle}(i)$. This is done by removing a prefix of $A_{\square}$, which can be done by cutting the syntax tree of A. Also, given a variable X of an FSLP G it is easy to compute $|[\![X]\!]_G|$. Furthermore, we can test if $A_{\triangle}(i) = [\![A']\!]_F$ because given two variables X, Y of an FSLP G we can test in polynomial time if $[\![X]\!]_G = [\![Y]\!]_G$ using a result from [7]. For a given $A \in V_0^{\square}$ we then take the $A' \in V_0^{\perp}$ with the smallest i such that $A_{\triangle}(i) = [\![A']\!]_F$ and set $\mathrm{sv}(A) = A'$ and $\mathrm{si}(A) = i$. If no such i exists then we set $\mathrm{sv}(A) = C$, where $\mathrm{rhs}(A) = B\langle C\rangle$, and $\mathrm{si}(A) = \ell(A) + 1$.

Example 4. Recall the definition of F_n from Example 1. Since $[\![C_i^{\ell}\langle A^{\ell}\rangle]\!]_{F_n} = c((ab)^{2^i}xb)[a] = c(ax(ab)^{2^i})[b] = [\![C_i^{\ell}\langle A^{\ell}\rangle]\!]_{F_n}$, we have $H_{i\,\triangle}^{\ell}(2^i+2) = H_{i\,\triangle}^{r}(2^i+2)$ for $0 \leq i \leq n$. We also have $H_{i\,\square}^{\ell}[j] = b\langle ExE\rangle = H_{i\,\square}^{r}[j]$ for all $2 \leq j \leq 2^i + 1$. This implies that $H_{i\,\triangle}^{\ell}(j) = H_{i\,\triangle}^{r}(j)$ for all $2 \leq j \leq 2^i + 1$. Therefore, $\mathrm{si}(H_i^{\ell}) = \mathrm{si}(H_i^{r}) = 2^i + 3$ and $\mathrm{sv}(H_i^{k}) = A^k$ for all $0 \leq i \leq n$ and $k \in \{\ell, r\}$. Since

$$[\![H_{i\,\square}^{\ell}[\mathrm{si}(H_i^{\ell}) - 1]\langle \mathrm{sv}(H_i^{\ell})\rangle]\!]_{F_n} = [\![H_{i\,\square}^{r}[\mathrm{si}(H_i^{r}) - 1]\langle \mathrm{sv}(H_i^{r})\rangle]\!]_{F_n}$$

we have $H_i^{\ell} \sim H_i^{r}$ for all $0 \leq i \leq n$, thus $V_0^{\perp}/{\sim} = \{\{H_i^{\ell}, H_i^{r}\} \mid 0 \leq i \leq n\}$, and $\mathrm{suff}(H_i^{\ell}, H_i^{r}) = 2^i$.

Let us change the definition of $\mathrm{rhs}(D_i^k)$ from $\mathrm{rhs}(D_i^k) = B_i\langle C_i^k\rangle$ to $\mathrm{rhs}(D_i^k) = B_i\langle C_0^k\rangle$. We then have for $0 \leq i \leq n$ that

$$[\![H_i^{\ell}]\!]_{F_n} = c((ab)^{2^i} \underbrace{b(\cdots b(}_{2^i} c(abab) \underbrace{) \cdots)}_{2^i} b) \text{ and}$$

$$[\![H_i^{r}]\!]_{F_n} = c(a \underbrace{b(\cdots b(}_{2^i} c(abab) \underbrace{) \cdots)}_{2^i}(ab)^{2^i}).$$

Since $[\![C_0^r\langle A^r\rangle]\!]_{F_n} = c(axab)[b] = c(abxb)[a] = [\![C_0^\ell\langle A^\ell\rangle]\!]_{F_n}$, we have

$$H_{i\,\triangle}^\ell(2^i + 2) = H_{i\,\triangle}^r(2^i + 2) = H_{j\,\triangle}^\ell(2^j + 2) = H_{j\,\triangle}^r(2^j + 2)$$

for all $0 \le i, j \le n$. Thus $H_{i\,\triangle}^k(2^i + 2 - m) = H_{j\,\triangle}^{k'}(2^j + 2 - m)$ for all $k, k' \in \{\ell, r\}$, $1 \le i, j \le 2^n$ and $0 \le m \le \max\{2^i, 2^j\}$. Therefore, $V_0^\perp/\!\sim\, = \{M\}$ consists of the single equivalence class $M = \{H_i^k \mid 0 \le i \le n, k \in \{\ell, r\}\}$. See Fig. 1 for the tree t_M.

We now explain how we have to change our vertical navigation structure $\mathcal{V}(F)$ and the operations on it to support the subtree equality check eq. We change the $\mathcal{V}(F)$-part of our navigation structure to $\mathcal{V}(F) = (\mathcal{N}(F_\square) \times \mathbb{N})^+$, where we use the $\mathbb{N}$ component to count how many $\downarrow$ steps we made. The operations $\triangle$, $\uparrow$ and z are straight-forward to implement. Let $v \in (\mathcal{N}(F_\square) \times \mathbb{N})^*$, $\gamma \in \mathcal{N}(F_\square)$ and $i \in \mathbb{N}$. We set $\mathrm{z}(v(\gamma, i)) = \mathrm{z}(\gamma)$ and $\triangle(A) = (\triangleleft(A), 1)$ for $A \in V_0^\perp$ and

$$\uparrow(v(\gamma, i)) = \begin{cases} v(\leftarrow(\gamma), i - 1) & \text{if } \leftarrow(\gamma) \ne \perp, \\ v & \text{if } \leftarrow(\gamma) = \perp \text{ and } v \ne \varepsilon, \\ \perp & \text{if } \leftarrow(\gamma) = \perp \text{ and } v = \varepsilon, \end{cases}$$

For the implementation of $\downarrow$, let $v(\gamma, i) \in \mathcal{V}(F)$ be the current state, where $v \in (\mathcal{N}(F_\square) \times \mathbb{N})^*$, $\gamma \in \mathcal{N}(F_\square)$ and $i \in \mathbb{N}$. Suppose the navigation γ started in $\triangleleft(A)$. This means that we are currently on $A_\square[i]$ and want to navigate to $A_\square[i+1]$. If $A \notin V_0^\square$, so rhs(A) is of the form $a\langle C\rangle$, there is nowhere to go. Now let $A \in V_0^\square$. In case si$(A) > i+1$, we can stay on $A_\square$, so we replace γ with $\to(\gamma)$ and i with $i+1$. In case si$(A) = i+1$, we have to continue to navigate in sv$(A)_\square$, since $A_\triangle(i+1) = [\![\mathrm{sv}(A)]\!]_F$. Therefore, we append $(\triangleleft(\mathrm{sv}(A)), 1)$. Formally, we have

$$\downarrow(v(\gamma, i)) = \begin{cases} v(\to(\gamma), i+1) & \text{if } S_\gamma \in V_0^\square \text{ and si}(S_\gamma) > i+1, \\ v(\gamma, i)\,\triangle(\mathrm{sv}(S_\gamma)) & \text{if } S_\gamma \in V_0^\square \text{ and si}(S_\gamma) = i+1, \\ \perp & \text{if } S_\gamma \notin V_0^\square. \end{cases}$$

With the new definition of $\mathcal{N}(F)$, the subtree equality check eq can easily be implemented. Suppose the rightmost elements from $\mathcal{N}(F)$ are $v(\gamma, i)$ and $v'(\gamma', i')$, where γ started with $\triangleleft(A)$ and γ' with $\triangleleft(A')$ for $A, A' \in V_0^\perp$. In case rhs$(A) = a\langle C\rangle$, so $A \in V_0^\perp \setminus V_0^\square$, then $A_\triangle(i) = A'_\triangle(i')$ if and only if rhs$(A') = a\langle C\rangle$. Now let $A, A' \in V_0^\square$, in which case we use Lemma 4 to test whether $A_\triangle(i) = A'_\triangle(i')$.

5   Discussion

We first implemented a data structure that can be precomputed in linear time with which we can do navigation steps in constant time. Later we added the ability to do subtree equality checks, again by precomputing a data structure of linear size. However, the precomputation time required is polynomial in this case. It would be interesting to show a lower bound for the exponent. Since the preprocessing requires equality checks for which the best known algorithm is quadratic (see [10]), it would be surprising if this exponent was lower than 2. Implementing all the algorithms of this work would also be interesting. In [6] it was shown that using linear time we can transform an FSLP into an equivalent one such that the height $h(F)$ of its syntax tree is logarithmic in $|F|$. If we can show that $h(F)$ does not increase when using Lemma 1, and that the size $|X|$ of elements $X \in \mathcal{N}(F)$ is bounded by $h(F)$, then we would obtain that $|X| \in \mathcal{O}(\log |F|)$.

References

1. Bille, P., Gørtz, I.L., Landau, G.M., Weimann, O.: Tree compression with top trees. Inf. Comput. **243**, 166–177 (2015). https://doi.org/10.1016/j.ic.2014.12.012
2. Bille, P., Landau, G.M., Raman, R., Sadakane, K., Satti, S.R., Weimann, O.: Random access to grammar-compressed strings and trees. SIAM J. Comput. **44**(3), 513–539 (2015). https://doi.org/10.1137/130936889
3. Bojanczyk, M., Walukiewicz, I.: Forest algebras. In: Flum, J., Grädel, E., Wilke, T. (eds.) Logic and Automata: History and Perspectives [in Honor of Wolfgang Thomas]. Texts in Logic and Games, vol. 2, pp. 107–132. Amsterdam University Press (2008)
4. Boneva, I., Niehren, J., Sakho, M.: Regular matching and inclusion on compressed tree patterns with context variables. In: Martín-Vide, C., Okhotin, A., Shapira, D. (eds.) LATA 2019. LNCS, vol. 11417, pp. 343–355. Springer, Cham (2019). https://doi.org/10.1007/978-3-030-13435-8_25
5. Cai, J., Paige, R.: Using multiset discrimination to solve language processing problems without hashing. Theor. Comput. Sci. **145**(1&2), 189–228 (1995). https://doi.org/10.1016/0304-3975(94)00183-J
6. Ganardi, M., Jez, A., Lohrey, M.: Balancing straight-line programs. In: Zuckerman, D. (ed.) 60th IEEE Annual Symposium on Foundations of Computer Science, FOCS 2019, Baltimore, Maryland, USA, 9–12 November 2019, pp. 1169–1183. IEEE Computer Society (2019). https://doi.org/10.1109/FOCS.2019.00073
7. Gascón, A., Lohrey, M., Maneth, S., Reh, C.P., Sieber, K.: Grammar-based compression of unranked trees. In: Fomin, F.V., Podolskii, V.V. (eds.) CSR 2018. LNCS, vol. 10846, pp. 118–131. Springer, Cham (2018). https://doi.org/10.1007/978-3-319-90530-3_11
8. Gasieniec, L., Kolpakov, R.M., Potapov, I., Sant, P.: Real-time traversal in grammar-based compressed files. In: 2005 Data Compression Conference (DCC 2005), Snowbird, UT, USA, 29–31 March 2005, p. 458. IEEE Computer Society (2005). https://doi.org/10.1109/DCC.2005.78
9. Hucke, D., Lohrey, M., Reh, C.P.: The smallest grammar problem revisited. In: Inenaga, S., Sadakane, K., Sakai, T. (eds.) SPIRE 2016. LNCS, vol. 9954, pp. 35–49. Springer, Cham (2016). https://doi.org/10.1007/978-3-319-46049-9_4
10. Jez, A.: Faster fully compressed pattern matching by recompression. ACM Trans. Algorithms **11**(3), 20:1–20:43 (2015). https://doi.org/10.1145/2631920
11. Lohrey, M.: Grammar-based tree compression. In: Potapov, I. (ed.) DLT 2015. LNCS, vol. 9168, pp. 46–57. Springer, Cham (2015). https://doi.org/10.1007/978-3-319-21500-6_3

12. Lohrey, M., Maneth, S., Reh, C.P.: Constant-time tree traversal and subtree equality check for grammar-compressed trees. Algorithmica **80**(7), 2082–2105 (2018). https://doi.org/10.1007/s00453-017-0331-3
13. Maneth, S., Peternek, F.: Constant delay traversal of compressed graphs. In: Bilgin, A., Marcellin, M.W., Serra-Sagristà, J., Storer, J.A. (eds.) 2018 Data Compression Conference, DCC 2018, Snowbird, UT, USA, 27–30 March 2018, pp. 32–41. IEEE (2018). https://doi.org/10.1109/DCC.2018.00011
14. Matsubara, W., Inenaga, S., Ishino, A., Shinohara, A., Nakamura, T., Hashimoto, K.: Efficient algorithms to compute compressed longest common substrings and compressed palindromes. Theor. Comput. Sci. **410**(8–10), 900–913 (2009). https://doi.org/10.1016/j.tcs.2008.12.016
15. Schieber, B., Vishkin, U.: On finding lowest common ancestors: simplification and parallelization. SIAM J. Comput. **17**(6), 1253–1262 (1988). https://doi.org/10.1137/0217079

Towards Efficient Interactive Computation of Dynamic Time Warping Distance

Akihiro Nishi[1], Yuto Nakashima[1] (ID), Shunsuke Inenaga[1,2](✉) (ID),
Hideo Bannai[3] (ID), and Masayuki Takeda[1] (ID)

[1] Department of Informatics, Kyushu University, Fukuoka, Japan
{yuto.nakashima,inenaga,takeda}@inf.kyushu-u.ac.jp
[2] PRESTO, Japan Science and Technology Agency, Kawaguchi, Japan
[3] M&D Data Science Center, Tokyo Medical and Dental University, Tokyo, Japan
hdbn.dsc@tmd.ac.jp

Abstract. The *dynamic time warping* (*DTW*) is a widely-used method that allows us to efficiently compare two time series that can vary in speed. Given two strings A and B of respective lengths m and n, there is a fundamental dynamic programming algorithm that computes the DTW distance $\mathrm{dtw}(A, B)$ for A and B together with an optimal alignment in $\Theta(mn)$ time and space. In this paper, we tackle the problem of interactive computation of the DTW distance for dynamic strings, denoted $\mathbf{D^2TW}$, where character-wise edit operation (insertion, deletion, substitution) can be performed at an *arbitrary* position of the strings. Let M and N be the sizes of the *run-length encoding* (*RLE*) of A and B, respectively. We present an algorithm for $\mathbf{D^2TW}$ that occupies $\Theta(mN + nM)$ space and uses $O(m + n + \#_{\mathrm{chg}}) \subseteq O(mN + nM)$ time to update a compact differential representation DS of the DP table per edit operation, where $\#_{\mathrm{chg}}$ denotes the number of cells in DS whose values change after the edit operation. Our method is at least as efficient as the algorithm recently proposed by Froese et al. running in $\Theta(mN + nM)$ time, and is faster when $\#_{\mathrm{chg}}$ is smaller than $O(mN + nM)$ which, as our preliminary experiments suggest, is likely to be the case in the majority of instances.

1 Introduction

The *dynamic time warping* (*DTW*) is a classical and widely-used method that allows us to efficiently compare two temporal sequences or time series that can vary in speed. A fundamental dynamic programming algorithm computes the DTW distance $\mathrm{dtw}(A, B)$ for two strings A and B together with an optimal alignment in $\Theta(mn)$ time and space [12], where $|A| = m$ and $|B| = n$. This algorithm allows one to update the DP table D for $\mathrm{dtw}(A, B)$ in $O(m)$ time (resp. $O(n)$ time) when a new character is appended to B (resp. to A).

In this paper, we introduce the "dynamic" DTW problem, denoted $\mathbf{D^2TW}$, where character-wise edit operation (insertion, deletion, substitution) can be

C. Boucher and S. V. Thankachan (Eds.): SPIRE 2020, LNCS 12303, pp. 27–41, 2020.
https://doi.org/10.1007/978-3-030-59212-7_3

performed at an *arbitrary* position of the strings. More formally, we wish to maintain a (space-efficient) representation of D that can dynamically be modified according to a given operation. This representation should be able to quickly answer the value of $D[m, n] = \mathsf{dtw}(A, B)$ upon query, together with an optimal alignment achieving $\mathsf{dtw}(A, B)$. This kind of interactive computation for (a representation of) D can be of practical merits, e.g. when simulating stock charts, or editing musical sequences. Another example of applications of **$\mathbf{D^2TW}$** is a sliding window version of DTW which computes $\mathsf{dtw}(A, B[j..j + d - 1])$ between A and every substring $B[j..j + d - 1]$ of B of arbitrarily fixed length d.

Incremental/decremental computation of a DP table is a restricted version of the aforementioned interactive computation, which allows for prepending a new character to B, and/or deleting the leftmost character from B. A number of incremental/decremental computation algorithms have been proposed for the *unit-cost edit distance* and *weighted edit distance*: Kim and Park [9] showed an incremental/decremental algorithm for the unit-cost edit distance that occupies $\Theta(mn)$ space and runs in $O(m + n)$ time per operation. Hyyrö et al. [7] proposed an algorithm for the edit distance with integer weights which uses $\Theta(mn)$ space and runs in $O(\min\{c(m + n), mn\})$ time per operation, where c is the maximum weight in the cost function. This translates into $O(m + n)$ time under constant weights. Schmidt [13] gave an algorithm that uses $\Theta(mn)$ space and runs in $O(n \log m)$ time per operation for a general weighted edit distance. Hyyrö and Inenaga [5] presented a space efficient alternative to incremental/decremental unit-cost edit distance computation which runs in $O(m + n)$ time per operation but uses only $\Theta(mN + nM)$ space, where M and N are the sizes of *run-length encoding* (RLE) of A and B, respectively. Since $M \leq m$ and $N \leq n$ always hold, the $mN + nM$ terms can be much smaller than the mn term for strings that contain many long character runs. Later, Hyyrö and Inenaga [6] presented a space-efficient alternative for edit distance with integer weights, which runs in $O(\min\{c(m + n), mn\})$ time per operation and requires $\Theta(mN + nM)$ space.

Fully-dynamic interactive computation for the (weighted) edit distance was also considered: Let j^* be the position in B where the modification has been performed. For the unit cost edit distance, Hyyrö et al. [8] presented a representation of the DP table which uses $\Theta(mn)$ space and can be updated in $O(\min\{rc(m + n), mn\})$ time per operation, where $r = \min\{j^*, n - j^* + 1\}$ and c is the maximum weight. They also showed that there exist instances that require $\Omega(\min\{rc(m + n), mn\})$ time to update their data structure per operation. Very recently, Charalampopoulos et al. [3] showed how to maintain an optimal (weighted) alignment of two fully-dynamic strings in $\tilde{O}(n \min\{\sqrt{n}, c\})$ time per operation, where $m = n$.

While computing *longest common subsequence* (*LCS*) and weighted edit distance of strings of length n can both be reduced to computing DTW of strings of length $O(n)$ [1,10], a reduction to the other direction is not known. It thus seems difficult to directly apply any of the aforementioned algorithms to our **$\mathbf{D^2TW}$** problem. Also, a conditional lower bound suggests that strongly sub-quadratic

DTW algorithms are unlikely to exist [1,2]. Thus, any method that recomputes the naïve DP table D from scratch should take almost quadratic time per update.

Our Contribution. This paper takes the first step towards an efficient solution to **D²TW**. Namely, we present an algorithm for **D²TW** that occupies $\Theta(mN + nM)$ space and uses $O(m + n + \#_{chg})$ time to update a compact differential representation DS for the DP table D per edit operation, where $\#_{chg}$ denotes the number of cells in DS whose values change after the edit operation. Since $\#_{chg} = O(mN + nM)$ always holds, our method is always at least as efficient as the naïve method that recomputes the full DP table D in $\Theta(mn)$ time, or the algorithm of Froese et al. [4] that recomputes another sparse representation of D in $\Theta(mN + nM)$ time. While there exist worst-case instances that give $\#_{chg} = \Omega(mN + nM)$, our preliminary experiments suggest that, in many cases, $\#_{chg}$ can be much smaller than the size of DS which is $\Theta(mN + nM)$.

Technically our algorithm is most related to Hyyrö et al.'s method [7,8] and Froese et al.'s method [4], but our algorithm is not straightforward from these.

Omitted proofs can be found in a full version of this paper [11].

2 Preliminaries

We consider sequences (strings) of characters from an alphabet Σ of real numbers. Let $A = a_1, \ldots, a_m$ be a string consisting of m characters from Σ. The *run-length encoding* rle(A) of string A is a compact representation of A such that each maximal run of the same characters in A is represented by a pair of the character and the length of the run. More formally, let $\mathbb{N}$ denote the set of positive integers. For any non-empty string A, rle(A) $= a_1^{e_1} \cdots a_M^{e_M}$, where $a_I \in \Sigma$ and $e_I \in \mathbb{N}$ for any $1 \leq I \leq M$, and $a_I \neq a_{I+1}$ for any $1 \leq I < M$. Each $a_I^{e_I}$ in rle(A) is called a (character) *run*, and e_I is called the exponent of this run. The *size* of rle(A) is the number M of runs in rle(A). E.g., for string $A = $ aaccccccccbbabbbb of length 16, rle(A) $= $ a²c⁷b²a¹b⁴ and its size is 5.

Dynamic time warping (DTW) is a commonly used method to compare two temporal sequences that may vary in speed. Consider two strings $A = a_1, \ldots, a_m$ and $B = b_1, \ldots, b_n$. To formally define the DTW for A and B, we consider an $m \times n$ grid graph $\mathcal{G}_{m,n}$ such that each vertex (i, j) has (at most) three directed edges; one to the lower neighbor $(i + 1, j)$ (if it exists), one to the right neighbor $(i, j + 1)$ (if it exists), and one to the lower-right neighbor $(i + 1, j + 1)$ (if it exists). A path in $\mathcal{G}_{m,n}$ that starts from vertex $(1, 1)$ and ends at vertex (m, n) is called a *warping path*, and is denoted by a sequence $(1, 1), \ldots, (i, j), \ldots, (m, n)$ of adjacent vertices. Let $\mathcal{P}_{m,n}$ be the set of all warping paths in $\mathcal{G}_{m,n}$. Note that each warping path in $\mathcal{P}_{m,n}$ corresponds to an alignment of A and B. The DTW for strings A and B, denoted dtw(A, B), is defined by dtw(A, B) $= \min_{p \in \mathcal{P}_{m,n}} \sqrt{\sum_{(i,j) \in p} (a_i - b_j)^2}$.

The fundamental $\Theta(mn)$-time and space solution for computing dtw(A, B), given in [12], fills an $m \times n$ dynamic programming table D such that $D[i, j] = $ dtw($A[1..i], B[1..j]$)² for $1 \leq i \leq m$ and $1 \leq j \leq n$. Therefore, after all the cells

	a	c	b	e	e	a	a	d
d	9	10	14	15	16	25	34	34
c	13	9	10	14	18	20	24	25
b	14	10	9	18	23	19	20	24
b	15	11	9	18	27	20	20	24
c	19	11	10	13	17	21	24	21
c	23	11	11	14	17	21	25	22
d	32	12	15	12	13	22	30	22
a	32	16	13	28	28	13	13	22

		c	b	e	e	a	a	d
d		1	5	6	7	16	25	25
c		1	2	6	10	11	15	16
b		2	1	10	15	11	12	16
b		3	1	10	19	12	12	16
c		3	2	5	9	13	16	13
c		3	3	6	9	13	17	14
d		4	7	4	5	14	22	14
a		8	5	20	20	5	5	14

Fig. 1. In this example where $A = $ dcbbccda and $B = $ acbeeaad, the values of $\Theta(mn)$ cells of the DP table for $\mathsf{dtw}(A, B)$ change after the edit operation on B (here, the first character $B[1] = $ a of B was deleted).

of D are filled, the desired result $\mathsf{dtw}(A, B)$ can be obtained by $\sqrt{D[m, n]}$. The value for each cell $D[i, j]$ is computed by the following well-known recurrence:

$$
\begin{aligned}
D[1, 1] &= (a_1 - b_1)^2, \\
D[i, 1] &= D[i - 1, 1] + (a_i - b_1)^2 \quad \text{for } 1 < i \leq m, \\
D[1, j] &= D[1, j - 1] + (a_1 - b_j)^2 \quad \text{for } 1 < j \leq n, \\
D[i, j] &= \min\{D[i, j - 1], D[i - 1, j], D[i - 1, j - 1]\} + (a_i - b_j)^2 \\
&\qquad \text{for } 1 < i \leq m \text{ and } 1 < j \leq n.
\end{aligned}
\tag{1}
$$

In the rest of this paper, we will consider the problem of maintaining a representation for D, each time one of the strings, B, is dynamically modified by an edit operation (i.e. single character insertion, deletion, or substitution) on an *arbitrary* position in B. We call this kind of interactive computation of $\mathsf{dtw}(A, B)$ as the *dynamic* DTW computation, denoted by $\mathbf{D^2TW}$.

Let B' denote the string after an edit operation is performed on B, and D' denote the dynamic programming table D after it has been updated to correspond to $\mathsf{dtw}(A, B')$. In a special case where the edit operation is performed at the right end of B, where we have $B' = Bc$ (insertion), $B' = B[1..n - 1]$ (deletion) or $B' = B[1..n - 1]c$ (substitution) with a character $c \in \Sigma$, then D can easily be updated to D' in $O(m)$ time by simply computing a single column at index $j = n$ or $j = n + 1$ using recurrence (1).

As in Fig. 1, in the worst case, the values of $\Theta(mn)$ cells of the DP table for $\mathsf{dtw}(A, B)$ can change after an edit on B. The following lemma gives a stronger statement that updating D to D' in our $\mathbf{D^2TW}$ scenario cannot be amortized:

Lemma 1. *There are strings A, B and a sequence of k edits on B such that $\Theta(kmn)$ cells in D' have different values in the corresponding cells in D.*

3   Our $\mathbf{D^2TW}$ Algorithm Based on RLE

We first explain the data structures which are used in our algorithm.

Differential Representation DR of D. The first idea of our algorithm is to use a differential representation DR of D: Each cell of DR contains two fields that respectively store the horizontal difference and the vertical difference, namely, $DR[i,j].U = D[i,j] - D[i-1,j]$ and $DR[i,j].L = D[i,j] - D[i,j-1]$. We let $DR[i,1].L = 0$ for any $1 \leq i \leq m$ and $DR[1,j].U = 0$ for any $1 \leq j \leq n$. The diagonal difference $D[i,j] - D[i-1,j-1]$ can easily be computed from $DR[i,j].U$ and $DR[i,j].L$ and thus is not explicitly stored in $DR[i,j]$.

In our algorithm we make heavy use of the following lemma:

Lemma 2. *For any $1 < i \leq m$,*

$$DR[i,j].U = \begin{cases} (a_i - b_1)^2 & \text{if } j = 1, \\ z - DR[i-1,j].L & \text{if } 2 \leq j \leq n, \end{cases}$$

and for any $1 < j \leq n$,

$$DR[i,j].L = \begin{cases} (a_1 - b_j)^2 & \text{if } i = 1, \\ z - DR[i,j-1].U & \text{if } 2 \leq i \leq m, \end{cases}$$

where $z = \min\{DR[i-1,j].L,\ DR[i,j-1].U,\ 0\} + (a_i - b_j)^2$.

Proof. $DR[i,1].U = (a_i - b_1)^2$ and $DR[1,j].L = (a_1 - b_j)^2$ are clear from recurrence (1). Now we consider $1 < i \leq m$ and $1 < j \leq n$, and let $d = D[i-1,j-1]$, $x = DR[i-1,j].L$, $y = DR[i,j-1].U$, and $d + z = D[i,j]$. Then we have $D[i-1,j] = d + x$ and $D[i,j-1] = d + y$ (see Fig. 2). It follows from the definition of DR that $DR[i,j].U = D[i,j] - D[i-1,j] = z - x$ and $DR[i,j].L = D[i,j] - D[i,j-1] = z - y$. Since $D[i,j] = \min\{D[i-1,j-1], D[i-1,j], D[i,j-1]\} + (a_i - b_j)^2$ by recurrence (1), we obtain $d + z = \min\{d, d+x, d+y\} + (a_i - b_j)^2$ which leads to $z = \min\{x, y, 0\} + (a_i - b_j)^2$. $\qquad\square$

RLE-Based Sparse Differential Representation DS. The second key idea of our algorithm is to divide the dynamic programming table D into "boxes" that are defined by intersections of maximal runs of A and B. Note that D contains $M \times N$ such boxes. Let $\mathrm{rle}(A) = A_1^{k_1} \ldots A_M^{k_M}$ and $\mathrm{rle}(B) = B_1^{l_1} \ldots B_N^{l_N}$ be the RLEs of A and B. Let $i_T^I = \sum_i^{I-1} k_i + 1$, $i_B^I = \sum_i^I k_i$, $j_L^J = \sum_j^{J-1} l_j + 1$, and $j_R^J = \sum_j^J l_j$. We define a sparse table DS for DR that consists only of the rows and columns on the borders of the maximal runs in A and B. Namely, DS is a sparse table that only stores the rows i_T^I, i_B^I ($1 \leq I \leq M$) and the columns j_L^J, j_R^J ($1 \leq J \leq N$), of DR (see Fig. 3). Each row and column of DS is implemented by a linked list as follows: each cell $DS[i,j]$ has four links to the upper, lower, left, and right neighbors in DS (if these neighbors exist), plus a diagonal link to the right-lower direction. This diagonal link from $DS[i,j]$ points to the first cell $DS[i+h, j+h]$ that is reached by following the right-lower diagonal path from $DS[i,j]$, namely, $h \geq 0$ is the smallest integer such that $i + h = i_B^I$ or $j + h = j_L^J$. Clearly DS occupies $\Theta(mN + nM)$ space. DS can answer $\mathrm{dtw}(A, B) = D[m,n]$ in $O(m+n)$ time by tracing $O(m+n)$ cells of DS from $(1,1)$ to (m,n).

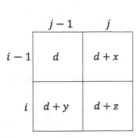

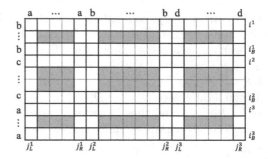

Fig. 2. Illustration for Lemma 2 which depicts the corresponding cells of the dynamic programming table D, where $D[i-1, j-1] = d$, $D[i-1, j] = d+x$, $D[i, j-1] = d+y$, and $D[i, j] = d+z$.

Fig. 3. Illustration for DS that consists only of the cells of DR corresponding to the maximal run boundaries of A and B (white rows and columns). The gray regions that are surrounded by the box boundaries are not stored in DS.

For each $1 \leq I < M$ and $1 \leq J < N$, we consider the region of DR that is surrounded by the borders of the Ith and $(I+1)$th runs of A, and the Jth and $(J+1)$th runs of B. This region is called a *box* for I, J, and is denoted by $\mathcal{B}^{I,J}$. For ease of description, we will sometimes refer to a box $\mathcal{B}^{I,J}$ also in D and DS.

3.1   Updating DS After an Edit Operation

Suppose that an edit operation has been performed at position j^* of string B and let B' denote the edited string. Let D' denote the dynamic programming table for $\text{dtw}(A, B')$, and DR' the difference representation for D'. As Fig. 4 shows, the number of changed cells in DR' can be much smaller than that of changed cells in D' (see also Fig. 1).

	a	c	b	e	e	a	a	d
d	-,-	1,-	4,-	1,-	1,-	9,-	9,-	0,-
c	-,4	-4,-1	1,4	4,-1	4,2	2,-5	4,-10	1,-9
b	-,1	-4,1	-1,-1	9,4	5,5	-4,-1	1,-4	4,-1
b	-,1	-4,1	-2,0	9,0	9,4	-7,1	0,0	4,0
c	-,4	-8,0	-1,1	3,-5	4,-10	4,1	3,4	-3,-3
c	-,4	-12,0	0,1	3,1	3,0	4,0	4,1	-3,1
d	-,9	-20,1	3,4	-3,-2	1,-4	9,1	8,5	-8,0
a	-,0	-16,4	-3,-2	15,16	0,15	-15,-9	0,-17	9,0

		c	b	e	e	a	a	d
d		-,-	4,-	1,-	1,-	9,-	9,-	0,-
c		-,0	1,-3	4,0	4,3	1,-5	4,-10	1,-9
b		-,1	-1,-1	9,4	5,5	-4,0	1,-3	4,0
b		-,1	-2,0	9,0	9,4	-7,1	0,0	4,0
c		-,0	-1,1	3,-5	4,-10	4,1	3,4	-3,-3
c		-,0	0,1	3,1	3,0	4,0	4,1	-3,1
d		-,1	3,4	-3,-2	1,-4	9,1	8,5	-8,0
a		-,4	-3,-2	15,16	0,15	-15,-9	0,-17	9,0

Fig. 4. For the running example from Fig. 1, only the gray cells have different values in the difference representations DR (left) and DR' (right).

Let DS' denote the sparse table for DR'. Since DS consists only of the boundary cells, the number of changed cells in DS' can even be much smaller. In what follows, we show how to efficiently update DS to DS'.

Because the prefix $B[1..j^*-1]$ remains unchanged after the edit operation, for any $j < j^*$ we have $DR[i,j] = DR'[i,j]$ by Lemma 2 and recurrence (1). Hence, we can restrict ourselves to the indices $j \geq j^*$. We define ℓ as a correcting offset of string indices before and after the update: $\ell = -1$ if a character has been inserted at position j^* of B, $\ell = 1$ if a character has been deleted from position j^* of B, and $\ell = 0$ otherwise. Now, for any $j \geq j^*$, $B'[j] = B[j+\ell]$ and column j in DR' corresponds to column $j+\ell$ in DR.

Let $\mathcal{B}^{I,J}$ be any box on DS'. For the the top row i_{T}^I of $\mathcal{B}^{I,J}$, we use a linked list $\Delta_{\mathrm{T}}^{I,J}$ that stores the column indices j ($j_{\mathrm{L}}^J \leq j \leq j_{\mathrm{R}}^J$) such that $DS[i_{\mathrm{T}}^I, j+\ell] \neq DS'[i_{\mathrm{T}}^I, j]$, in increasing order. We also compute, in each element of the list, the value for $D'[i_{\mathrm{T}}^I, j]$ of the corresponding column index j. We use similar lists $\Delta_{\mathrm{B}}^{I,J}$, $\Delta_{\mathrm{L}}^{I,J}$, and $\Delta_{\mathrm{R}}^{I,J}$ for the bottom row, left column, and right column of $\mathcal{B}^{I,J}$, respectively. We compute these lists when an edit operation is performed to string B, and use them to update DS to DS' efficiently.

Let $\#_{\mathrm{chg}}$ denote the number of cells in our sparse representation such that $DS[i+\ell, j] \neq DS'[i,j]$. In the sequel, we prove:

Theorem 1. *Our* $\mathbf{D^2TW}$ *algorithm updates DS to DS' in $O(m+n+\#_{\mathrm{chg}})$ time.*

Initial Step. Suppose that j^* is in the Jth run of string B. Let $\mathcal{B}^{I,J}$ be any of the M boxes of DR that contain column j^*, where $j_{\mathrm{L}}^J \leq j^* \leq j_{\mathrm{R}}^J$. Due to Lemma 2, $(1, j^*)$ is the only cell in the first row where we may have $DS'[1, j^*] \neq DS[1, j^*+\ell]$. $DS'[1, j^*]$ can be easily computed in $O(1)$ time by Lemma 2. Then, $D'[1, j^*]$ can be computed in $O(j^*) \subseteq O(n)$ time by tracing the first row and using $DS'[1,j].L$ for increasing $j = 1, \ldots, j^*$. The list $\Delta_{\mathrm{T}}^{I,J}$ only contains j^* (coupled with $D'[1, j^*]$) if $DS'[1, j^*] \neq DS[1, j^* + \ell]$, and it is empty otherwise.

Editing string B at position j^* incurs some structural changes to DS: (a) $\mathcal{B}^{I,J}$ gets wider by one (insertion of the same character to a run), (b) $\mathcal{B}^{I,J}$ gets narrower by one (deletion of a character), (c) $\mathcal{B}^{I,J}$ is divided into $2M$ or $3M$ boxes (insertion of a different character to a run, or character substitution).

In cases (a) and (b), the diagonal links of $\mathcal{B}^{I,J}$ need to be updated. A crucial observation is that the total number of such diagonal links to update is bounded by m for all the M boxes $\mathcal{B}^{1,J}, \ldots, \mathcal{B}^{M,J}$, since the destinations of such diagonal links are within the same column of DS' ($j_{\mathrm{R}}^J + 1$ in case (a), and $j_{\mathrm{R}}^J - 1$ in case (b)). For each box $\mathcal{B}^{I,J}$, if $j_{\mathrm{R}}^J - j_{\mathrm{L}}^J \geq i_{\mathrm{T}}^I - i_{\mathrm{B}}^I$ (i.e. $\mathcal{B}^{I,J}$ is a square or a horizontal rectangle), then we scan the top row i_{T}^I from right to left and fix the diagonal links until encountering the first cell in i_{T}^I whose diagonal link needs no updates (see Fig. 5). The case with $j_{\mathrm{R}}^J - j_{\mathrm{L}}^J < i_{\mathrm{T}}^I - i_{\mathrm{B}}^I$ (i.e. $\mathcal{B}^{I,J}$ is a vertical rectangle) can be treated similarly. By the above observation, these costs for all boxes $\mathcal{B}^{I,J}$ that contain the edit position j^* sum up to $O(m)$.

In case (a), we shift the right column j_{R}^J of DS to the right by one position, and reuse it as the right column $j_{\mathrm{R}}^J + 1$ of DS'. This incurs two new cells $(i_{\mathrm{T}}^I, j_{\mathrm{R}}^J)$ and $(i_{\mathrm{B}}^I, j_{\mathrm{R}}^J)$ in DS' (the gray cells in Fig. 5). We can compute $DS'[i_{\mathrm{T}}^I, j_{\mathrm{R}}^J]$ in $O(1)$

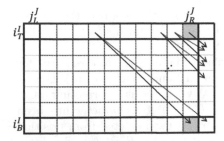

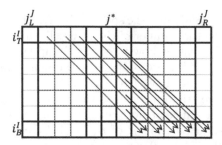

Fig. 5. Case (a) of the initial step. The dashed arcs are the old diagonal links in DS, and the sold arcs are the modified diagonal links in DS'. The gray cells depict cells $(i_\mathrm{T}^I, j_\mathrm{R}^J)$ and $(i_\mathrm{B}^I, j_\mathrm{R}^J)$.

Fig. 6. Case (c) of the initial step, where character substitution has been performed at position j^*. The dashed arcs are the old diagonal links in DS from row i_T^I up to j^*, and the sold arcs are the modified diagonal links from new column j^* in DS'.

time using Lemma 2. Now consider to compute $DS'[i, j_\mathrm{R}^J + 1]$ for the new right column. Since this right column initially stores $DS[i, j_\mathrm{R}^J]$ for the old DS, using Lemma 2, we can compute $DS'[i, j_\mathrm{R}^J + 1]$ in increasing order of $i = 1, \ldots, m$, from top to bottom, in $O(1)$ time each. We can compute $D'[1, j_\mathrm{R}^J + 1]$ in $O(j_\mathrm{R}^J)$ time by simply scanning the first row. Then, we can compute $D'[i, j_\mathrm{R}^J + 1]$ for increasing $i = 2, \ldots, m$, using $DS'[i, j_\mathrm{R}^J + 1]$, and construct $\Delta_\mathrm{R}^{I,J}$. This takes a total of $O(j_\mathrm{R}^J + m) \subseteq O(m + n)$ time. Finally, $DS'[i_\mathrm{B}^I, j_\mathrm{R}^J]$ is computed from $D'[i_\mathrm{B}^I, j_\mathrm{R}^J + 1]$ and $DS'[i_\mathrm{B}^I, j_\mathrm{R}^J + 1].L$ in $O(1)$ time. Case (b) can be treated similarly.

For case (c), we consider a sub-case where a character substitution was performed completely inside a run of B, at position j^*. This divides an existing box $\mathcal{B}^{I,J}$ into three boxes $\mathcal{B}^{I,J}$, $\mathcal{B}^{I,J+1}$, and $\mathcal{B}^{I,J+2}$. Thus, there appear three new columns $j^* - 1$, j^*, and $j^* + 1$ in DS'. Then, the diagonal links for these new columns can be computed in $O(1)$ time each, by scanning row i_T^I from $j^* + 1$, from right to left (see Fig. 6). The DS' values for the cells in these new columns, as well as the D' values for column $j^* + 1$, can also be computed in similar ways to cases (a) and (b). The other sub-cases of (c) can be treated similarly.

Updating Cells on Row i_T^I and Column j_L^J. In what follows, suppose that we are given a box $\mathcal{B}^{I,J}$ to the right of the edit position j^*, in which some boundary cell values may have to be updated. For ease of exposition, we will discuss the simplest case with substitution where the column indices do not change between DS and DS'. The cases with insertion/deletion can be treated similarly by considering the offset value ℓ appropriately.

Now our task is to quickly detect the boundary cells (i, j) of $\mathcal{B}^{I,J}$ such that $DS[i, j] \neq DS'[i, j]$, and to update them. We assume that the boundary cell values of the preceding boxes $\mathcal{B}^{I-1,J}$ and $\mathcal{B}^{I,J-1}$ have already been computed.

We consider how to detect the cells on the top boundary row i_T^I and the cells on the left boundary column j_L^J of box $\mathcal{B}^{I,J}$ that need to be updated, and how to update them. For this sake, we use the following lemma on the values of DR, which is immediate from Lemma 2:

Lemma 3. *Let $1 \le i \le m$ and $1 \le j \le n$. Suppose that for any cell (i', j') with $i' < i$ or $j' < j$, the value of $DR'[i', j']$ has already been computed. If $DR[i, j] \ne DR'[i, j]$, then $DR[i, j-1].U \ne DR'[i, j-1].U$ or $DR[i-1, j].L \ne DR'[i-1, j].L$.*

Intuitively, Lemma 3 states that the cell (i, j) such that $DR[i, j] \ne DR'[i, j]$ must be propagated from its left neighbor or its top neighbor. We use this lemma for updating the boundaries of each box $\mathcal{B}^{I,J}$ stored in DS. Recall that the values on the preceding row $i_{\mathrm{T}}^I - 1 = i_{\mathrm{B}}^{I-1}$ and on the preceding column $j_{\mathrm{L}}^J - 1 = j_{\mathrm{R}}^{J-1}$ have already been updated. Then, the cells on i_{T}^I and j_{L}^J of box $\mathcal{B}^{I,J}$ with $DS[i, j] \ne DS'[i', j']$ can be found in constant time each, from the lists $\Delta_{\mathrm{B}}^{I-1,J}$ and $\Delta_{\mathrm{R}}^{I,J-1}$ maintained for the preceding row $i_{\mathrm{T}}^I - 1 = i_{\mathrm{B}}^{I-1}$ and preceding column $j_{\mathrm{L}}^J - 1 = j_{\mathrm{R}}^{J-1}$, respectively.

We process column indices $\Delta_{\mathrm{B}}^{I-1,J}$ in increasing order, and suppose that we are currently processing column index $\hat{j} \in \Delta_{\mathrm{B}}^{I-1,J}$ in the bottom row i_{B}^{I-1} of the preceding box $\mathcal{B}^{I-1,J}$. According to the above arguments, this indicates that the cells (i_{T}^I, j) in the top row i_{T}^I of $\mathcal{B}^{I,J}$ that need to be updated (i.e., $DS[i_{\mathrm{T}}^I, j] \ne DS'[i_{\mathrm{T}}^I, j]$). We assume that, for any j' with $j_{\mathrm{L}}^J \le j' < \hat{j}$, the value of $DS'[i_{\mathrm{T}}^I, j']$ has already been computed. Also, we have maintained a partial list for $\Delta_{\mathrm{T}}^{I,J}$ where the last element of this partial list stores the largest j'' such that $j_{\mathrm{L}}^J \le j'' < \hat{j}$ and $DS[i_{\mathrm{T}}^I, j''] \ne DS'[i_{\mathrm{T}}^I, j'']$, together with the value of $D'[i_{\mathrm{T}}^I, j'']$. Now it follows from Lemma 2 that both $DS'[i_{\mathrm{T}}^I, \hat{j}].U$ and $DS'[i_{\mathrm{T}}^I, \hat{j}].L$ can be respectively computed in constant time from $DS'[i_{\mathrm{T}}^I - 1, \hat{j}].L$ and $DS'[i_{\mathrm{T}}^I, \hat{j}-1].U$, and thus we can check whether $DS[i_{\mathrm{T}}^I, \hat{j}] \ne DS'[i_{\mathrm{T}}^I, \hat{j}]$ in constant time as well. In case $DS[i_{\mathrm{T}}^I, \hat{j}] \ne DS'[i_{\mathrm{T}}^I, \hat{j}]$, we append $\hat{j}$ to the partial list for $\Delta_{\mathrm{T}}^{I,J}$. By the definition of DS, we have $D'[i_{\mathrm{T}}^I, \hat{j}] = D'[i_{\mathrm{T}}^I - 1, \hat{j}] - DS'[i_{\mathrm{T}}^I, \hat{j}].U$. Since $D'[i_{\mathrm{T}}^I - 1, \hat{j}] = D'[i_{\mathrm{B}}^{I-1}, \hat{j}]$ is stored with the current column index $\hat{j}$ in the list $\Delta_{\mathrm{B}}^{I-1,J}$, $D'[i_{\mathrm{T}}^I, \hat{j}]$ can also be computed in constant time.

Suppose we have processed cell $(i_{\mathrm{T}}^I, \hat{j})$. We perform the same procedure as above for the right-neighbor cells $(i_{\mathrm{T}}^I, \hat{j} + p)$ with $p = 1$ and increasing p, until encountering the first cell $(i_{\mathrm{T}}^I, \hat{j} + p)$ such that (1) $DS[i_{\mathrm{T}}^I, \hat{j} + p] = DS'[i_{\mathrm{T}}^I, \hat{j} + p]$, (2) $\hat{j} + p \in \Delta_{\mathrm{B}}^{I-1,J}$, or (3) $\hat{j} + p = j_{\mathrm{R}}^J + 1$. In cases (1) and (2), we move on to the next element of in $\Delta_{\mathrm{B}}^{I-1,J}$, and perform the same procedure as above. We are done when we encounter case (3) or $\Delta_{\mathrm{B}}^{I-1,J}$ becomes empty. The total number of cells $(i_{\mathrm{T}}^I, \hat{j} + p)$ for all boxes in DS' is bounded by $\#_{\mathrm{chg}}$.

In a similar way, we process row indices $\Delta_{\mathrm{R}}^{I,J-1}$ in increasing order, update the cells on the left column j_{L}^J, and maintain another partial list for $\Delta_{\mathrm{L}}^{I,J}$.

Updating Cells on Row i_{B}^I and Column j_{R}^J. Let us consider how to detect the cells on the bottom row i_{B}^I and the cells on the right column j_{R}^J of box $\mathcal{B}^{I,J}$ that need to be updated, and how to update them.

The next lemma shows monotonicity on the values of D inside each $\mathcal{B}^{I,J}$.

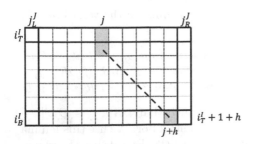

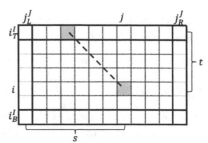

Fig. 7. Diagonal propagation of $DR[i, j] \neq DR'[i, j]$ inside box $\mathcal{B}^{I,J}$.

Fig. 8. Illustration for the case where $s > t$ in Lemma 6.

Lemma 4 [4]. *For any (i, j) with $1 \leq i \leq m$ and $j_L^J < j \leq j_R^J$, $D[i, j] \geq D[i, j-1]$. For any (i, j) with $i_T^I < i \leq i_B^I$ and $1 \leq j \leq n$, $D[i, j] \geq D[i-1, j]$.*

The next corollary is immediate from Lemma 4.

Corollary 1. *For any cell (i, j) with $1 \leq i \leq m$ and $j_L^J < j \leq j_R^J$, $DR[i, j].L \geq 0$. Also, for any cell (i, j) with $i_T^I < i \leq i_B^I$ and $1 \leq j \leq n$, $DR[i, j].U \geq 0$.*

Now we obtain the next lemma, which is a key to our algorithm.

Lemma 5. *For any cell (i, j) with $i_T^I + 1 < i \leq i_B^I$ and $j_L^J + 1 < j \leq j_R^J$, $DR[i, j] = DR[i-1, j-1]$.*

Proof. By Corollary 1, $DR[i-1, j].L \geq 0$ and $DR[i, j-1].U \geq 0$ for $i_T^I+1 < i \leq i_B^I$ and $j_L^J + 1 < j \leq j_R^J$. Thus clearly $\min\{DR[i-1, j].L, DR[i, j-1].U, 0\} = 0$. Therefore, for the value of z in Lemma 2, we have $z = (a_i - b_j)^2$, which leads to

$$DR[i, j].U = (a_i - b_j)^2 - DR[i-1, j].L \tag{2}$$
$$DR[i, j].L = (a_i - b_j)^2 - DR[i, j-1].U \tag{3}$$

By applying Eq. (3) to the $DR[i-1, j].L$ term of Eq. (2), we get

$$DR[i, j].U = (a_i - b_j)^2 - ((a_{i-1} - b_j)^2 - DR[i-1, j-1].U).$$

Recall that $a_i = a_{i-1}$, since we are considering cells in the same box $\mathcal{B}^{I,J}$. Thus $DR[i, j].U = DR[i-1, j-1].U$. By applying Eq. (2) to the $DR[i, j-1].U$ term of Eq. (3), we similarly obtain $DR[i, j].L = DR[i-1, j-1].L$.   □

For any $i_T^I + 1 < i \leq i_B^I$ and $j_L^J + 1 < j \leq j_R^J$, let ℓ be the smallest positive integer that satisfies $i - \ell = i_T^I + 1$ or $j - \ell = j_L^J + 1$. By Lemma 5, for any cell (i, j) on the bottom row i_B^I or on the right column j_R^J, we have $DS[i, j] = DR[i-\ell, j-\ell]$ and $DS'[i, j] = DR'[i-\ell, j-\ell]$. This means that $DS[i, j] \neq DS'[i, j]$ iff $DR[i-\ell, j-\ell] \neq DR'[i-\ell, j-\ell]$. Thus, finding cells (i, j) with $DS[i, j] \neq DS'[i, j]$ on the bottom row i_B^I or on the right column j_R^J reduces to finding cells (i', j') with $DR[i', j'] \neq DR'[i', j']$ on the row $i_T^I + 1$ or on the column $j_L^J + 1$. See Fig. 7.

We have shown how to compute $\Delta_T^{I,J}$ for the top row i_T^I and $\Delta_L^{I,J}$ for the left column j_L^J. We here explain how to use $\Delta_T^{I,J}$ (we can use $\Delta_L^{I,J}$ in a symmetric manner). We process column indices in $\Delta_T^{I,J}$ in increasing order, and suppose that we are currently processing column index $\hat{j} \in \Delta_T^{I,J}$ in the top row i_T^I of the current box $\mathcal{B}^{I,J}$. We check whether $DR[i_T^I + 1, \hat{j}] \neq DR'[i_T^I + 1, \hat{j}]$. For this sake, we need to know the values of $DR[i_T^I + 1, \hat{j}]$ and $DR'[i_T^I + 1, \hat{j}]$. Recall that, by Lemma 5, $DR[i_T^I+1, \hat{j}]$ is equal to $DR[i_T^I+1+h, \hat{j}+h]$ ($= DS[i_T^I+1+h, \hat{j}+h]$) on the bottom row i_B^I (if $i_T^I + 1 + h = i_B^I$) or on the right column j_R^J (if $\hat{j} + h = j_R^J$), where $h > 0$. Since the cell $(i_T^I + 1 + h, \hat{j} + h)$ can be retrieved in constant time by the diagonal link from the cell $(i_T^I, \hat{j} - 1)$ on the top row i_T^I, we can compute $DR[i_T^I + 1, \hat{j}]$ in constant time, applying Lemma 5 to the upper-left direction.

Computing $DR'[i_T^I + 1, \hat{j}]$ is more involved. By Lemma 2, we can compute $DR'[i_T^I + 1, \hat{j}]$ from $DR'[i_T^I, \hat{j}].L$ and $DR'[i_T^I + 1, \hat{j} - 1].U$. Since $(i_T^I, \hat{j})$ is on the top row i_T^I, $DR'[i_T^I, \hat{j}].L = DS'[i_T^I, \hat{j}].L$ has already been computed. Consider to compute $DR'[i_T^I + 1, \hat{j} - 1].U$. Since $DR'[i_T^I + 1, \hat{j} - 1].U = D'[i_T^I + 1, \hat{j} - 1] - D'[i_T^I, \hat{j} - 1]$, it suffices to compute $D'[i_T^I, \hat{j} - 1]$ and $D'[i_T^I + 1, \hat{j} - 1]$. By definition, $D'[i_T^I, \hat{j}-1] = D'[i_T^I, \hat{j}] - DR'[i_T^I, \hat{j}].L$. Since $\hat{j} \in \Delta_T^{I,J}$, we can retrieve the value of $D'[i_T^I, \hat{j}]$ from the current element of the list $\Delta_T^{I,J}$, in $O(1)$ time. Since $DR'[i_T^I, \hat{j}].L = DS'[i_T^I, \hat{j}].L$, we can compute $D'[i_T^I, \hat{j} - 1]$ in $O(1)$ time. What remains is how to compute $D'[i_T^I + 1, \hat{j} - 1]$. We use the next lemma.

Lemma 6. *For any cell (i, j) with $i_T^I + 1 < i \leq i_B^I$ and $j_L^J + 1 < j \leq j_R^J$, let $s = j - j_L^J$ and $t = i - i_T^I$. Then,*

$$D[i, j] = D[i_T^I + \max\{t - s, 0\}, j_L^J + \max\{s - t, 0\}] + \min\{s, t\} \cdot (a_i - b_j)^2.$$

Proof. Consider the case where $s > t$. By applying Lemma 4 to recurrence (1), we obtain $D[i, j] = D[i - 1, j - 1] + (a_i - b_j)^2$. Since $a_i = a_{i'}$ and $b_j = b_{j'}$ for $i_T^I < i' < i$ and $j_L^J < j' < j$, by repeatedly applying Lemma 4 to the above equation, we get $D[i, j] = D[i_T^I, j_L^J + (s - t)] + t \cdot (a_i - b_j)^2$. See also Fig. 8. The case $s \leq t$ is similar and we obtain $D[i, j] = D[i_T^I + (t - s), j_L^J] + s \cdot (a_i - b_j)^2$. By merging the two equations for $s > t$ and $s \leq t$, we obtain the desired equation. $\square$

Let $k = \hat{j} - j_L^J$. Since $j_L^J + 1 < \hat{j}$, $k \geq 2$. Since $s = \hat{j} - 1 - j_L^J = k - 1$, $t = i_T^I + 1 - i_T^I = 1$, and $k \geq 2$, we get $s \geq t$. Thus it follows from Lemma 6 that

$$D'[i_T^I+1, \hat{j}-1] = D'[i_T^I, j_L^J+(k-2)] + (A[i_T^I]-B[\hat{j}])^2 = D'[i_T^I, \hat{j}-2] + (A[i_T^I]-B[\hat{j}])^2.$$

Since the value $D'[i_T^I, \hat{j}]$ is already computed and stored in the corresponding element of $\Delta_T^{I,J}$, we can compute, in $O(1)$ time, $D'[i_T^I, \hat{j} - 2]$ by

$$
\begin{aligned}
D'[i_T^I, \hat{j} - 2] &= D'[i_T^I, \hat{j}] - DR'[i_T^I, \hat{j}].L - DR'[i_T^I, \hat{j} - 1].L \\
&= D'[i_T^I, \hat{j}] - DS'[i_T^I, \hat{j}].L - DS'[i_T^I, \hat{j} - 1].L.
\end{aligned}
$$

Thus, we can determine in $O(1)$ time whether $DR[i_T^I + 1, \hat{j}] \neq DR'[i_T^I + 1, \hat{j}]$, and hence whether $DS[i_T^I + 1 + h, \hat{j} + h] \neq DS'[i_T^I + 1 + h, \hat{j} + h]$.

Suppose $DS[i_{\mathrm{T}}^I + 1 + h, \hat{j} + h] \neq DS'[i_{\mathrm{T}}^I + 1 + h, \hat{j} + h]$. Then we need to compute $D'[i_{\mathrm{T}}^I + 1 + h, \hat{j} + h]$. This can be computed in constant time using Lemma 6, by $D'[i_{\mathrm{T}}^I + 1 + h, \hat{j} + h] = D'[i_{\mathrm{T}}^I, \hat{j} - 1] + (h + 1) \cdot (A[i_{\mathrm{T}}^I] - B[\hat{j}])^2$, where $D'[i_{\mathrm{T}}^I, \hat{j} - 1] = D'[i_{\mathrm{T}}^I, \hat{j}] - DR'[i_{\mathrm{T}}^I, \hat{j}].L$. We add the column index $\hat{j} + h$ to list $\Delta_{\mathrm{B}}^{I,J}$ if $i_{\mathrm{T}}^I + 1 + h = i_{\mathrm{B}}^I$, and/or add the row index $i_{\mathrm{T}}^I + 1 + h$ to list $\Delta_{\mathrm{R}}^{I,J}$ if $\hat{j} + h = j_{\mathrm{R}}^J$, together with the value of $D'[i_{\mathrm{T}}^I + 1 + h, \hat{j} + h]$.

The above process of computing $DR'[i_{\mathrm{T}}^I + 1, \hat{j}]$ is illustrated in Fig. 9. Suppose we have processed cell $(i_{\mathrm{T}}^I + 1, \hat{j})$. We perform the same procedure as above for the right-neighbor cells $(i_{\mathrm{T}}^I + 1, \hat{j} + q)$ with $q = 1$ and increasing q, until encountering the first cell $(i_{\mathrm{T}}^I + 1, \hat{j} + q)$ such that (1) $DR[i_{\mathrm{T}}^I + 1, \hat{j} + q] = DR'[i_{\mathrm{T}}^I + 1, \hat{j} + q]$, (2) $\hat{j} + q \in \Delta_{\mathrm{T}}^{I,J}$, or (3) $\hat{j} + q = j_{\mathrm{R}}^J + 1$. In cases (1) and (2), we remove $\hat{j}$ from $\Delta_{\mathrm{T}}^{I,J}$ and move to the next element of in $\Delta_{\mathrm{T}}^{I,J}$. We are done when we encounter case (3) or $\Delta_{\mathrm{T}}^{I,J}$ becomes empty. By Lemma 5, the total number of cells $(i_{\mathrm{T}}^I + 1, \hat{j} + q)$ for all boxes in DS' is $O(\#_{\mathrm{chg}})$.

Batched Updates. Our algorithm can efficiently support *batched updates* for insertion, deletion, substitution of a run of characters.

Theorem 2. *Let B' be the string after a run-wise edit operation on B, and let $n' = |B'|$. DS can be updated to DS' in $O(m + \max\{n, n'\} + \#'_{\mathrm{chg}})$ time where $\#'_{\mathrm{chg}}$ denotes the number of cells where the values differ between DS and DS'.*

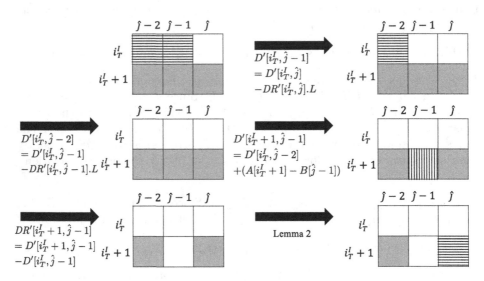

Fig. 9. Illustration for the process of computing $DR'[i_{\mathrm{T}}^I + 1, \hat{j}]$. The gray cells show those for which both values of D' and DR' are unknown, the vertically striped cells show those for which only the value of D' is known, the horizontally striped cells show those for which only the value of DR' is known, and the white cells show those for which both values of D' and DR' are known. At the final step (lower-right), the desired value $DR'[i_{\mathrm{T}}^I + 1, \hat{j}]$ has been computed.

Since n' is the length of the string $|B'|$ after modification, $\#'_{chg}$ in Theorem 2 is bounded by $O(mN + \max\{n', n\}M)$. Thus, we can perform a batched run-wise update on our sparse table DS in worst-case $O(m + \max\{n, n'\} + \#'_{chg}) \subseteq O(mN + \max\{n, n'\}M)$ time. Let k be the total number of characters that are involved in a run-wise batched edit operation from B to B' (namely, a run of k characters is inserted, a run of k characters is deleted, or a run of k_1 characters is substituted for a run of k_2 characters with $k = k_1 + k_2$). Then a naïve k-time applications of Theorem 1 to the run-wise batched edit operation requires $O(k(m + n + \#_{chg})) \subseteq O(k(mN + nM))$ time. Since $n' \leq n + k$, the batched update of Theorem 2 is faster than the naïve method by a factor of k whenever $k \in O(n)$. We also remark that our batched update algorithm is at least as efficient as building the sparse DP table of Froese et al.'s algorithm [4] from scratch using $\Theta(mN + \max\{n, n'\}M)$ time and space.

3.2   Evaluation of $\#_{chg}$

As was proven previously, our **D²TW** algorithm works in $O(m + n + \#_{chg})$ time per edit operation on one of the strings. In this subsection, we analyze how large

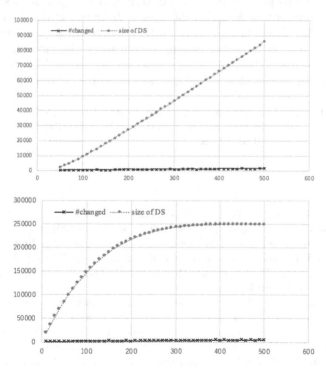

Fig. 10. Comparisons of the values of $\#_{chg}$ and the sizes of the sparse table DS on two randomly generated strings A and B. Upper: With fixed RLE size $N = M = 50$ and varying lengths $n = m$ from 50 to 500 (horizontal axis). Lower: With fixed length $n = m = 500$ and varying RLE sizes $N = M$ from 10 to 500 (horizontal axis).

the $\#_{\mathrm{chg}}$ would be in theory and practice. Although $\#_{\mathrm{chg}} = \Theta(mN + nM)$ in the worst case for some strings (Theorem 3), our preliminary experiments shown below suggest that $\#_{\mathrm{chg}}$ can be much smaller than $mN + nM$ in many cases.

Theorem 3. *Consider strings $A = A_1^k \cdots A_M^k$ and $B = B_1^l \cdots B_N^l$ of RLE sizes M and N, respectively, where $|A| = m = kM$ and $|B| = n = lN$. We assume lexicographical orders of characters as $A_{I-1} > A_I$ for $1 < I \le M$, $B_{J-1} < B_J$ for $1 < J \le N$, and $A_M > B_N$. If we delete $B[1]$ from B, then $\#_{\mathrm{chg}} = \Omega(mN+nM)$.*

We have also conducted preliminary experiments to estimate practical values of $\#_{\mathrm{chg}}$, using randomly generated strings. For simplicity, we set $m = n$ and $M = N$ for all experiments. We fixed the alphabet size $|\Sigma| = 26$ throughout our experiments. In the first experiment, we fixed the RLE size $M = N = 50$, randomly generated two strings A and B of varying lengths $m = n$ from 50 to 500, and compared the values of $\#_{\mathrm{chg}}$ and the sizes of DS. For each m, we randomly generated 50 pairs of strings A and B of length m each, and took the average values for $\#_{\mathrm{chg}}$ and the sizes of DS when $B[1]$ was deleted from B. In the second experiment, we fixed the string length $m = n = 500$ and randomly generated two strings A and B of varying RLE sizes $M = N$ from 10 to 500. For each M, we randomly generated 50 pairs of strings A and B of RLE size M, and took the average values for $\#_{\mathrm{chg}}$ and the sizes of DS when $B[1]$ was deleted from B. The results are shown in Fig. 10. In both experiments, $\#_{\mathrm{chg}}$ is much smaller than the size of DS. It is noteworthy that even when the values of M $(= N)$ and m $(= n)$ are close, the value of $\#_{\mathrm{chg}}$ stayed very small. This suggests that our algorithm can be fast also on strings that are *not* RLE-compressible.

Acknowledgments. This work was supported by JSPS KAKENHI Grant Numbers JP18K18002 (YN), JP17H01697 (SI), JP20H04141 (HB), JP18H04098 (MT), and JST PRESTO Grant Number JPMJPR1922 (SI).

References

1. Abboud, A., Backurs, A., Williams, V.V.: Tight hardness results for LCS and other sequence similarity measures. In: FOCS 2015, pp. 59–78 (2015)
2. Bringmann, K., Künnemann, M.: Quadratic conditional lower bounds for string problems and dynamic time warping. In: FOCS 2015, pp. 79–97 (2015)
3. Charalampopoulos, P., Kociumaka, T., Mozes, S.: Dynamic string alignment. In: CPM 2020, pp. 9:1–9:13 (2020)
4. Froese, V., Jain, B.J., Rymar, M., Weller, M.: Fast exact dynamic time warping on run-length encoded time series. CoRR abs/1903.03003 (2020)
5. Hyyrö, H., Inenaga, S.: Compacting a dynamic edit distance table by RLE compression. In: Freivalds, R.M., Engels, G., Catania, B. (eds.) SOFSEM 2016. LNCS, vol. 9587, pp. 302–313. Springer, Heidelberg (2016). https://doi.org/10.1007/978-3-662-49192-8_25
6. Hyyrö, H., Inenaga, S.: Dynamic RLE-compressed edit distance tables under general weighted cost functions. Int. J. Found. Comput. Sci. **29**(4), 623–645 (2018)

7. Hyyrö, H., Narisawa, K., Inenaga, S.: Dynamic edit distance table under a general weighted cost function. In: van Leeuwen, J., Muscholl, A., Peleg, D., Pokorný, J., Rumpe, B. (eds.) SOFSEM 2010. LNCS, vol. 5901, pp. 515–527. Springer, Heidelberg (2010). https://doi.org/10.1007/978-3-642-11266-9_43
8. Hyyrö, H., Narisawa, K., Inenaga, S.: Dynamic edit distance table under a general weighted cost function. J. Discret. Algorithms **34**, 2–17 (2015)
9. Kim, S.R., Park, K.: A dynamic edit distance table. J. Discret. Algorithms **2**, 302–312 (2004)
10. Kuszmaul, W.: Dynamic time warping in strongly subquadratic time: algorithms for the low-distance regime and approximate evaluation. In: ICALP 2019, pp. 80:1–80:15 (2019)
11. Nishi, A., Nakashima, Y., Inenaga, S., Bannai, H., Takeda, M.: Towards efficient interactive computation of dynamic time warping distance. CoRR abs/2005.08190 (2020). https://arxiv.org/abs/2005.08190
12. Sakoe, H., Chiba, S.: Dynamic programming algorithm optimization for spoken word recognition. IEEE Trans. Acoust. Speech Signal Process. **26**(1), 43–49 (1978)
13. Schmidt, J.P.: All highest scoring paths in weighted grid graphs and their application in finding all approximate repeats in strings. SIAM J. Comput. **27**(4), 972–992 (1998)

Smaller Fully-Functional Bidirectional BWT Indexes

Djamal Belazzougui[1] and Fabio Cunial[2,3(✉)]

[1] CAPA, DTISI, Centre de Recherche sur l'Information Scientifique et Technique, Algiers, Algeria
dbelazzougui@cerist.dz
[2] Max Planck Institute for Molecular Cell Biology and Genetics (MPI-CBG), Dresden, Germany
cunial@mpi-cbg.de
[3] Center for Systems Biology Dresden (CSBD), Dresden, Germany

Abstract. Burrows-Wheeler indexes that support both extending and contracting any substring of the text T of length n on which they are built, in any direction, provide substantial flexibility in traversing the text and can be used to implement several algorithms. The practical appeal of such indexes is contingent on them being compact, and current designs that are sensitive to the compressibility of the input take either $O(e + \bar{e})$ words of space, where e and $\bar{e}$ are the number of right and left extensions of the maximal repeats of T, or $O(r \log(n/r) + \bar{r} \log(n/\bar{r}))$ words, where r and $\bar{r}$ are the number of runs in the Burrows-Wheeler transform of T and of its reverse. In this paper we describe a fully-functional bidirectional index that takes $O(m + r + \bar{r})$ words, where m is the number of maximal repeats of T, as well as a variant that takes $O(r + \bar{r})$ words.

Keywords: BWT · Suffix tree · Suffix-link tree · BWT runs · Maximal repeats · Bidirectional index

1 Introduction

Data structures that allow appending characters both to the left and to the right side of any substring of a text are called *bidirectional indexes*, and have been used extensively in bioinformatics. Such indexes are called *fully-functional* if they also support removing characters from both sides: this enables applications that slide over the text a window whose size can change dynamically, and whose position can move in both directions, like variable-order Markov models and bidirectional, variable-order de Bruijn graphs (see [3] for more details on applications). It is well-known that several classes of sequence datasets in post-genome bioinformatics are highly compressible, and that it is desirable to have indexes whose size is sensitive to some measure of compressibility. To date, the only fully-functional bidirectional index that is sensitive to such a measure takes

© Springer Nature Switzerland AG 2020
C. Boucher and S. V. Thankachan (Eds.): SPIRE 2020, LNCS 12303, pp. 42–59, 2020.
https://doi.org/10.1007/978-3-030-59212-7_4

$O(e + \overline{e})$ words of space, where e (respectively, $\overline{e}$) is the number of right (respectively, left) extensions of the *maximal repeats* of the input string T of length n (defined in Sect. 2), and it implements all operations in $O(\log \log n)$ time [3]. Since Theorem 1 in [3] can be applied to any suffix tree representation that provides the necessary operations, one could also set up a fully-functional bidirectional index that takes $O(r \log(n/r) + \overline{r} \log(n/\overline{r}))$ words and answers queries in $O(\log(n/\min\{r, \overline{r}\}) \log \sigma)$ time, where r (respectively, $\overline{r}$) is the number of *runs in the Burrows-Wheeler transform* of T (respectively, of its reverse) and σ is the size of the alphabet, by using two synchronized instances of the run-length compressed suffix tree described in [10].

In this paper we describe a fully-functional bidirectional index that takes $O(m + r + \overline{r})$ words of space, where m is the number of maximal repeats of the input, and that is much simpler than the data structures in [10]. It is well known that $\max\{m, r\} \leq e$, that m/r can be $\Theta(n)$ in families of very compressible strings (like $a^n b$), and that r/m can be $\Theta(n)$ in other families of strings (e.g. when $\sigma \in \Theta(n)$), so there are no dominance relations between m and r that hold for all strings [4,10]. In practice m if often smaller than r, and both are often significantly smaller than e [4]. The query time of such index is $O(\log \log n + t)$, where t can be $O(\max\{h, \overline{h}\} \log \log \sigma)$, $O(\sigma)$, $O(r\sigma/m)$, or $O(\log \sigma / \log(m/(r \log \sigma)))$, depending on design choices (see Sect. 3 for details), and where σ is the size of the alphabet and h (respectively, $\overline{h}$) is the height of the *maximal repeat subgraph* of T (defined in Sect. 2). Such heights are themselves related to the compressibility of the input, and they can be smaller than the height of the full suffix trees in practice. Our index is practical, and a similar setup has already been implemented e.g. in [8]. We then remove the space dependency on m, describing a fully-functional bidirectional index that takes just $O(r + \overline{r})$ words, and that supports queries in $O(H^2 \log \log n)$ time, where $H \geq \max\{h, \overline{h}\}$ is the length of a longest maximal repeat of the text.

Our data structures combine the run-length encoded Burrows-Wheeler transform with pruned versions of the suffix tree topology that are similar to those introduced in [8]. We use runs in the Burrows-Wheeler transform to define a new category of redundant maximal repeats, whose information can be reconstructed from partial isomorphisms between subtrees of the suffix tree connected by sequences of Weiner links.

2   Preliminaries

Let $\Sigma = [1..\sigma]$ be an integer alphabet, let $\# = 0$ be a separator not in Σ, and let $T \in [1..\sigma]^{n-1}$ be a string. We denote with $\overline{W}$ the reverse of a string $W \in [0..\sigma]^*$, i.e. string W written from right to left. A *repeat* is a string that occurs at least twice in T. We call the set of characters $a \in [0..\sigma]$ such that Wa occurs in $T\#$ the *right-extensions of W*, and we call *right-maximal* a repeat with at least two distinct right-extensions. The *right-saturation* of a substring W of T is the shortest right-maximal substring WV of T (where $V \in [1..\sigma]^*$). *Left-extensions*, *left-maximality* and *left-saturation* are defined symmetrically.

It is well-known that T can have at most $n - 1$ right-maximal substrings and at most $n - 1$ left-maximal substrings. A *maximal repeat* is a repeat that is both left- and right-maximal. We denote with m_T the number of maximal repeats of T. A maximal repeat W is called *right-frontier* if it has at least one right-extension Wa that is not left-maximal. A right-frontier maximal repeat W is called *rightmost* if no right-extension Wa with $a \in [0..\sigma]$ is left-maximal. *Left-frontier* and *leftmost* maximal repeats are defined symmetrically.

We denote with ST_T the *suffix tree* of $T\#$, and with $\overline{\mathsf{ST}}_T$ the suffix tree of $\overline{T}\#$. We assume the reader to be already familiar with the basics of suffix trees, which we do not further describe here. We denote by $\ell(v)$ the label of a node v of a suffix tree, and we say that v is the *locus* of all substrings $W[1..k]$ of T where $|\ell(u)| < k \le |\ell(v)|$, u is the parent of v, and $W = \ell(v)$. We use $\mathsf{ST}_T(v)$ (respectively, $\mathsf{ST}_T(W)$) to denote the subtree of ST_T rooted at node v (respectively, rooted at the node w with $\ell(w) = W$), and we indicate that two subtrees of ST_T are isomorphic by writing $\mathsf{ST}_T(v) \simeq \mathsf{ST}_T(w)$ (respectively, $\mathsf{ST}_T(V) \simeq \mathsf{ST}_T(W)$). It is well-known that a substring W of T is right-maximal (respectively, left-maximal) iff $W = \ell(v)$ for some internal node v of ST_T (respectively, for some internal node v of $\overline{\mathsf{ST}}_T$). A *suffix link* is an arc (v, v') such that v and v' are nodes of ST_T and $\ell(v) = a \cdot \ell(v')$ for some $a \in [0..\sigma]$. Suffix links and internal nodes of ST_T form a tree, called the *suffix-link tree* of T and denoted by SLT_T, and inverting the direction of all suffix links yields the so-called *explicit Weiner links*, whose label is the character that each suffix link removed. Given an internal node v and a character $a \in [0..\sigma]$, it might also happen that string $a\ell(v)$ occurs in T but is not right-maximal, i.e. it is not the label of any internal node of ST_T: every such left extension of an internal node that ends in the middle of an edge is called *implicit Weiner link*, and its label is character a. An internal node v of ST_T can have more than one outgoing Weiner link, and all such Weiner links have distinct labels: in this case, $\ell(v)$ is a maximal repeat, as well as the label of a node in $\overline{\mathsf{ST}}_T$. Since left-maximality is closed under prefix operation, the maximal repeats of T are all and only the nodes of ST_T that lie on paths that start from the root and that end at nodes labelled by rightmost maximal repeats: we call this the *maximal repeat subgraph of* ST_T.

We assume the reader to be familiar with the Burrows-Wheeler transform of $T\#$, which we denote with BWT_T (we use $\overline{\mathsf{BWT}}_T$ to denote the BWT of $\overline{T}\#$) and we do not further describe here. A *run of* BWT_T is a maximal substring of BWT_T that contains exactly one distinct character. We denote by r_T and $\overline{r}_T$ the number of runs in BWT_T and $\overline{\mathsf{BWT}}_T$, respectively, and we call *run-length encoded BWT* (RLBWT_T) any representation of BWT_T that takes $O(r_T)$ words of space and that supports the well-known rank and select operations (see e.g. [13,14,18]). It is easy to implement a version of RLBWT_T that supports rank and select in $O(\log \log n)$ time (see e.g. [4] and references therein). Note that the maximal repeat subgraph of ST_T can have unary paths, but a run-length encoding of its balanced parentheses representation takes $O(r_T)$ space, since such unary paths are compressed and every rightmost maximal repeat can be charged to a distinct BWT run. We call *fully-functional bidirectional index* [3] a data structure that, given a constant-space descriptor $\mathrm{id}(W)$ of any substring W of T, supports the following operations: $\mathtt{extendRight}(\mathrm{id}(W), a) = \mathrm{id}(Wa)$ if Wa occurs in T for $a \in [0..\sigma]$, or an error otherwise; $\mathtt{contractRight}(\mathrm{id}(W)) = \mathrm{id}(V)$ if $W = Va$ occurs in T, or an error otherwise; $\mathtt{extendLeft}$ and $\mathtt{contractLeft}$ are defined symmetrically. We consider bidirectional indexes based on the BWT, and we use $\mathrm{id}(W) = (\mathbb{I}(W, T), \mathbb{I}(\overline{W}, \overline{T}), |W|)$, where $\mathbb{I}(W, T)$ is the function that maps a substring W of T to the interval of W in BWT, i.e. to the interval of all suffixes of $T\#$ that start with W. A bidirectional index is called *synchronous* if it updates both $\mathbb{I}(W, T)$ and $\mathbb{I}(\overline{W}, \overline{T})$ after every operation, and *asynchronous* if it updates just one such interval after every operation, but allows to reconstruct the other interval when needed [6]. We call *fully-functional unidirectional index* a data structure that supports extend and contract operations on just one side of W: in this case $\mathrm{id}(W)$ is either $(\mathbb{I}(W, T), |W|)$ or $(\mathbb{I}(\overline{W}, \overline{T}), |W|)$. See e.g. [12,17,19,20] for a sampler of bidirectional and unidirectional indexes that support extension.

A $\mathtt{contractLeft}$ operation from a right-maximal substring of T (i.e. a suffix link) can be implemented using just BWT_T and a compact representation of the topology of ST_T, as follows (see e.g. [5,15]). Let $[i..j]$ be the interval in BWT_T of the source node v of the suffix link, and let $\ell(v) = aW$ where $a \in [0..\sigma]$ and $W \in [0..\sigma]^*$. We convert v to $[i..j]$ by accessing the leftmost and rightmost leaves in the subtree of v. Let aWX and aWY be the suffixes of T that correspond to positions i and j in BWT_T, respectively, for some strings X and Y. Note that the position i' of WX in BWT_T is $\mathtt{select}_a(\mathsf{BWT}_T, i - C[a])$, the position j' of WY in BWT_T is $\mathtt{select}_a(\mathsf{BWT}, j - C[a])$, and W is the longest prefix of the suffixes that correspond to positions i' and j' in $\mathsf{BWT}_T{}^1$. We convert i' and j' to identifiers of leaves in ST_T by selecting the i'-th and j'-th leaves in the topology, and we compute w by taking the lowest common ancestor (LCA) of such leaves. Clearly $j' - i' \geq j - i$, since the $j - i + 1$ contiguous suffixes of interval $[i..j]$ are projected to a set of not necessarily contiguous suffixes inside interval $[i'..j']$, such projected suffixes correspond to all and only the occurrences of character a in $\mathbb{I}(W)$, and $[i'..j']$ either coincides with $\mathbb{I}(W, T)$ or is a subinterval of it (and in the latter case $[i'..j']$ does not coincide with the interval of any node of ST_T). If $j' - i' > j - i$ then W is left-maximal; otherwise, W is not necessarily left-maximal and $[i'..j']$ does not necessarily coincide with a BWT run.

In the rest of the paper we drop T from all subscripts whenever it is clear from the context.

3 Bidirectional Indexes in $O(m + r + \overline{r})$ Space

In this section we describe bidirectional indexes whose space complexity depends on the number of BWT runs, rather than on the number of extensions of maximal repeats. We achieve this by using the following pruning of the suffix tree topology of T, which we denote with $\mathcal{R}_T$ in what follows: (1) For every node v of ST that

---

[1] $\mathtt{select}_a(S, i)$ is the well-known select operation on string S with character a and rank i, and $C[a]$ for $a \in [0..\sigma]$ contains the number of occurrences of all characters smaller than a in lexicographic order.

corresponds to a maximal repeat, we compact into a single node every maximal run of consecutive children of v, in lexicographic order, that are not left-maximal, and such that all their BWT intervals contain the same character (see e.g. Fig. 1). We call *red* such a compacted node w, and we store in field $w.\mathtt{cardinality} \geq 1$ the number of nodes of ST that were compacted into it. (2) We remove from the topology of ST_T all internal nodes that are not left-maximal and that were not compacted into a red node. We denote with $\rho(v)$ the map that projects a node v of ST onto a (possibly red) node v' of $\mathcal{R}$; we set $\rho(v) = v$ for every maximal repeat of T, and for every leaf v that was not compacted into a red node; we set $\rho(v) = \mathtt{null}$ for every internal node that we removed. (3) We connect every leaf v, which was not compacted into a red node, to the deepest node $u' \notin \{v, \mathtt{null}\}$ of $\mathcal{R}$ such that $u' = \rho(u)$ and u is an ancestor of v in ST.

Note that a node v of ST can have multiple, non-consecutive red children whose BWT intervals contain all the same character, and that a rightmost maximal repeat has at least two red children. If T has no maximal repeat of length at least one, the pruning does not change the topology of ST. We denote with $\overline{\mathcal{R}}_T$ the pruned topology built from $\overline{\mathsf{ST}}_T$, and we remove subscripts whenever T is clear from the context. It is easy to see that the suffix link algorithm in Sect. 2 can be implemented using $\mathcal{R}$ rather than the full suffix tree topology:

Lemma 1. *Let T be a string of length n, with m maximal repeats and r BWT runs. There is an index that takes $O(m + r)$ words of space and that implements the unidirectional $\mathtt{contractLeft}(\mathtt{id}(aW)) = \mathtt{id}(W)$ operation on BWT in $O(\log \log n)$ time for any substring W of T.*

Proof. Assume first that aW is right-maximal: we implement $\mathtt{suffixLink}(v)$ from the locus v of aW in ST, using $\mathcal{R}$ rather than ST and replacing BWT with RLBWT. As in Sect. 2, we use $[i..j]$ to denote $\mathbb{I}(aW, T)$, and $[i'..j']$ to denote the projected interval that results from the select operations on i and j. If $j' - i' > j - i$, then W is a maximal repeat, and we return the interval of the LCA between the i'-th and j'-th leaf in $\mathcal{R}$. If $j' - i' = j - i$, we know that all and only the occurrences of a in the BWT interval of W belong to $[i'..j']$, thus we take again the LCA v of the i'-th and j'-th leaves: if v is not red, we return its interval, since this occurs only when $[i'..j']$ straddles the intervals of two children of W, in which case v is the locus of W (Fig. 1a). If v is red, we compute its BWT interval $[i^*..j^*]$: if $[i'..j'] \subset [i^*..j^*]$, then W is not left-maximal, and its locus is either one of the non left-maximal nodes of ST that were merged into v, or it is a descendant of one such node: in both cases, the interval of the locus of W is $[i'..j']$ itself. If $[i'..j'] = [i^*..j^*]$, we return $[i'..j']$ if $v.\mathtt{cardinality} = 1$, otherwise we return the interval of the parent of v in $\mathcal{R}$ (Fig. 1b).

If aW is not right-maximal, we just run the algorithm described in [3] on $\mathcal{R}$ rather than on ST: specifically, we take the suffix link (v, v') from the locus v of aW, as described above; we check whether v' is a maximal repeat and, in the positive case, we issue a weighted level ancestor query from it, where weights are string depths. A data structure that supports such queries on the maximal repeat subgraph of ST takes $O(m)$ words. If v' is not a maximal repeat, we

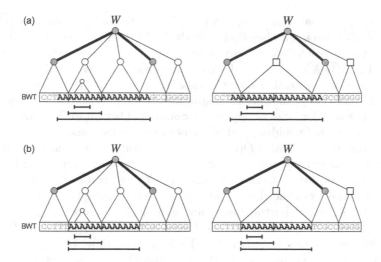

Fig. 1. Illustrating Lemma 1. Circles: nodes in the full suffix tree; gray circles: maximal repeats; white circles: nodes that are not left-maximal; squares: red nodes in $\mathcal{R}$. Intervals below the BWT: possible values of $[i'..j']$ in the lemma. Left: full topology; right: pruned topology $\mathcal{R}$. The occurrences of A in the BWT interval of W can straddle the intervals of left-maximal children of W (a), or they can coincide with the union of the intervals of children of W that are not left-maximal (b).

move to the lowest ancestor v'' of v' that is a maximal repeat, by just taking the parent of the corresponding red node of $\mathcal{R}$, we measure $|\ell(v'')|$ (which we store in every maximal repeat node of $\mathcal{R}$), and if $|\ell(v'')| \geq |W|$, we issue a weighted level ancestor query; otherwise, the locus of W is v'.

Note that a BWT run can be fragmented into multiple red nodes of $\mathcal{R}$, but every red node can be mapped either to a BWT run or to a frontier maximal repeat, in such a way that every frontier maximal repeat and every run is used at most twice in the mapping: thus we can represent $\mathcal{R}$ in $O(m+r)$ words by run-length encoding its balanced parentheses representation. We leave the details of how to represent each data structure to the reader. Here we just mention that we build a predecessor data structure on the sequence of first positions of the intervals in which the BWT is partitioned by red nodes and leaves: such a data structure takes $O(m+r)$ words. The claimed time complexity derives from known bounds on each data structure: see e.g. [2, 4, 9, 11, 16, 22].   □

To implement `contractLeft` in a synchronous index, we also need to compute $\mathbb{I}(\overline{W}, \overline{T})$ from $\mathbb{I}(W, T)$ and $\mathbb{I}(\overline{aW}, \overline{T})$. Note that we already know whether W is left-maximal or right-maximal from Lemma 1. If W is not left-maximal, $\mathbb{I}(\overline{W}, \overline{T}) = \mathbb{I}(\overline{aW}, \overline{T})$. Otherwise, if W is a maximal repeat, we can just move to the lowest maximal-repeat ancestor of $\overline{aW}$ in $\overline{\mathcal{R}}$ and derive its interval. If W is left-maximal but not right-maximal, its locus WX in ST is a maximal repeat, thus its locus in $\mathcal{R}$ is a maximal repeat as well, and we could store $\mathbb{I}(\overline{WX}, \overline{T})$ in such a maximal repeat. Since $\overline{\mathsf{ST}}(\overline{WX}) \simeq \overline{\mathsf{ST}}(\overline{W})$, we could reconstruct $\mathbb{I}(\overline{W}, \overline{T})$ from $\mathbb{I}(\overline{aW}, \overline{T})$ if we knew the offset of $\mathbb{I}(\overline{aWX}, \overline{T})$ with respect to $\mathbb{I}(\overline{WX}, \overline{T})$.

To compute such offset, we would need a unidirectional operation $\texttt{extendLeft}(\text{id}(\overline{WX}), a) = \text{id}(\overline{aWX})$ that works on $\overline{\mathsf{BWT}}$: this is typically implemented using operation $\texttt{countSmaller}(i, j, a)$ on BWT, which returns all characters in $\mathsf{BWT}[i..j]$ that are lexicographically smaller than a, for any choice of i and j. The simplest way to implement $\texttt{countSmaller}$ on RLBWT rather than on BWT is probably by adding partial counts to the runs encoded as σ predecessor data structures: this takes $O(r)$ words of additional space and $O(\sigma \log \log n)$ query time. Alternatively, one could store σ words for every run, containing the result of $\texttt{countSmaller}$ up to that run: this takes $O(r\sigma)$ words of additional space and $O(\log \log n)$ query time. This approach can be generalized to blocks of B consecutive runs, storing σ partial counts for each block and answering a query by scanning a block: this takes $O(\sigma r / B)$ additional words and $O(\log \log n + B)$ query time, which become $O(r)$ words and $O(\log \log n + \sigma)$ time, or $O(r\sigma / \log \log n)$ words and $O(\log \log n)$ time, or $O(m)$ words and $O(\log \log n + r\sigma/m)$ time, by suitable choices of B.

Alternatively, one could create an instance of the weighted 2D orthogonal range counting problem, in which the horizontal dimension is the sequence of runs, the vertical dimension is the alphabet, and the weight of a point is the length of a run: then, a $\texttt{countSmaller}$ query is a rectangle with a given range in the horizontal dimension and with a lower half-space in the vertical dimension. This can be implemented with a range tree on the sequence of runs, in which we additionally store a word for every bit of every bitvector: this takes $O(r \log \sigma)$ additional words, and answers queries in $O(\log \log n + \log \sigma)$ time [1,23]. This approach can be generalized by building a range tree of height i and node degree $2^{\log \sigma / i} = \sigma^{1/i}$: this yields $O(ri\sigma^{1/i})$ words and $O(\log \log n + i)$ time, which become $O(m)$ words and $O(\log \log n + \log \sigma / (\log \frac{m}{r \log \sigma}))$ time (when $m \geq 2r \log \sigma$), or $O(r \cdot \log \log n \cdot 2^{\log \sigma / \log \log n})$ words and $O(\log \log n)$ time.

In the following lemma we show yet another tradeoff that is specific to intervals of maximal repeats, and that uses a pruning strategy similar to $\mathcal{R}$:

Lemma 2. *Let T be a string of length n on alphabet $[1..\sigma]$, with m maximal repeats and $\bar{r}$ runs in $\overline{\mathsf{BWT}}$. There is an index that takes $O(m + \bar{r})$ words of space, and that implements the unidirectional $\texttt{extendLeft}(\text{id}(\overline{W}), b) = \text{id}(\overline{bW})$ operation on $\overline{\mathsf{BWT}}$ in $O(h \log \log \sigma)$ time from any maximal repeat W of T and any $b \in [0..\sigma]$, where h is the height of the maximal repeat subgraph of ST.*

Proof. Since $\overline{W}$ labels a node of $\overline{\mathsf{ST}}$, to compute the interval of $\overline{W}b$ in $\overline{\mathsf{BWT}}$ we could just store a map from character to offset inside the interval of $\overline{W}$ in $\overline{\mathsf{BWT}}$: however, this would take space proportional to the number of left-extensions of all maximal repeats. Instead, we store a map from character to relative first and last position inside the interval of $\overline{W}$, but just for the children of $\overline{W}$ in $\overline{\mathsf{ST}}$ that are maximal repeats, or whose interval in $\overline{\mathsf{BWT}}$ is the beginning of a run: we call such nodes the *marked children* of $\overline{W}$. This takes $O(m + \bar{r})$ words overall. The only problematic case occurs when the intervals of at least two consecutive children of $\overline{W}$ in $\overline{\mathsf{ST}}$ are contained in the same run of $\overline{\mathsf{BWT}}$, since in this case we don't store all their offsets inside the interval of $\overline{W}$. Thus, let $\overline{W}b_0X_0, \overline{W}b_1X_1, \ldots, \overline{W}b_{k-1}X_{k-1}$ with $k \geq 2$ be a maximal sequence of children of $\overline{W}$, in lexicographic order, whose intervals in $\overline{\mathsf{BWT}}$ are all contained in the same run of character a: we call such a sequence a *run of children* of $\overline{W}$. Note that $\overline{W}$ might have other runs of children with the same character a, as well as with different characters, and that a run of children might be preceded or followed by another run of children (for a different character) or by a maximal repeat child of $\overline{W}$. Given a character b_q, we want to compute the sum of the sizes of the intervals of $\overline{W}b_iX_i$ for all $i < q$. We can determine the run of children of $\overline{W}$ to which b_q belongs, by maintaining a predecessor data structure on the marked children of $\overline{W}$.

Clearly $a\overline{W}$ labels a node of $\overline{\mathsf{ST}}$, characters b_i for all $i \in [0..k-1]$ label consecutive children $a\overline{W}b_0X_0', a\overline{W}b_1X_1', \ldots, a\overline{W}b_{k-1}X_{k-1}'$ of $a\overline{W}$ in lexicographic order (with $|X_i'| \geq |X_i|$ for each i), and the intervals of all such children have the same size and relative order as the corresponding children of $\overline{W}$ (see e.g. Fig. 2). However, the characters in the BWT intervals of such children of $a\overline{W}$ are arbitrary, so, for example, a child of $\overline{W}$ might not correspond to any child of $a\overline{W}$ (iff the interval of the child of $\overline{W}$ does not contain character a), $a\overline{W}b_iX_i'$ might become a maximal repeat for some i, and a left-maximal child of $\overline{W}$ might become a child of $a\overline{W}$ that is not left-maximal, so a run of children of $\overline{W}$ might become embedded into a longer run of children of $a\overline{W}$. Note that, if $a\overline{W}$ is not left-maximal, the set of children and their intervals do not change if we add one more character to the left of $a\overline{W}$, thus from now on we assume that we are in the left-saturation $V\overline{W}$ of $\overline{W}$ for some $V \in \Sigma^+$, i.e. that $V\overline{W}$ is a maximal repeat, and that we store pointers from $\overline{W}$ to $V\overline{W}$, which are clearly within our space budget.

We can also afford to store the offset of $V\overline{W}b_0X_0'$ inside the interval of $V\overline{W}$, since we can charge it to a marked child of $\overline{W}$. Thus, we query the map of $V\overline{W}$ for character b_q: if we find it, then we know the offset of the interval of $V\overline{W}b_qX_q'$ inside the interval of $V\overline{W}$, and we can derive its offset with respect to $V\overline{W}b_0X_0'$. Otherwise, we use the predecessor data structure of $V\overline{W}$ to find the closest marked child b_0' of $V\overline{W}$ that is lexicographically smaller than b_q; note that b_0' can be either lexicographically bigger or smaller than b_0, thus we might have to correct the current estimate of the offset with a negative value. Finally, we compute the offset of the interval of $V\overline{W}b_qX_q'$ with respect to the interval of $V\overline{W}b_0'Y'$, by recurring on the run of children of $V\overline{W}$ to which b_0' and b_q belong. Such a recursion corresponds to a descent from W along the maximal repeat subgraph of ST, thus it takes $O(h)$ steps, where h is the height of the subgraph. Recursion must eventually yield the relative offset of character b_q. Assume by contradiction that we are at a rightmost maximal repeat W^*, and that the interval of b_q is still inside a run of children of $\overline{W^*}$, say of character c: then, by left-saturating $c\overline{W^*}$ we would get a maximal repeat that is deeper than W^* in ST, a contradiction.

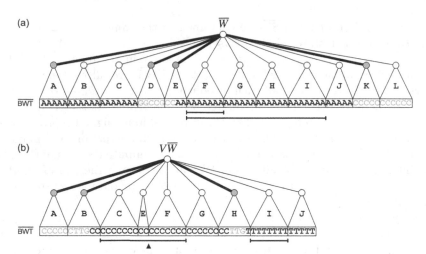

Fig. 2. Illustrating Lemma 2 on the topology of $\overline{\mathsf{ST}}$. Marked children are highlighted with gray nodes and thick edges. For clarity, the first character of a parent-child label is displayed inside the subtree of the child. (a) Assume that we want to know the offsets of children **G** and **J** inside the interval of $\overline{W}$: this reduces to computing the offsets of **G** and **J** with respect to the end of **E**. (b) Recursion on $V\overline{W}$, where the last character of V is **A**. The intervals at the bottom show the results of the recursion. The dark triangle is the position of the end of **E** inside the interval of $V\overline{W}$, which can be stored in the corresponding marked child of $\overline{W}$.

We store the characters of the marked children of every maximal repeat in a deterministic dictionary, which answers queries in constant time, and we build a predecessor data structure on the marked children of every maximal repeat, which answers queries in $O(\log \log \sigma)$ time. All such data structures take $O(m+\overline{r})$ words of space. □

The recursion along Weiner links of Lemma 2 is reminiscent of the recursion along suffix links to reconstruct the label of an edge in the suffix tree or CDAWG [7,21]. Lemmas 1 and 2 are clearly all we need to implement a synchronous, fully functional bidirectional index:

Theorem 1. *Let T be a string of length n on alphabet $[1..\sigma]$, with m maximal repeats, r runs in BWT and $\overline{r}$ runs in $\overline{\mathsf{BWT}}$. Let h (respectively, $\overline{h}$) be the height of the maximal repeat subgraph of the suffix tree of T (respectively, of the suffix tree of $\overline{T}$). There is a fully-functional bidirectional index that takes $O(m + r + \overline{r})$ words of space, and that supports all operations in $O(\log \log n + \max\{h, \overline{h}\} \log \log \sigma)$ time.*

In practice, if the target application never uses strings shorter than a threshold known during construction, one could even prune the top part of all topologies, as noted in [3]. Note that a sequence of k synchronous extendLeft operations can be computed in overall $O(k \log \log n + \min\{k, \overline{h}'\} \cdot h \log \log \sigma)$ time, where $\overline{h'}$ is the height of $\overline{\mathsf{ST}}$, since during such a sequence of operations we need to update the interval in $\overline{\mathsf{BWT}}$ only for left-maximal suffixes.

By symmetry, one might also want a unidirectional implementation of `contractRight` that works on BWT. This can be useful, for example, to implement an *asynchronous bidirectional de Bruijn graph*, i.e. a de Bruijn graph that the user needs to traverse in just one direction (specifically, from right to left) in a session, but which allows switching to the other direction in another session (and the user might afford to store just one BWT and related data structures per session).

Lemma 3. *Let T be a string of length n, with m maximal repeats and r BWT runs. There is an index that takes $O(m + r)$ words of space and that, given the interval in BWT of a string Ua and of its longest left-maximal suffix Va, computes the interval in BWT of all strings resulting from a sequence of c* `contractRight` *operations, and of their longest left-maximal suffixes, in overall $O(\max\{c, |U| - |V|\} \cdot \log\log n)$ time.*

Proof. We store RLBWT, the pruned topology $\mathcal{R}$ described above, and the maximal repeat subgraph of $\overline{\mathsf{ST}}$. In each node of the latter subgraph, we store a pointer to the corresponding node of $\mathcal{R}$, and a deterministic dictionary on the characters that lead to its maximal repeat children; such data structures answer queries in constant time and take $O(m)$ words of space overall. Symmetrically, in each maximal repeat node W of $\mathcal{R}$ we store a pointer to node $\overline{W}$ of the maximal repeat subgraph of $\overline{\mathsf{ST}}$.

Assume WLOG that Va is a proper suffix of Ua. Let W be the maximal repeat locus of Va in ST, and let b be the character that precedes Va in Ua. Clearly $\mathsf{ST}(Ua) \simeq \mathsf{ST}(bW)$. Since we can access the string depth of W and of its parent W' in ST using $\mathcal{R}$, we know whether the locus of V is W or W'. If the locus of V is W, then $\mathbb{I}(U, T) = \mathbb{I}(Ua, T)$, and the longest left-maximal suffix of U is V. Otherwise, we do a top-down traversal of the maximal repeat subgraph of $\overline{\mathsf{ST}}$, starting from $\overline{W'}$ and accessing the characters of U at specific string depths, and stopping at the longest maximal repeat suffix V' of U. Let b' be the character that precedes V' in U, and let $i \leq j$ be, respectively, the starting position of V' and of V inside Ua. We perform the same top-down traversal from $\overline{W}$ as well, stopping at position $i - 1$: in this way, we can compute the offset of $\mathbb{I}(b'U[i..j-1]Va)$ inside $\mathbb{I}(b'V')$, which is the same as the offset of $\mathbb{I}(Ua)$ inside $\mathbb{I}(U)$ since $\mathsf{ST}(b'V') \simeq \mathsf{ST}(U)$. Clearly every operation can be charged to a distinct position in $[i..j-1]$. □

It is also easy to keep up to date the BWT interval of the longest left-maximal suffix W of the current string UbW after every unidirectional `extendLeft` operation on BWT. In a unidirectional de Bruijn graph of fixed order k, the time complexity of moving from one k-mer to the next according to Lemma 3 becomes thus $O(k \cdot \log\log n)$. However, when traversing a large number $N \gg k$ of k-mers, the characters added to the left side are consumed on the right side, thus the cost of moving from one k-mer to the next amortizes to $O(\log\log n)$.

Having the BWT interval of the longest left-maximal suffix W of the current string UbW available at all times, enables also switching to the other direction after any sequence of extendLeft and contractRight operations, by following a pointer from the locus WV of W in ST (a maximal repeat) to the locus of $\overline{WV}$ in $\overline{\text{ST}}$, and taking the child of $\overline{WV}$ labeled with character b. This child might not be left-maximal, so it might have been compressed into a red node of $\overline{\mathcal{R}}$: in this case, we reconstruct its offset inside the interval of $\overline{WV}$ using the technique in Lemma 2. This offset is the same as the offset of $\mathbb{I}(\overline{UbW}, \overline{T})$ inside $\mathbb{I}(\overline{W}, \overline{T})$, thus we just need to compute $\mathbb{I}(\overline{W}, \overline{T})$ by issuing $|V|$ select operations from each end of $\mathbb{I}(\overline{WV}, \overline{T})$. The whole process takes $O(h \log \log \sigma + \lambda \log \log n)$ time, where λ is the maximum string length of an edge of ST that connects two maximal repeats.

4 Bidirectional Indexes in $O(r + \bar{r})$ Space

The pruned topology $\mathcal{R}$ described in Sect. 3 can be compressed further. We call *blue* a maximal repeat W, with $\mathbb{I}(W, T) = [i..j]$, such that: (1) exactly one child WV of W in ST is a maximal repeat: let $\mathbb{I}(WV, T) = [i'..j'] \subset [i..j]$; (2) either $i' = i$, or $\text{BWT}[i..x']$ contains just one distinct character for some $x' \in [i'..j']$ (let x be the largest such x'); and symmetrically (3) either $j' = j$ or $\text{BWT}[y'..j]$ contains just one distinct character for some $y' \in [i'..j']$ (let y be the smallest such y'). We call *black* every maximal repeat that is not blue. Note that all children of a blue node in $\mathcal{R}$ are red, except for one child which is either blue or black. We call $\mathcal{B}$ a version of $\mathcal{R}$ in which: (1) we compact every maximal (unary) path of blue nodes $v_1, \ldots, v_k$ into a single node v^*, which we mark as blue in $\mathcal{B}$, and in which we store $\mathbb{I}(v_1, T)$, $\mathbb{I}(v_k, T)$, $|\ell(v_1)|$, $|\ell(v_k)|$, and the x and y values of v_k defined above; (2) we compact into a single red node, the red children of every v_i that are located on the same side of path $v_1, \ldots, v_k$, and we connect to such a red child of v^* all the leaves that were connected to a corresponding red child of $\mathcal{R}$. We denote with $\beta(v)$ the map that projects a node v of ST onto a (possibly null) node of $\mathcal{B}$.

Note that the procedure that constructs $\mathcal{B}$ does not necessarily compact every unary path of the maximal repeat subgraph of ST, i.e. the subgraph of $\mathcal{B}$ induced by blue and black nodes can still contain unary paths of black nodes. However, every black node in the union of all such unary paths can be charged to a distinct boundary between runs of BWT. Rightmost maximal repeats are mapped to deepest black nodes of $\mathcal{B}$, and they can all be charged to distinct boundaries between BWT runs as well. Thus, the number of black nodes of $\mathcal{B}$ with at least two maximal repeat children is $O(r)$. Every blue node of $\mathcal{B}$ can be charged to a distinct black node of $\mathcal{B}$, thus the total number of internal nodes of $\mathcal{B}$ is $O(r)$. Finally, it is still true that every red node of $\mathcal{B}$ can be mapped either to a BWT run, or to a blue or black node of $\mathcal{B}$, in such a way that every blue and black node and every run is used at most twice in the mapping. Thus, $\mathcal{B}$ with run-length-encoded leaves takes $O(r)$ words of space.

Lemma 1 can be adapted to work on $\mathcal{B}$ rather than on $\mathcal{R}$, by using a recursion similar to Lemma 2:

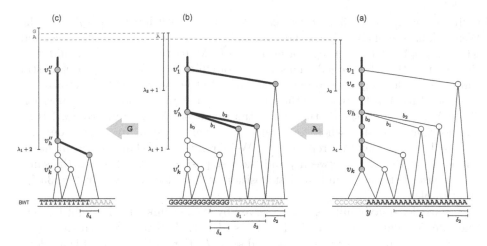

Fig. 3. Illustrating Lemma 4 on the topology of ST. Circles: left-maximal (gray) and non-left-maximal (white) nodes of ST. Thick gray arrows: Weiner links. (a) We want to compute the BWT interval of the highest node with string depth at least λ_1 inside the blue path $v_1, \ldots, v_k$, i.e. we want to compute offset δ_1. (b) We take a Weiner link with character A, we find the lowest left-maximal ancestor v'_h of v'_k, and we read its interval and string depth. Since $\lambda + 1$ is bigger, we add δ_2 to the output and we want to compute δ_3. To do so, we add to the output the number of non-G characters between the end of v'_k and the end of v'_h, and (c) we compute δ_4 by taking another Weiner link from v'_h and v'_k with character G. If our query length were λ_2, we would stop at panel (b).

Lemma 4. *Let T be a string of length n with r BWT runs. There is an index that takes $O(r)$ words of space and that implements the unidirectional* `contractLeft(id(aW))` $=$ `id(W)` *operation on* BWT *in $O(H \log \log n)$ time for any substring W of T, where H is the length of a longest maximal repeat of T.*

Proof. Assume that, after an LCA query of Lemma 1 on $\mathcal{B}$, we reach a blue node v of $\mathcal{B}$: then we have to compute the BWT interval of one of the (unknown) blue nodes $v_1, \ldots, v_k$ of ST such that $\beta(v_i) = v$ for $i \in [1..k]$, and the blue node we want is the lowest ancestor of v_k in ST with string depth at most λ, where λ is known from the contract algorithm. We denote this instance of the problem with the triplet $(\mathbb{I}(v_1), \mathbb{I}(v_k), \lambda)$, which we fully know from $\mathcal{B}$.

Let x, y be the values defined above and stored in v. If none of the nodes $v_1, \ldots, v_k$ has a right-extension whose BWT interval is fully contained inside $[y..j]$, where j is the last position of $\mathbb{I}(v_1)$, then every node in the path has the same last position as v_1 (we can check this by comparing the last position of $\mathbb{I}(v_1)$ to the last position of $\mathbb{I}(v_k)$). Otherwise, consider a generic node v_h in the blue path: let $\mathbb{I}(v_h) = [i..j]$, let b_0 be its only right-extension that leads to another maximal repeat node v_{h+1}, let its right-extensions with characters $b_1, \ldots, b_k$ have BWT intervals fully contained inside $[y..j]$, with $k \geq 1$, and let $\mathrm{BWT}[p] = a$ for all $p \in [y..j]$. If we take the Weiner link in ST from v_h

with character a (Fig. 3a, b), we reach a node v'_h with interval $[i'..j']$ (i.e. the Weiner link is explicit), v'_h has the same right-extensions $b_0, b_1, \ldots, b_k$ as v_h, and $\mathsf{ST}(\ell(v'_h) \cdot b_p) \simeq \mathsf{ST}(\ell(v_h) \cdot b_p)$ for every $p > 0$. Moreover, since we are adding a character to the left of $\ell(v_h)$, the sequence of children $b_0, b_1, \ldots, b_k$ of v'_h is still contiguous in lexicographic order. Thus, the offset between j and the known last position of $\mathbb{I}(v_1)$, is the same as the offset between j' and the known last position of $\mathbb{I}(v'_1)$, where v'_1 is the destination of the (explicit) Weiner link from v_1. If v_h has no right-extension whose interval is fully contained in $[y..j]$ (e.g. because v_h is a leftmost maximal repeat), the Weiner link with character a might be implicit, but the offset between j and the last position of $\mathbb{I}(v_1)$ is still the same as the offset between j' and the last position of $\mathbb{I}(v'_1)$ (see node v_e in Fig. 3a for an example).

Clearly left-maximality is not preserved by Weiner links, i.e. it might happen that the destination nodes of Weiner links $v'_1, \ldots, v'_k$ in ST are such that v'_q is not left-maximal for all $q \in [h+1..k]$ and some h (Fig. 3b). If v'_1 itself is not left-maximal, we recur on instance $(\mathbb{I}(v'_1), \mathbb{I}(v'_k), \lambda + 1)$. By taking more Weiner links from this instance, some left-maximality will eventually appear in the projected path: indeed, there is at least one $p \in [1..k]$ such that v_p has a child whose interval is fully contained in $[y..q]$, where q is the last position of $\mathbb{I}(v_1)$; this implies that v'_p is right-maximal, thus a sequence of Weiner links will eventually produce a maximal repeat when reaching the left-saturation of v'_p. Otherwise, we find v'_h by issuing an LCA query in $\mathcal{B}$ from the ends of $\mathbb{I}(v'_k)$, and by moving to the lowest maximal repeat ancestor w in $\mathcal{B}$ of the node returned by the LCA query (note that the node returned by the LCA query might be red, since v'_k might not be left-maximal). Assume that w is black: we access its string depth (which we store inside every black node) and, if $|\ell(w)| > \lambda + 1$, we move to the highest ancestor u of w in $\mathcal{B}$ with $|\ell(u)| \geq \lambda + 1$ (we set the string depth of a blue node to the string depth of the deepest node of ST that was compacted onto it). If u is black, we know its BWT interval and we return the offset between the end of $\mathbb{I}(u)$ and the end of $\mathbb{I}(v'_1)$. If u is blue, let $u_1, \ldots, u_p$ be the nodes of ST such that $\beta(u_q) = u$ for $q \in [1..p]$. We check whether u_1 is a descendant of node $\beta(v'_1)$ in $\mathcal{B}$, using known BWT intervals: if this is the case, we add to the output the offset between the last position of $\mathbb{I}(u_1)$ and the last position of $\mathbb{I}(v'_1)$, and we recur on instance $(\mathbb{I}(u_1), \mathbb{I}(u_p), \lambda + 1)$. If u is the blue node that contains v'_1, we just recur on instance $(\mathbb{I}(v'_1), \mathbb{I}(u_p), \lambda + 1)$. If $|\ell(w)| < \lambda + 1$, we add to the output the offset between the end of $\mathbb{I}(w)$ and the end of $\mathbb{I}(v'_1)$, as well as the number of occurrences of characters different from c in $\mathsf{BWT}[p + 1..q]$, where p (respectively, q) is the last position of $\mathbb{I}(v'_k)$ (respectively, of $\mathbb{I}(w)$), and c is the only character in $\mathbb{I}(v'_k)$; then, we recur on instance $(\mathbb{I}(w), \mathbb{I}(v'_k), \lambda + 1)$. If w is blue and corresponds to the blue path $w_1, \ldots, w_p$ in ST, we run a similar algorithm using $\mathbb{I}(w_1), \mathbb{I}(w_p), |\ell(w_1)|, |\ell(w_p)|$, as well as a descriptor of the deepest w_q that is an ancestor of v'_k: this information can be stored in the blue node v of $\mathcal{B}$ we are coming from, without affecting the space budget.

It could also happen that the Weiner link by character a projects the blue path $v_1, \ldots, v_k$ of ST onto a subpath of a longer blue path $w'_1, \ldots, v'_1, \ldots, v'_k, \ldots, w'_p$ of ST: this can be easily detected with $\mathcal{B}$, and in this

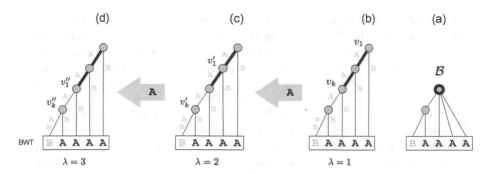

Fig. 4. Running Lemma 4 on string $T =$ AAAAB. (a) Compacted topology. The internal node with thick border is blue, the other is black. (b-d) The suffix tree of T. Gray nodes are maximal repeats; the nodes connected by thick edges are those compacted onto the blue node of $\mathcal{B}$. We want to reconstruct the BWT interval of substring A.

case we just recur on instance $(\mathbb{I}(v_1'), \mathbb{I}(w_p'), \lambda + 1)$. The projection of $v_1, \ldots, v_k$ might also straddle $v_1, \ldots, v_k$ itself (see e.g. Fig. 4), or it might straddle multiple, distinct blue paths; the former case is already handled by our algorithm, and the second can be detected and handled with $\mathcal{B}$ as well: we leave the details to the reader.

Note that the initial blue path $v_1, \ldots, v_k$ cannot be projected onto another blue path by every step of the algorithm: indeed, after at most H Weiner links, the projection of any maximal repeat must have lost its right-maximality, and the fact that a node in some projected path loses its right-maximality after a Weiner link, implies that the projected path did not entirely consist of blue nodes. More generally, let v_p be the node we are trying to compute the interval of, inside the initial blue path $v_1, \ldots, v_k$ of ST. The algorithm considers a sequence of instances $(\mathbb{I}(u), \mathbb{I}(w), q)$ such that the length of the suffix tree path from node u to the projection of v_p by Weiner links is monotonically non-increasing at every step. After at most H Weiner links, the projection of any node in the path $v_1, \ldots, v_p$ of ST is not right-maximal any more, thus the algorithm eventually ends and yields the interval of v_p.

Assume instead that, after an LCA query of Lemma 1 on $\mathcal{B}$ from the initial interval $[i..j]$, we reach a red node of $\mathcal{B}$. If the red node is the child of a black maximal repeat, we already have all the information we need to implement Lemma 1. If the red node is the child of a blue node v, we represent this instance of the problem with the triplet $(\mathbb{I}(v_1), \mathbb{I}(v_k), [i..j])$, where $v_1, \ldots, v_k$ is the maximal path of blue nodes of ST that were merged onto v, and where the node of ST that we want to reconstruct is the deepest node v_p in $v_1, \ldots, v_k$ such that $\mathbb{I}(v_p)$ contains both $\mathbb{I}(v_k)$ and $[i..j]$. We take a backward step with the only character a in BWT$[i..j]$, reaching intervals $\mathbb{I}(v_1')$, $\mathbb{I}(v_k')$ and $[i'..j']$, where v_1' and v_k' are nodes of ST. We compute the LCA in $\mathcal{B}$ of the ends of $\mathbb{I}(v_k')$, and we find its lowest maximal repeat ancestor w in $\mathcal{B}$: such node can be either blue or black.

If w is black, we check whether $[i'..j'] \subset \mathbb{I}(w)$: if this is not the case, $\beta(v'_p)$ is an ancestor of w in $\mathcal{B}$, where v'_p is the projection of v_p by the Weiner link, thus we compute the LCA in $\mathcal{B}$ of the i'-th and j'-th leaves (this might be a maximal repeat), and then we take the LCA u in $\mathcal{B}$ between the resulting node and w. Clearly u is a maximal repeat: if u is blue, we recur on problem instance $([e..f], \mathbb{I}(u_h), [i'..j'])$, where $u_1, \dots, u_h$ are the blue nodes of ST that were merged onto u, and $[e..f]$ is the smallest of $\mathbb{I}(v'_1)$ and $\mathbb{I}(u_1)$. If u is black, let b_0 be the right-extension of u that contains $\mathbb{I}(v'_k)$: we know that the subtrees rooted at all the children of u in ST that are lexicographically greater than b_0 are isomorphic to the corresponding subtrees of v_p, thus we just need to know the total number of such children and the sum of the lengths of all their BWT intervals. To access such values, it suffices to store in u a predecessor data structure on the starting positions of the BWT intervals of all its children in $\mathcal{B}$, and to query such structure with the last position of $\mathbb{I}(v'_k)$. If $[i'..j'] \subset \mathbb{I}(w)$, we check whether $[i'..j']$ is also contained in the interval of the red child of w in $\mathcal{B}$ whose interval contains v'_k: if so, both v'_k and $[i'..j']$ lie in a subtree of ST that is not left-maximal, thus we take another backward step from $\mathbb{I}(w)$, $\mathbb{I}(v'_k)$ and $[i'..j']$ with the only character in BWT$[i'..j']$, and we recur on the resulting problem instance. Otherwise $w = v'_p$ and all the information we need is stored in w.

If w is blue, let $w_1, \dots, w_h$ be the maximal path of blue nodes of ST that were merged onto w. If $[i'..j'] \not\subset \mathbb{I}(w_1)$ we proceed as above. Otherwise we store, in the red child of v we are coming from, the following information: a descriptor of the lowest blue ancestor w_q of v'_k in ST; a descriptor of the non-left-maximal child z of w_q in ST that is an ancestor of v'_k; the sum of the lengths of all BWT intervals of children of w_q that are lexicographically larger than z, and the number of such children. If $[i'..j'] \not\subset \mathbb{I}(w_q)$, we recur on problem instance $([e..f], \mathbb{I}(w_q), [i'..j'])$, where $[e..f]$ is the smallest of $\mathbb{I}(v'_1)$ and $\mathbb{I}(w_1)$, since the projection of v_p by the Weiner link is an ancestor of w_q in ST. If $[i'..j'] \subset \mathbb{I}(z)$, we take a backward step with the only character in $[i'..j']$, as described above. Otherwise, w_q is the projection of v_p by the Weiner link, and all the information we need is stored in the red node of $\mathcal{B}$ we are coming from.

Lemma 1 needs also weighted level ancestor queries on the string depth of maximal repeats: this might require computing the interval of the highest blue node v_p of ST, in a blue path $v_1, \dots, v_k$ that was compressed onto a blue node of $\mathcal{B}$, and that is the locus of a string of known length: this can be done with the methods already described. $\square$

As mentioned in Sect. 3, to build a synchronous bidirectional index we also need to compute the interval of $\overline{W}$ in $\overline{\text{BWT}}$ after a left-contraction from aW in BWT. We still know from Lemma 4 whether W is left-maximal or right-maximal: if W is not left-maximal, the interval of $\overline{W}$ is the same as the interval of $\overline{aW}$, and if W is a maximal repeat we can just move to the deepest maximal repeat ancestor of $\overline{aW}$ in $\mathcal{B}$ using the methods described above. If W is left-maximal but not right-maximal, however, we cannot proceed as in Sect. 3, since we cannot afford to store the mapping between all maximal repeats in the two directions. Instead, we proceed as follows. If the locus of W in ST is mapped to a black node of $\mathcal{B}$ with label W^*, we store in such a black node the interval $[e..f]$ of $\overline{W^*}$ in $\overline{\mathsf{BWT}}$. Since $\overline{\mathsf{ST}}(\overline{W}) \simeq \overline{\mathsf{ST}}(\overline{W^*})$, we call $\mathtt{extendRight}(\mathrm{id}(\overline{W^*}), a)$ on $\overline{\mathsf{BWT}}$ and we apply the offsets of the interval we get to the interval of $\overline{aW}$ we started from. If the locus of W in ST is mapped to a blue node v of $\mathcal{B}$, with corresponding blue path $v_1, \ldots, v_k$ in ST, we store in v the number $|\ell(v_k)|$ and the interval $[i..j]$ of $\overline{\ell(v_k)}$ in $\overline{\mathsf{BWT}}$ and, since we know $|W|$, we issue $|\ell(v_k)| - |W|$ $\mathtt{contractRight}$ operations on $\overline{\mathsf{BWT}}$ from $\overline{\ell(v_k)}$: this takes $O(H^2 \log \log n)$ time.

Lemma 5. *Let T be a string of length n on alphabet $[1..\sigma]$ with $\bar{r}$ runs in $\overline{\mathsf{BWT}}$. There is an index that takes $O(\bar{r})$ words of space, and that implements the unidirectional $\mathtt{extendLeft}(\mathrm{id}(\overline{W}), b) = \mathrm{id}(\overline{bW})$ operation on $\overline{\mathsf{BWT}}$ in $O(H \log \log n)$ time from any maximal repeat W of T and any $b \in [0..\sigma]$, where H is the length of a longest maximal repeat of T.*

Proof. We implement the algorithm in Lemma 2 on $\overline{\mathcal{B}}$, recurring on Weiner links as described in Lemma 4. Specifically, assume that the maximal repeat node v_p of $\overline{\mathsf{ST}}$ we start from, of known length λ, is compacted onto a blue node v of $\overline{\mathcal{B}}$ with path $v_1, \ldots, v_k$, and let x and y be the properties of v defined at the beginning of Sect. 4. We don't know whether the extension character b we need lies on the left side of $v_1, \ldots, v_k$, on the right side of the path, or along the path itself. We assume first that it lies on the left side, i.e. we try to compute the offset of the first position of $\mathbb{I}(\ell(v_p)b)$ with respect to x, and we recur as described in Lemma 4.

If the path $v'_1, \ldots, v'_k$ projected by the Weiner link in $\overline{\mathsf{ST}}$ is compacted onto another blue node of $\overline{\mathcal{B}}$, character b still lies on the left side of the projected path, and we recur on that problem instance. Assume instead that path $v'_1, \ldots, v'_k$ in $\overline{\mathsf{ST}}$ is right-maximal up to some node v'_h. We can find $w = \overline{\beta}(v'_h)$ using $\overline{\mathcal{B}}$ and $\mathbb{I}(v'_k)$, as described previously. If w is black, we access its string depth: if $|\ell(w)| > \lambda + 1$, we find the locus u of v'_p in $\overline{\mathcal{B}}$. If u is black, we continue as in Lemma 2. Otherwise, b still lies on the left side of the path of u, and we recur on this problem instance. If $|\ell(w)| < \lambda + 1$, v'_p is a non-right-maximal descendant of v'_h, and we recur on problem instance $(\mathbb{I}(w), \mathbb{I}(v'_k), \mathtt{left})$. If $|\ell(w)| = \lambda + 1$, we continue from w as in Lemma 2. If w is blue, we perform similar operations using the descriptor of node v'_h in $\overline{\mathsf{ST}}$, which we store in the blue node v we came from. In addition to the information about v'_h described in Lemma 4, we also store the right-extension of v'_h along which the blue path of $\overline{\mathsf{ST}}$ compacted to w continues: this is enough to recur on the correct instance of the problem. While running Lemma 2, we might be in a black node z of $\overline{\mathcal{B}}$ and we might take a Weiner link from one of its marked children, ending up in a node z' of $\overline{\mathsf{ST}}$ that has been compacted onto a blue node of $\overline{\mathcal{B}}$: in every such case, we store a descriptor of z', and the right-extension of z' along which the path of $\overline{\beta}(z')$ continues. This is enough to recur on the correct instance of the problem.

The initial assumption that b lies on the left side of the path of v might be wrong. Assume that $\mathbb{I}(v_1) = [i..j]$, that $\mathsf{BWT}[i..x]$ contains just character a, and that $\mathsf{BWT}[y..j]$ contains just character $c \neq a$. If b actually lies on the right side of the path, the first Weiner link we took added to the left of v_p character

a, but $a\ell(v_p)$ is never followed by b in T. This might be detected as soon as the recursion reaches a black node, in which case we restart the whole process assuming that b lies on the right side of the path. Detecting the mistake might still not be possible at the first black node, since b might fall inside a run of its children. By the end of the recursion, however, the mistake will be detected: this holds by the same criterion used in Lemma 2 to prove that the offset of b can eventually be reconstructed. If b lies on the right side of the path and $c = a$, and if this configuration continues to be true after several Weiner links, until v_p is eventually projected to a black node of $\overline{\mathcal{B}}$, then such node is the same that we would have reached assuming that b lies on the right side of the path: thus, making the wrong assumption does not affect the outcome of the recursion in this case. If, after some Weiner links in the same configuration, we reach a blue node in which the character c' on the right side is different from the character a' on the left side, then recursion will eventually reach a maximal repeat $Va'W\ell(v_p)$, for some $V, W \in [1..\sigma]^*$, such that $Va'W\ell(v_p)b$ does not occur in T, and this will be detected by the end of the process. Finally, the case in which b lies along the path of v is identical to the case in which b lies fully on the left side of the path.   □

Lemmas 4 and 5 are clearly all we need to prove the main result of this section:

Theorem 2. *Let T be a string of length n, with r runs in* BWT *and $\bar{r}$ runs in* $\overline{\text{BWT}}$. *Let H be the length of a longest maximal repeat of T. There is a fully-functional bidirectional index that takes $O(r + \bar{r})$ words of space, and that supports all operations in $O(H^2 \log \log n)$ time.*

Acknowledgements. We thank Timothy Chan for insights on static weighted 2D orthogonal range counting, and Gene Myers and the Myers' lab for hosting and fruitful discussions.

References

1. Alstrup, S., Stolting Brodal, G., Rauhe, T.: New data structures for orthogonal range searching. In: Proceedings of the 41st Annual Symposium on Foundations of Computer Science, pp. 198–207 (2000)
2. Amir, A., Landau, G.M., Lewenstein, M., Sokol, D.: Dynamic text and static pattern matching. ACM Trans. Algorithms (TALG) **3**(2), 19 (2007)
3. Belazzougui, D., Cunial, F.: Fully-functional bidirectional Burrows-Wheeler indexes and infinite-order de Bruijn graphs. In: 30th Annual Symposium on Combinatorial Pattern Matching (CPM 2019). Schloss Dagstuhl-Leibniz-Zentrum fuer Informatik (2019)
4. Belazzougui, D., Cunial, F., Gagie, T., Prezza, N., Raffinot, M.: Composite repetition-aware data structures. In: Cicalese, F., Porat, E., Vaccaro, U. (eds.) CPM 2015. LNCS, vol. 9133, pp. 26–39. Springer, Cham (2015). https://doi.org/10.1007/978-3-319-19929-0_3

5. Belazzougui, D., Cunial, F., Kärkkäinen, J., Mäkinen, V.: Versatile succinct representations of the bidirectional Burrows-Wheeler transform. In: Bodlaender, H.L., Italiano, G.F. (eds.) ESA 2013. LNCS, vol. 8125, pp. 133–144. Springer, Heidelberg (2013). https://doi.org/10.1007/978-3-642-40450-4_12
6. Cánovas, R., Rivals, E.: Full compressed affix tree representations. In: Data Compression Conference (DCC 2017), pp. 102–111. IEEE (2017)
7. Crochemore, M., Epifanio, C., Grossi, R., Mignosi, F.: Linear-size suffix tries. Theoret. Comput. Sci. **638**, 171–178 (2016)
8. Cunial, F., Alanko, J., Belazzougui, D.: A framework for space-efficient variable-order Markov models. Bioinformatics **35**(22), 4607–4616 (2019)
9. Farach, M., Muthukrishnan, S.: Perfect hashing for strings: formalization and algorithms. In: Hirschberg, D., Myers, G. (eds.) CPM 1996. LNCS, vol. 1075, pp. 130–140. Springer, Heidelberg (1996). https://doi.org/10.1007/3-540-61258-0_11
10. Gagie, T., Navarro, G., Prezza, N.: Fully functional suffix trees and optimal text searching in BWT-runs bounded space. J. ACM **67**(1), 1–54 (2020)
11. Hagerup, T., Miltersen, P.B., Pagh, R.: Deterministic dictionaries. J. Algorithms **41**(1), 69–85 (2001)
12. Maaß, M.G.: Linear bidirectional on-line construction of affix trees. In: Giancarlo, R., Sankoff, D. (eds.) CPM 2000. LNCS, vol. 1848, pp. 320–334. Springer, Heidelberg (2000). https://doi.org/10.1007/3-540-45123-4_27
13. Mäkinen, V., Navarro, G.: Succinct suffix arrays based on run-length encoding. In: Apostolico, A., Crochemore, M., Park, K. (eds.) CPM 2005. LNCS, vol. 3537, pp. 45–56. Springer, Heidelberg (2005). https://doi.org/10.1007/11496656_5
14. Mäkinen, V., Navarro, G., Sirén, J., Välimäki, N.: Storage and retrieval of highly repetitive sequence collections. J. Comput. Biol. **17**(3), 281–308 (2010)
15. Munro, J.I., Navarro, G., Nekrich, Y.: Space-efficient construction of compressed indexes in deterministic linear time. In: Proceedings of the Twenty-Eighth Annual ACM-SIAM Symposium on Discrete Algorithms, pp. 408–424. SIAM (2017)
16. Okanohara, D., Sadakane, K.: Practical entropy-compressed rank/select dictionary. In: 2007 Proceedings of the Ninth Workshop on Algorithm Engineering and Experiments (ALENEX), pp. 60–70. SIAM (2007)
17. Schnattinger, T., Ohlebusch, E., Gog, S.: Bidirectional search in a string with wavelet trees and bidirectional matching statistics. Inf. Comput. **213**, 13–22 (2012)
18. Sirén, J., Välimäki, N., Mäkinen, V., Navarro, G.: Run-length compressed indexes are superior for highly repetitive sequence collections. In: Amir, A., Turpin, A., Moffat, A. (eds.) SPIRE 2008. LNCS, vol. 5280, pp. 164–175. Springer, Heidelberg (2008). https://doi.org/10.1007/978-3-540-89097-3_17
19. Stoye, J.: Affix trees. Master's thesis, Universität Bielefeld (2000)
20. Strothmann, D.: The affix array data structure and its applications to RNA secondary structure analysis. Theoret. Comput. Sci. **389**(1–2), 278–294 (2007)
21. Takagi, T., Goto, K., Fujishige, Y., Inenaga, S., Arimura, H.: Linear-size CDAWG: new repetition-aware indexing and grammar compression. In: Fici, G., Sciortino, M., Venturini, R. (eds.) SPIRE 2017. LNCS, vol. 10508, pp. 304–316. Springer, Cham (2017). https://doi.org/10.1007/978-3-319-67428-5_26
22. Willard, D.E.: Log-logarithmic worst-case range queries are possible in space θ (n). Inf. Process. Lett. **17**(2), 81–84 (1983)
23. Willard, D.E.: New data structures for orthogonal range queries. SIAM J. Comput. **14**(1), 232–253 (1985)

Internal Quasiperiod Queries

Maxime Crochemore[1] , Costas S. Iliopoulos[1] , Jakub Radoszewski[2] ,
Wojciech Rytter[2] , Juliusz Straszyński[2] , Tomasz Waleń[2] ,
and Wiktor Zuba[2]([✉])

[1] Department of Informatics, King's College London, London, UK
{maxime.crochemore,c.iliopoulos}@kcl.ac.uk
[2] Institute of Informatics, University of Warsaw, Warsaw, Poland
{jrad,rytter,jks,walen,w.zuba}@mimuw.edu.pl

Abstract. Internal pattern matching requires one to answer queries
about factors of a given string. Many results are known on answering
internal period queries, asking for the periods of a given factor. In this
paper we investigate (for the first time) internal queries asking for covers
(also known as quasiperiods) of a given factor. We propose a data structure that answers such queries in $\mathcal{O}(\log n \log \log n)$ time for the shortest
cover and in $\mathcal{O}(\log n (\log \log n)^2)$ time for a representation of all the covers, after $\mathcal{O}(n \log n)$ time and space preprocessing.

Keywords: Cover · Quasiperiodicity · Internal pattern matching ·
Seed · Run (maximal repetition)

1 Introduction

A *cover* (also known as a quasiperiod) is a weak version of a period. It is a
factor of a text T whose occurrences cover all positions in T; see Fig. 1. The
notion of cover is well-studied in the off-line model. Linear-time algorithms for
computing the shortest cover and all the covers of a string of length n were
proposed in [2] and [23,24], respectively. Moreover, linear-time algorithms for
computing shortest and longest covers of all prefixes of a string are known; see [6]
and [22], respectively. Covers were also studied in parallel [5,7] and streaming [13]
models of computation. Definitions of other variants of quasiperiodicity can be
found in the survey [12]. In this work we introduce covers to the internal pattern
matching model [20].

In the internal pattern matching model, a text T of length n is given in
advance and the goal is to answer queries related to factors of the text. One
of the basic internal queries in texts are *period queries*, that were introduced
in [19] (actually, internal primitivity queries were considered even earlier [9,10]).

J. Radoszewski, T. Waleń and W. Zuba—Supported by the Polish National Science
Center, grant no. 2018/31/D/ST6/03991.
J. Straszyński—Partially supported by ERC Consolidator Grant 772346 TUgbOAT
and by the Polish National Science Center, grant no. 2018/31/D/ST6/03991.

C. Boucher and S. V. Thankachan (Eds.): SPIRE 2020, LNCS 12303, pp. 60–75, 2020.
https://doi.org/10.1007/978-3-030-59212-7_5

$$T: \quad \text{a b a a b a b a a b a b a}$$

Fig. 1. MINCOVER(T) = aba is the shortest cover of T and MINCOVER($T[2 . . 13]$) = *baababa* is the shortest cover of its suffix of length 12.

A period query requires one to compute all the periods of a given factor of T. It is known that they can be expressed as $\mathcal{O}(\log n)$ arithmetic sequences. The fastest known algorithm answering period queries is from [20]. It uses a data structure of $\mathcal{O}(n)$ size that can be constructed in $\mathcal{O}(n)$ expected time and answers period queries in $\mathcal{O}(\log n)$ time (a deterministic construction of this data structure was given in [16]). A special case of period queries are *two-period queries*, which ask for the shortest period of a factor that is known to be periodic. In [20] it was shown that two-period queries can be answered in constant time after $\mathcal{O}(n)$-time preprocessing. Another algorithm for answering such queries was proposed in [3].

Let us denote by MINCOVER(S) and ALLCOVERS(S), respectively, the length of the shortest cover and the lengths of all covers of a string S. Similarly as in the case of periods, it can be shown that the set ALLCOVERS(S) can be expressed as a union of $\mathcal{O}(\log |S|)$ pairwise disjoint arithmetic sequences. We consider data structures that allow to efficiently answer these queries in the internal model.

> INTERNAL QUASIPERIOD QUERIES
>
> **Input:** A text T of length n
>
> **Query:** For any factor S of T, compute MINCOVER(S) or
> ALLCOVERS(S) after efficient preprocessing of the text T

Recently [11] we have shown how to compute the shortest cover of each cyclic shift of a string T of length n, that is, the shortest cover of each length-$|T|$ factor of T^2, in $\mathcal{O}(n \log n)$ total time. This work can be viewed as a generalization of [11] to computing covers of any factor of a string. It also generalizes the earlier works on computing covers of prefixes of a string [6,22].

Our Results. We show that MINCOVER and ALLCOVERS queries can be answered in $\mathcal{O}(\log n \log \log n)$ time and $\mathcal{O}(\log n (\log \log n)^2)$ time, respectively, with a data structure that uses $\mathcal{O}(n \log n)$ space and can be constructed in $\mathcal{O}(n \log n)$ time. In particular, the time required to answer an ALLCOVERS query is slower by only a poly log log n factor from optimal. Moreover, we show that any m MINCOVER or ALLCOVERS queries can be answered off-line in $\mathcal{O}((n+m) \log n)$ and $\mathcal{O}((n+m) \log n \log \log n)$ time, respectively, and $\mathcal{O}(n+m)$ space. In particular, the former matches the complexity of the best known solution for computing shortest covers of all cyclic shifts of a string [11], despite being far more general. We assume the word RAM model of computation with word size $\Omega(\log n)$.

Our Approach. Our main tool are *seeds*, a known generalization of the notion of cover. A seed is defined as a cover of a superstring of the text [14]. A representation of all seeds of a string T, denoted here *SeedSet*(T), can be computed in linear time [17]. We will frequently extract individual seeds from $SeedSet(T)$; each time such an auxiliary query needs $\mathcal{O}(\log\log n)$ time. Consequently, $\log\log n$ is a frequent factor in our query times related to internal covers.

We construct a tree-structure (static range tree) of so-called *basic factors* of a string. For each basic factor F we store a compact representation of the set $SeedSet(F)$. The crucial point is that the total length of all these factors is $\mathcal{O}(n\log n)$ and every other factor can be represented, using the tree-structure, as a concatenation of $\mathcal{O}(\log n)$ basic factors. Representations of seed-sets of basic factors are precomputed. Then, upon an internal query related to a specific factor S, we decompose S into concatenation of basic factors $F_1, F_2, \ldots, F_k$. Intuitively, the representation of the set of covers or (in easier queries) the shortest cover will be computed as a "composition" of $SeedSet(F_1), SeedSet(F_2), \ldots, SeedSet(F_k)$, followed by adjusting it to border conditions using internal pattern matching. To get efficiency, when querying about covers of a factor S, we do not compute the whole representation of $SeedSet(S)$ (these representations are only precomputed for basic factors).

Finally, several stringology tools related to properties of covers and string periodicity are used to improve `polylog`n-factors in the query time that would result from a direct application this approach.

2   Preliminaries

We consider a text T of length n over an integer alphabet $\{0, \ldots, n^{\mathcal{O}(1)}\}$. If this is not the case, its letters can be sorted and renumbered in $\mathcal{O}(n\log n)$ time, which does not influence the preprocessing time of our data structure.

For a string S, by $|S|$ we denote its length and by $S[i]$ we denote its ith letter $(i = 1, \ldots, |S|)$. By $S[i\mathbin{..}j]$ we denote the string $S[i] \ldots S[j]$ called a factor of S; it is a prefix if $i = 1$ and a suffix if $j = |S|$. A factor that occurs both as a prefix and as a suffix of S is called a border of S. A factor is proper if it is shorter than the string itself. A positive integer p is called a period of S if $S[i] = S[i + p]$ holds for all $i = 1, \ldots, |S| - p$. By $\mathsf{per}(S)$ we denote the smallest period of S. A string S is called periodic if $|S| \geq 2\mathsf{per}(S)$ and aperiodic otherwise. If $S = XY$, then any string of the form YX is called a cyclic shift of S. We use the following simple fact related to covers.

Observation 1. *Let A, B, C be strings such that $|A| < |B| < |C|$.*

(a) If A is a cover of B and B is a cover of C, then A is a cover of C.
(b) If B is a border of C and A is a cover of C, then A is a cover of B.

Below we list several algorithmic tools used later in the paper.

Queries Related to Suffix Trees and Arrays
A range minimum query on array $A[1\mathbin{..}n]$ requires to compute $\min\{A[i], \ldots, A[j]\}$.

Lemma 2 ([4]). *Range minimum queries on an array of size n can be answered in $\mathcal{O}(1)$ time after $\mathcal{O}(n)$-time preprocessing.*

By $\mathsf{lcp}(i,j)$ ($\mathsf{lcs}(i,j)$) we denote the length of the longest common prefix of $T[i \mathinner{..} n]$ and $T[j \mathinner{..} n]$ (longest common suffix of $T[1 \mathinner{..} i]$ and $T[1 \mathinner{..} j]$, respectively). Such queries are called longest common extension (LCE) queries. The following lemma is obtained by using range minimum queries on suffix arrays.

Lemma 3 ([4,15]). *After $\mathcal{O}(n)$-time preprocessing, one can answer LCE queries for T in $\mathcal{O}(1)$ time.*

The suffix tree of T, denoted as $\mathcal{T}(T)$, is a compact trie of all suffixes of T. Each implicit or explicit node of $\mathcal{T}(T)$ corresponds to a factor of T, called its *string label*. The *string depth* of a node of $\mathcal{T}(T)$ is the length of its string label.

We use *weighted ancestor (WA) queries* on a suffix tree. Such queries, given an explicit node v and an integer value ℓ that does not exceed the string depth of v, ask for the highest explicit ancestor u of v with string depth at least ℓ.

Lemma 4 ([1,17]). *Let $\mathcal{T}(T)$ be the suffix tree of T. WA queries on $\mathcal{T}(T)$ can be answered in $\gamma_n = \mathcal{O}(\log \log n)$ time after $\mathcal{O}(n)$-time preprocessing. Moreover, any m WA queries on $\mathcal{T}(T)$ can be answered off-line in $\mathcal{O}(n+m)$ time.*

Internal Pattern Matching (IPM)

The data structure for IPM queries is built upon a text T and allows efficient location of all occurrences of one factor X of T inside another factor Y of T, where $|Y| \leq 2|X|$.

Lemma 5 ([20]). *The result of an IPM query is a single arithmetic sequence. After linear-time preprocessing one can answer IPM queries for T in $\mathcal{O}(1)$ time.*

A period query, for a given factor X of text T, returns a compact representation of all the periods of X (as a set of $\mathcal{O}(\log n)$ arithmetic sequences).

Lemma 6 ([20]). *After $\mathcal{O}(n)$ time and space preprocessing, for any factor of T we can answer a period query in $\mathcal{O}(\log n)$ time.*

The data structures of Lemmas 5 and 6 are constructed in $\mathcal{O}(n)$ expected time. These constructions were made worst-case in [16].

Static Range Trees

A *basic interval* is an interval $[a \mathinner{..} a + 2^i)$ such that 2^i divides $a - 1$. We assume w.l.o.g. that n is a power of two. We consider a static range tree structure whose nodes correspond to basic subintervals of $[1 \mathinner{..} n]$ and a non-leaf node has children corresponding to the two halves of the interval. (See e.g. [18]). The total number of basic intervals is $\mathcal{O}(n)$. Using the tree, every interval $[i \mathinner{..} j]$ can be decomposed into $\mathcal{O}(\log n)$ pairwise disjoint basic intervals. The decomposition can be computed in $\mathcal{O}(\log n)$ time by inspecting the paths from the leaves corresponding to i and j to their lowest common ancestor. A *basic factor* of T is a factor that corresponds to positions from a basic interval.

Seeds

We say that a string S is a *seed* of a string U if S is a factor of U and S is a cover of a string U' such that U is a factor of U'; see Fig. 2. The second point of the lemma below follows from Lemma 4.

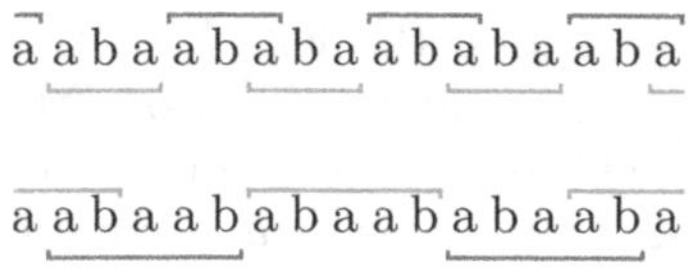

$$\overline{a\;a}\,\overline{b\;a\;a}\,\overline{b\;a\;b}\,\overline{a\;a}\,\overline{b\;a\;b}\,\overline{a\;a}\,\overline{b\;a}$$

$$\overline{a\;a}\,\overline{b\;a\;a}\,\overline{b\;a\;b}\,\overline{a\;a}\,\overline{b\;a\;b}\,\overline{a\;a}\,\overline{b\;a}$$

Fig. 2. The strings *aba, abaab* are seeds of the given string (as well as strings *abaaba, abaababa, abaababaa*).

Lemma 7 ([17]).

(a) All the seeds of T can be represented as a collection of a linear number of disjoint paths in the suffix tree $T(T)$. Moreover, this representation can be computed in $\mathcal{O}(n)$ time if T is over an integer alphabet.

(b) After $\mathcal{O}(n)$ time preprocessing we can check if a given factor of T is a seed of T in $\mathcal{O}(\gamma_n)$ time.

Our main data structure is a static range tree *SeedSets*(T) which stores all seeds of every basic factor of T represented as a collection of paths in its suffix tree. Actually, only seeds of length at most half of a string will be of interest; see Fig. 3.

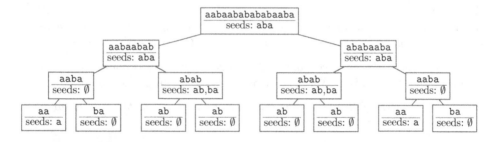

Fig. 3. A schematic view of tree *SeedSets* of T (in the real data structure, seeds are stored on suffix trees of basic factors). For example, **ba** is a seed of $T[5..12]$ since it is a seed of basic factors $T[5..8]$ and $T[9..12]$ and its occurrence covers $T[8..9]$ (Lemma 10).

The sum of lengths of basic factors in T is $\mathcal{O}(n \log n)$. Consequently, due to Lemma 7, the tree *SeedSets*(T) has total size $\mathcal{O}(n \log n)$ and can be computed in $\mathcal{O}(n \log n)$ time. (To use Lemma 7(a) we renumber letters in basic factors of T via bucket sort so that the letters of a basic factor S are from $\{0, \ldots, |S|^{\mathcal{O}(1)}\}$.)

3   Internal Cover of a Given Length

In this section we show how to use *SeedSets*(T) to answer internal queries related to computing the longest prefix of a factor S of T that is covered by its length-ℓ prefix. We start with the following, easier queries.

> COVER OF A GIVEN LENGTH QUERY (ISCOVER(ℓ, S))
> **Input:** A factor S of T and a positive integer ℓ
> **Query:** Does S have a cover of length ℓ?

The following three lemmas provide the building blocks of the data structure for answering ISCOVER queries.

Lemma 8 (Seed of a basic factor). *After $\mathcal{O}(n \log n)$-time preprocessing, for any factor C and basic factor B of T such that $2|C| \leq |B|$, we can check if C is a seed of B in $\mathcal{O}(\gamma_n)$ time.*

Proof. Let $|C| = c$ and $B = T[a..b]$. We first ask an IPM query to find an occurrence of C inside $T[a..a+2c-1]$. If such an occurrence does not exist, then C cannot be a seed of $T[a..b]$ as it is already not a seed of $T[a..a+2c-1]$ (there must be a full occurrence to cover the middle letter, and $a+2c-1 \leq b$). Otherwise, we can use the occurrence to check if C is a seed of B with Lemma 7(b). □

For strings C and S, by $Cov(C, S)$ we denote the set of positions of S that are covered by occurrences of C.

Lemma 9 (Covering short factors). *After $\mathcal{O}(n)$-time preprocessing, for any two factors C and F of T such that $|F|/|C| = \mathcal{O}(1)$, the set $Cov(C, F)$, represented as a union of maximal intervals, can be computed in $\mathcal{O}(1)$ time.*

Proof. We ask IPM queries for pattern C on length-$2|C|$ factors of F with step $|C|$. Each IPM query returns an arithmetic sequence of occurrences that corresponds to an interval of covered positions (possibly empty). It suffices to compute the union of these intervals. □

Lemma 10 (Seeds of strings concatenation). *After $\mathcal{O}(n)$-time preprocessing, for any three factors C, $F_1 = T[i..j]$ and $F_2 = T[j+1..k]$ of T such that $2|C| \leq |F_1|, |F_2|$ and C is a seed of both F_1 and F_2, we can check if C is also a seed of $F_1 F_2$ in constant time.*

Proof. For a string C of length c being a seed of both $T[i..j]$ and $T[j+1..k]$ to be a seed of $T[i..k]$, it is enough if its occurrences cover the string $U = T[j-c+1..j+c]$. We can check this condition if we apply Lemma 9 for C and $F = T[j-2c+1..j+2c]$. □

Lemma 11. *After $\mathcal{O}(n \log n)$ time and space preprocessing of T, a query* ISCOVER(ℓ, S) *can be answered in $\mathcal{O}(\log(|S|/\ell)\gamma_n + 1)$ time.*

Proof. Let $S = T[i..j]$, $|S| = s$ and $C = T[i..i+\ell-1]$.

We consider a decomposition of S into basic factors, but we are only interested in basic factors of length at least 2ℓ in the decomposition. Let $F_1, \ldots, F_k$ be those factors and $T[i..i'], T[j'..j]$ be the remaining prefix and suffix of length $\mathcal{O}(\ell)$. Note that $k = \mathcal{O}(\log(s/\ell))$. Moreover, this decomposition can be computed in $\mathcal{O}(k+1)$ time by starting from the leftmost and rightmost basic factors of length 2^b, where $b = \lceil \log \ell \rceil + 1$, that are contained in S.

If C is a cover of S, it must be a seed of each of the basic factors $F_1, \ldots, F_k$. We can check this condition by using Lemma 8 in $\mathcal{O}(k\gamma_n)$ total time.

Next we check if C is a seed of $F_1 \cdots F_k$ in $\mathcal{O}(k)$ total time using Lemma 10. Finally, we use IPM queries to check if occurrences of C cover all positions in each of the strings $T[i \mathinner{.\,.} i'+c-1]$, $T[j'-c+1 \mathinner{.\,.} j]$ and if C is a suffix of $T[i \mathinner{.\,.} j]$, using Lemma 9. This takes $\mathcal{O}(1)$ time.

The total time complexity is $\mathcal{O}(k\gamma_n + 1)$. $\square$

As we will see in the next section, IsCover queries immediately imply a slower, $\mathcal{O}(\log^2 n \, \gamma_n)$-time algorithm for answering MinCover queries. However, they are also used in our algorithm for answering AllCovers queries. In the efficient algorithm for MinCover queries we use the following generalization of IsCover queries.

Longest Covered Prefix Query (CoveredPref(ℓ, S))

Input: A factor S of T and a positive integer ℓ

Query: The longest prefix P of S that is covered by $S[1 \mathinner{.\,.} \ell]$

To answer these queries, we introduce an intermediate problem that is more directly related to the range tree containing seeds representations.

SeededBasicPref(C, ℓ, S) query

Input: A length-ℓ factor C of T and a factor S being a concatenation of basic factors of T of length 2^p, where $p = \min\{q \in \mathbb{Z} : 2^q \geq 2\ell\}$

Output: The length m of the longest prefix of S which is a concatenation of basic factors of length 2^p such that C is a seed of this prefix

---

In other words, we consider only blocks of S which are basic factors of length $2^p = \Theta(\ell)$. Everything starts and ends in the beginning/end of a basic factor of length 2^p. The number of such blocks in the prefix returned by SeededBasicPref is $\mathcal{O}(\mathsf{result}'/\ell)$, where $\mathsf{result}' = \text{SeededBasicPref}(C, \ell, S)$, and, as we show in Lemma 13, it can be computed in $\mathcal{O}(\log(\mathsf{result}'/\ell)\gamma_n + 1)$ time. This is how we achieve $\mathcal{O}(\log(\mathsf{result}/\ell) \, \gamma_n + 1)$ time for CoveredPref(ℓ, S) queries. In a certain sense the computations behind Lemma 13 can work in a pruned range tree $SeedSets(T)$.

Lemma 12. *After $\mathcal{O}(n)$-time preprocessing, a CoveredPref(ℓ, S) query reduces in $\mathcal{O}(1)$ time to a SeededBasicPref(C, ℓ, S') query with $|S'| \leq |S|$.*

Proof. First, let us check if the answer to CoveredPref(ℓ, S) is small, i.e. at most 4ℓ, using Lemma 9. Otherwise, let p be defined as in a SeededBasicPref query, $C = S[1 \mathinner{.\,.} \ell]$ and S' be the maximal factor of S that is composed of basic factors of length 2^p (S' can be the empty string, if $|S| < 3 \cdot 2^p$). Let $S = T[i \mathinner{.\,.} j]$ and $S' = T[i' \mathinner{.\,.} j']$. Then

$$|(i' + \textsc{SeededBasicPref}(C, \ell, S')) - (i + \textsc{CoveredPref}(\ell, S))| < 2^p;$$

see Fig. 4. Hence, knowing $d = \textsc{SeededBasicPref}(C, \ell, S')$, we check in $\mathcal{O}(1)$ time, using Lemma 9 in a factor $T[i' + d - 2^p \mathinner{.\,.} i' + d + 2^p - 1]$ of length 2^{p+1}, what is the exact value of $\textsc{CoveredPref}(\ell, S)$.

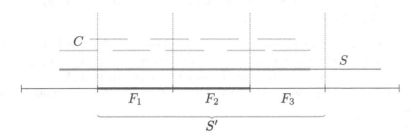

Fig. 4. F_1, F_2, F_3 are basic factors of length 2^p. The answers to $\textsc{CoveredPref}(\ell, S)$ and $\textsc{SeededBasicPref}(C, \ell, S')$ queries are shown in bold. Note that C is a seed of F_1 and F_2 and that it could be the case that C is also a seed of F_3, even though it has no further full occurrence.

We compute p using the formula $p = 1 + \lceil \log \ell \rceil$. Then the endpoints of S' can be computed from the endpoints of S in $\mathcal{O}(1)$ time using simple modular arithmetic. The $\mathcal{O}(n)$ preprocessing is due to Lemma 9. □

A proof of the following lemma is left for the full version.

Lemma 13. *After $\mathcal{O}(n \log n)$ time and space preprocessing of T, a query $\textsc{SeededBasicPref}(C, \ell, S)$ can be answered in $\mathcal{O}(\log(\mathsf{result}/\ell)\, \gamma_n + 1)$ time, where $\mathsf{result} = |\textsc{SeededBasicPref}(C, \ell, S)|$.*

As a corollary of Lemma 12 and 13, we obtain the following result.

Lemma 14. *After $\mathcal{O}(n \log n)$ time and space preprocessing of T, a query $\textsc{CoveredPref}(\ell, S)$ can be answered in $\mathcal{O}(\log(\mathsf{result}/\ell)\, \gamma_n + 1)$ time, where $\mathsf{result} = |\textsc{CoveredPref}(\ell, S)|$.*

4   Internal Shortest Cover Queries

For a string S, by *Borders*(S) we denote a decomposition of the set of all border lengths of S into $\mathcal{O}(\log |S|)$ arithmetic sequences $A_1, \ldots, A_k$ such that each sequence A_i is either a singleton or, if p is its difference, then the borders with lengths in $A_i \setminus \{\min(A_i)\}$ are periodic with the shortest period p. Moreover, $\max(A_i) < \min(A_{i+1})$ for every $i \in [1 \mathinner{.\,.} k - 1]$. See e.g. [8]. The following lemma is shown by applying a period query (Lemma 6).

Lemma 15 ([16,20]). *For any factor S of T, Borders(S) can be computed in $\mathcal{O}(\log n)$ time after $\mathcal{O}(n)$-time preprocessing.*

4.1   Simple Algorithm with $\mathcal{O}(\log^2 n\, \gamma_n)$ Query Time

Let us start with a much simpler but slower algorithm for answering MINCOVER queries using IsCOVER queries. We improve it in Theorem 17 by using COVEREDPREF queries and applying an algorithm for computing shortest covers that resembles, to some extent, computation of the shortest cover from [2].

Proposition 16. *Let T be a string of length n. After $\mathcal{O}(n \log n)$-time pre-processing, for any factor S of T we can answer a MINCOVER(S) query in $\mathcal{O}(\log^2 n \log \log n)$ time.*

Proof. Using Lemma 15 we compute the set $Borders(S) = A_1, \ldots, A_k$ in $\mathcal{O}(\log n)$ time. Let us observe that the shortest cover of a string is aperiodic. This implies that from each progression A_i only the border of length $\min(A_i)$ can be the shortest cover of S. We use Lemma 11 to test each of the $\mathcal{O}(\log n)$ candidates in $\mathcal{O}(\log n\, \gamma_n)$ time. $\square$

4.2   Faster Queries

Theorem 17. *Let T be a string of length n. After $\mathcal{O}(n \log n)$-time preprocessing, for any factor S of T we can answer a MINCOVER(S) query in $\mathcal{O}(\log n \log \log n)$ time.*

Proof. Again we use Lemma 15 we compute the set $Borders(S) = A_1, \ldots, A_k$, in $\mathcal{O}(\log n)$ time. Let us denote the border of length $\min(A_i)$ by C_i and $C_{k+1} = S$. We assume that C_i's are sorted in increasing order of lengths. Then we proceed as shown in Algorithm 1. See also Fig. 5.

Algorithm 1: MINCOVER(S) query.

$i := 1$;
while true do
    // Invariant: $C_1, \ldots, C_{i-1}$ are not covers of S
    // C_i is an *active* border
    $P := \text{COVEREDPREF}(|C_i|, S)$;
    if $P = S$ **then return** $|C_i|$;
    while $|C_i| \le |P|$ **do**
        $i := i + 1$;

To argue for the correctness of the algorithm it suffices to show the invariant. The proof goes by induction.

The base case is trivial. Let us consider the value of i at the beginning of a step of the while-loop. If $P = S$, then by the inductive assumption C_i is the shortest cover of S and can be returned. Otherwise, C_i is not a cover of S.

Moreover, for each j such that $|C_i| < |C_j| \le |P|$, since C_j is a prefix of P, C_i is a seed of C_j. Moreover, both C_i and C_j are borders of S, so C_i is a border of C_j. Consequently, C_j cannot be a cover of T, as then C_i would also be a cover of T by Observation 1. This shows that the inner while-loop correctly increases i.

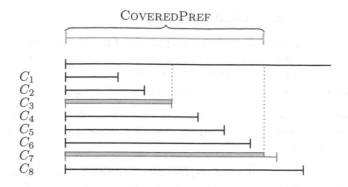

Fig. 5. If C_3 is an active border, then the next active one is C_7. We skip C_4, C_5, C_6 as candidates for the shortest cover.

The algorithm stops because at each point $|P| \geq |C_i|$ and i is increased.

Let $c_1, \ldots, c_p$ be equal to the length of an active border in the algorithm at the start of subsequent outer while-loop iterations and let $c_{p+1} = |S|$.

Let us note that, for all $j = 1, \ldots, p$, $|\text{COVEREDPREF}(c_j, S)| \leq c_{j+1}$. By Lemma 14, the total complexity of answering longest covered prefix queries in the algorithm is at most

$$\mathcal{O}\left(p + \gamma_n \sum_{j=1}^{p} \log \tfrac{c_{j+1}}{c_j}\right) = \mathcal{O}(\log n + \gamma_n(\log c_{p+1} - \log c_1)) = \mathcal{O}(\log n \, \gamma_n).$$

The preprocessing of Lemmas 14 and 15 takes $\mathcal{O}(n \log n)$ time. The conclusion follows.   □

If MINCOVER queries are to be answered in a batch, we can use off-line WA queries of Lemma 4 to save the γ_n-factor. We can also avoid storing the whole data structure *SeedSets* by using an approximate version of COVEREDPREF queries. Details are left for the full version.

Theorem 18. *For a string T of length n, any m queries $\text{MINCOVER}(T[i \mathinner{.\,.} j])$ can be answered in $\mathcal{O}((n + m) \log n)$ time and $\mathcal{O}(n + m)$ space.*

5   Internal All Covers Queries

In this section we refer to ALLCOVERS(S) as to the set of lengths of all covers of S. This set consists of a logarithmic number of arithmetic sequences since the same is true for all borders. In each sequence of borders we show that it is needed only to check $\mathcal{O}(1)$ borders to be a cover of S. Hence we start with an algorithm testing any sequence of $\mathcal{O}(\log n)$ candidate borders.

5.1 Verifying $\mathcal{O}(\log n)$ Candidates

Assume that B is an increasing sequence $b_1, \ldots, b_k$ of lengths of borders of a given factor S (not necessarily all borders), with $b_k = |S|$. A *chain* in B is a maximal subsequence $b_i, \ldots, b_j$ of consecutive elements of B such that $S[1 .. b_t]$ is a cover of $S[1 .. b_{t+1}]$ for each $t \in [i .. j)$. From Observation 1 we get the following.

Observation 19. *The set of elements of a chain that belong to* $\text{ALLCOVERS}(S)$ *is a prefix of this chain. Moreover, if the last element of a chain is not* $|S|$, *then it is not a cover of* S.

We denote by $chains(B)$ and $covers(B)$, respectively, the partition of B into chains and the set of elements $b \in B$ such that $S[1 .. b]$ is a cover of S. For $b \in B$ by $prev(b)$ we denote the previous element in its chain (if it exists). Moreover, for $C \subseteq B$ by $next_C(b)$ we denote the smallest $c \in C$ such that $c > b$.

Lemma 20. *Let T be a string of length n. After $\mathcal{O}(n \log n)$-time preprocessing, for any factor S of T and a sequence B of $\mathcal{O}(\log n)$ borders of S we can compute* $covers(B)$ *in $\mathcal{O}(\log n \log \log n \, \gamma_n)$ time.*

Proof. We introduce two operations and use them in a recursive Algorithm 2.

refine(B): removes the last element of each chain in B and every second element of each chain, except $|S|$ (see Fig. 6). Note that $|refine(B)| \leq |B|/2 + 1$.

computeUsing(B, C): Assuming that we know the set C of all covers of S among $refine(B)$, for each element b of $B \setminus refine(B)$ we add it to C if $prev(b) \in C$ and $S[1 .. b]$ is a cover of $S[1 .. next_C(b)]$. The set of all elements that satisfy this condition together with C is returned as $covers(B)$.

$$
\begin{array}{ccc}
\overset{\textstyle |S| \atop \textstyle \|}{} & & \overset{\textstyle |S| \atop \textstyle \|}{} \\
b_1 \to b_2 \to b_3 \to b_4 \to b_5 \quad b_6 \to b_7 \to b_8 \to b_9 & \xrightarrow{\ refine\ } & b_1 \to b_3 \quad b_6 \to b_8 \to b_9
\end{array}
$$

Fig. 6. There is an arrow from b_i to b_{i+1} iff $S[1 .. b_i]$ is a cover of $S[1 .. b_{i+1}]$. Note that all elements in the last chain b_6, b_7, b_8, b_9 are cover lengths of S, b_5 is not, but some prefix of b_1, b_2, b_3, b_4 may be.

Algorithm 2: $covers(B)$

Compute $chains(B)$;
if B *is a single chain (ending with* $|S|$) **then return** B;
$B' := refine(B)$; $// |B'| \leq |B|/2 + 1$
$C := covers(B')$;
return $computeUsing(B, C)$;

If $B = (b_1, \ldots, b_k)$, then $chains(B)$ can be constructed in $\mathcal{O}(\sum_{i=1}^{k-1} (\log \frac{b_{i+1}}{b_i} \gamma_n + 1)) = \mathcal{O}(\log n \, \gamma_n)$ time using Lemma 11. Similarly, operation *computeUsing*(B, C) requires $\mathcal{O}(\log n \gamma_n)$ time since the intervals $[b, next_C(b)]$ for $b \in B \backslash refine(B)$ such that $prev(b) \in C$ are pairwise disjoint. The depth of recursion of Algorithm 2 is $\mathcal{O}(\log \log n)$. This implies the required complexity.

$\square$

5.2   Computing Periodic Covers

Our tool for periodic covers are (as usual) *runs*. A *run* (also known as a *maximal repetition*) is a periodic factor $R = T[a \mathinner{..} b]$ which can be extended neither to the left nor to the right without increasing the period $p = \mathsf{per}(R)$, i.e., $T[a-1] \neq T[a + p - 1]$ and $T[b - p + 1] \neq T[b + 1]$ provided that the respective positions exist. The following observation is well-known.

Observation 21. *Two runs in T with the same period p can overlap on at most $p - 1$ positions.*

The *exponent* $\mathsf{exp}(S)$ of a string S is $|S|/\mathsf{per}(S)$. The *Lyndon root* of a string S is the minimal cyclic shift of $S[1 \mathinner{..} \mathsf{per}(S)]$.

If $S = T[a \mathinner{..} b]$ is periodic, then by $\mathsf{run}(S)$ we denote the run R with the same period that contains S. We say that S is *induced* by R. A periodic factor of T is induced by exactly one run [10]. The run-queries are essentially equivalent to two-period queries. By $\mathcal{R}(T)$ we denote the set of all runs in a string T.

Lemma 22 ([3,10,21])**.**

(a) $|\mathcal{R}(T)| \leq n$ and $\mathcal{R}(T)$ can be computed in $\mathcal{O}(n)$ time.
(b) After $\mathcal{O}(n)$-time preprocessing, $\mathsf{run}(S)$ queries can be answered in $\mathcal{O}(1)$ time.
(c) The runs from $\mathcal{R}(T)$ can be grouped by their Lyndon roots in $\mathcal{O}(n)$ time.

The following lemma implies that indeed for any string S, ALLCOVERS(S) can be expressed as a union of $\mathcal{O}(\log |S|)$ arithmetic sequences. It also shows a relation between periodic covers and runs in S.

Lemma 23. *Let S be a string, $A \in Borders(S)$ be an arithmetic sequence with difference p, $A' = A \backslash \{\min(A)\}$ and $a' = \min(A')$. Moreover, let x be the minimal exponent of a run in S with Lyndon root being a cyclic shift of $S[1 \mathinner{..} p]$.*

(a) *If $a' \notin$ ALLCOVERS(S), then $A' \cap$ ALLCOVERS$(S) = \emptyset$.*
(b) *Otherwise, there exists $c \in ((x - 2)p, xp] \cap A'$ such that*
    $A' \cap$ ALLCOVERS$(S) = \{a', a' + p, \dots, c\}$.

Proof. Part (a) follows from Observation 1. Indeed, assume that S has a cover of length $b \in A'$, with $b > a'$. As $S[1 \mathinner{..} a']$ is a cover of $S[1 \mathinner{..} b]$, we would have $a' \in$ ALLCOVERS(S).

We proceed to the proof of part (b). Let c be the maximum element of A' such that $C := S[1 \mathinner{..} c]$ is a cover of S. By the same argument as before, we have that $A' \cap$ ALLCOVERS$(S) = A' \cap [1, c]$. It suffices to prove the bounds for c.

Let L be the minimum cyclic shift of $S[1 \mathinner{..} p]$. We consider all runs $R_1, \dots, R_k$ in S with Lyndon root L. Each occurrence of C in S is induced by one of them.

Each of the runs must hold an occurrence of C. Indeed, by Observation 21, no two of the runs overlap on more than $p-1$ positions, so the pth position of each run cannot be covered by occurrences of C that are induced by other runs. The shortest of the runs has length xp, so $c \le xp$.

Furthermore, let $C' = S[1..c']$ be a prefix of S of length $c' = c + p$. If $p \cdot \exp(R_i) \ge c' + p - 1$, then R_i induces an occurrence of C' and $Cov(C', R_i) = Cov(C, R_i)$. Hence, if $px \ge c' + p - 1$ would hold, C' would be a cover of S, which contradicts our assumption. Therefore, $px < c' + p - 1 = c + 2p - 1$, so $c > (x - 2)p$. $\qquad\square$

Lemma 25 transforms Lemma 23 into a data structure. We use static dictionaries.

Lemma 24 (Ružić [25]). *A static dictionary of n integers that supports $\mathcal{O}(1)$-time lookups can be stored in $\mathcal{O}(n)$ space and constructed in $\mathcal{O}(n(\log \log n)^2)$ time. The elements stored in the dictionary may be accompanied by satellite data.*

Lemma 25 (Computing $\mathcal{O}(\log n)$ Candidates).
For any factor S of T we can compute in $\mathcal{O}(\log n)$ time $\mathcal{O}(\log n)$ borders of S which are candidates for covers of S. After knowing which of these candidates are covers of S, we can in $\mathcal{O}(\log n)$ time represent (as $\mathcal{O}(\log n)$ arithmetic progressions) all borders which are covers of S. The preprocessing time is $\mathcal{O}(n(\log \log n)^2)$ and the space used is $\mathcal{O}(n)$.

Proof. It is enough to show that for any factor S of T and a single arithmetic sequence $A \in Borders(S)$ we can compute in $\mathcal{O}(1)$ time up to four candidate borders. Then, after knowing which of them are covers of S, we can in $\mathcal{O}(1)$ time represent (as a prefix subsequence of A) all borders in A which are covers of S. We first describe the data structure and then the query algorithm.

Data Structure. Let $T[a_1..b_1], \ldots, T[a_k..b_k]$ be the set of all runs in T with Lyndon root L, with $a_1 < \cdots < a_k$ (and $b_1 < \cdots < b_k$). The part of the data structure for this Lyndon root consists of an array A_L containing $a_1, \ldots, a_k$, an array E_L containing the exponents of the respective runs, as well as a dictionary on A_L and a range-minimum query data structure on E_L. Formally, to each Lyndon root we assign an integer identifier in $[1, n]$ that is retained with every run with this Lyndon root and use it to index the data structures. We also store a dictionary of all the runs. The data structure takes $\mathcal{O}(n)$ space and can be constructed in $\mathcal{O}(n(\log \log n)^2)$ time by Lemmas 2, 22 and 24. We also use LCE-queries on T (Lemma 3).

Queries. Let us consider a query for $S = T[i..j]$ and $A \in Borders(S)$. If $|A| = 1$, we have just one candidate. Otherwise, A is an arithmetic sequence with difference p. Let $a = \min(A)$, $A' = A \backslash \{a\}$, and $a' = \min(A')$. We select borders of length a and a' as candidates. If $a' \notin \text{ALLCOVERS}(S)$, then Lemma 23(a) implies that $A \cap \text{ALLCOVERS}(S) \subseteq \{a\}$. We also select borders of lengths in $A \cap ((x - 2)p, xp]$ as candidates, where x is defined as in Lemma 23. Note that there are at most two of them. Let c be the maximum candidate which turned out to be a cover of S. Then $A \cap \text{ALLCOVERS}(S) = A \cap [1, c]$ by Lemma 23(b).

What is left is to compute x, that is, the minimum exponent of a run in S with Lyndon root L that is a cyclic shift of $S[1 .. p]$. Since $|A| \geq 2$, S has a prefix run with Lyndon root L. Then $\ell = \min(p + d, |S|)$, where $d = \text{lcp}(i, i + p)$, is the length of the run. If $\ell = |S|$, then $x = \ell/p$ and we are done. Otherwise, let $i' = i + p + d$. We make the following observation.

Claim. If $a' \in \text{ALLCOVERS}(S)$, then $T[i' .. i' + p]$ is contained in a run in T with Lyndon root L.

Proof. We identify the run $T[a .. b]$ with period p containing $T[i' .. i' + p]$ by asking $\text{lcp}(i', i' + p)$ and $\text{lcs}(i', i' + p)$ queries. This lets us recover the identifier of its Lyndon root L. Similarly we compute the suffix run with Lyndon root L in S and the previous run $T[a' .. b']$ with Lyndon root L in T. Using the dictionary on A_L, we recover the range in the array that corresponds to elements from a to a'. This lets us use a range minimum query on this range in E_L and use it together with the exponents of the prefix and suffix runs of S to compute x. $\square$

5.3 Main Query Algorithm

The main result of this section follows from Lemma 20 and Lemma 25.

Theorem 26. *Let T be a string of length n. After $\mathcal{O}(n \log n)$-time preprocessing, for any factor S of T we can answer a query $\text{ALLCOVERS}(S)$, with output represented as a union of $\mathcal{O}(\log n)$ pairwise disjoint arithmetic sequences, in $\mathcal{O}(\log n (\log \log n)^2)$ time.*

The transformation to the off-line model is similar as in Theorem 18.

Corollary 27. *For a string T of length n, any m queries $\text{ALLCOVERS}(T[i .. j])$ can be answered in $\mathcal{O}((n + m) \log n \log \log n)$ time and $\mathcal{O}(n + m)$ space.*

6 Final Remarks

We showed an efficient data structure for computing internal covers. However, a similar problem for seeds, which are another well-studied notion in quasiperiodicity, seems to be much harder. We pose the following question.

Open Problem
Can one answer internal queries related to seeds in $\mathcal{O}(\text{polylog } n)$ time after $\mathcal{O}(n \cdot \text{polylog } n)$ time preprocessing?

References

1. Amir, A., Landau, G.M., Lewenstein, M., Sokol, D.: Dynamic text and static pattern matching. ACM Trans. Algorithms **3**(2), 19 (2007). https://doi.org/10.1145/1240233.1240242

2. Apostolico, A., Farach, M., Iliopoulos, C.S.: Optimal superprimitivity testing for strings. Inf. Process. Lett. **39**(1), 17–20 (1991). https://doi.org/10.1016/0020-0190(91)90056-N

3. Bannai, H., Tomohiro, I., Inenaga, S., Nakashima, Y., Takeda, M., Tsuruta, K.: The "runs" theorem. SIAM J. Comput. **46**(5), 1501–1514 (2017). https://doi.org/10.1137/15M1011032

4. Bender, M.A., Farach-Colton, M.: The LCA problem revisited. In: Gonnet, G.H., Viola, A. (eds.) LATIN 2000. LNCS, vol. 1776, pp. 88–94. Springer, Heidelberg (2000). https://doi.org/10.1007/10719839_9

5. Berkman, O., Iliopoulos, C.S., Park, K.: The subtree max gap problem with application to parallel string covering. Inf. Comput. **123**(1), 127–137 (1995). https://doi.org/10.1006/inco.1995.1162

6. Breslauer, D.: An on-line string superprimitivity test. Inf. Process. Lett. **44**(6), 345–347 (1992). https://doi.org/10.1016/0020-0190(92)90111-8

7. Breslauer, D.: Testing string superprimitivity in parallel. Inf. Process. Lett. **49**(5), 235–241 (1994). https://doi.org/10.1016/0020-0190(94)90060-4

8. Crochemore, M., et al.: The maximum number of squares in a tree. In: Kärkkäinen, J., Stoye, J. (eds.) CPM 2012. LNCS, vol. 7354, pp. 27–40. Springer, Heidelberg (2012). https://doi.org/10.1007/978-3-642-31265-6_3

9. Crochemore, M., Iliopoulos, C., Kubica, M., Radoszewski, J., Rytter, W., Waleń, T.: Extracting powers and periods in a string from its runs structure. In: Chavez, E., Lonardi, S. (eds.) SPIRE 2010. LNCS, vol. 6393, pp. 258–269. Springer, Heidelberg (2010). https://doi.org/10.1007/978-3-642-16321-0_27

10. Crochemore, M., Iliopoulos, C.S., Kubica, M., Radoszewski, J., Rytter, W., Waleń, T.: Extracting powers and periods in a word from its runs structure. Theor. Comput. Sci. **521**, 29–41 (2014). https://doi.org/10.1016/j.tcs.2013.11.018

11. Crochemore, M., et al.: Shortest covers of all cyclic shifts of a string. In: Rahman, M.S., Sadakane, K., Sung, W.-K. (eds.) WALCOM 2020. LNCS, vol. 12049, pp. 69–80. Springer, Cham (2020). https://doi.org/10.1007/978-3-030-39881-1_7

12. Czajka, P., Radoszewski, J.: Experimental evaluation of algorithms for computing quasiperiods. CoRR abs/1909.11336 (2019). http://arxiv.org/abs/1909.11336

13. Gawrychowski, P., Radoszewski, J., Starikovskaya, T.: Quasi-periodicity in streams. In: Pisanti, N., Pissis, S.P. (eds.) 30th Annual Symposium on Combinatorial Pattern Matching, CPM 2019. LIPIcs, Pisa, Italy, 18–20 June 2019, vol. 128, pp. 22:1–22:14. Schloss Dagstuhl - Leibniz-Zentrum für Informatik (2019). https://doi.org/10.4230/LIPIcs.CPM.2019.22

14. Iliopoulos, C.S., Moore, D.W.G., Park, K.: Covering a string. Algorithmica **16**(3), 288–297 (1996). https://doi.org/10.1007/BF01955677

15. Kärkkäinen, J., Sanders, P., Burkhardt, S.: Linear work suffix array construction. J. ACM **53**(6), 918–936 (2006). https://doi.org/10.1145/1217856.1217858

16. Kociumaka, T.: Efficient data structures for internal queries in texts. Ph.D. thesis, University of Warsaw (2018). https://mimuw.edu.pl/~kociumaka/files/phd.pdf

17. Kociumaka, T., Kubica, M., Radoszewski, J., Rytter, W., Waleń, T.: A linear-time algorithm for seeds computation. ACM Trans. Algorithms **16**(2) (2020). https://doi.org/10.1145/3386369

18. Kociumaka, T., Radoszewski, J., Rytter, W., Straszyński, J., Waleń, T., Zuba, W.:
    Efficient representation and counting of antipower factors in words. In: Martín-
    Vide, C., Okhotin, A., Shapira, D. (eds.) LATA 2019. LNCS, vol. 11417, pp. 421–
    433. Springer, Cham (2019). https://doi.org/10.1007/978-3-030-13435-8_31
19. Kociumaka, T., Radoszewski, J., Rytter, W., Waleń, T.: Efficient data structures
    for the factor periodicity problem. In: Calderón-Benavides, L., González-Caro, C.,
    Chávez, E., Ziviani, N. (eds.) SPIRE 2012. LNCS, vol. 7608, pp. 284–294. Springer,
    Heidelberg (2012). https://doi.org/10.1007/978-3-642-34109-0_30
20. Kociumaka, T., Radoszewski, J., Rytter, W., Waleń, T.: Internal pattern matching
    queries in a text and applications. In: Indyk, P. (ed.) Proceedings of the Twenty-
    Sixth Annual ACM-SIAM Symposium on Discrete Algorithms, SODA 2015, San
    Diego, CA, USA, 4–6 January 2015, pp. 532–551. SIAM (2015). https://doi.org/
    10.1137/1.9781611973730.36
21. Kolpakov, R.M., Kucherov, G.: Finding maximal repetitions in a word in linear
    time. In: 40th Annual Symposium on Foundations of Computer Science, FOCS
    1999, New York, NY, USA, 17–18 October 1999, pp. 596–604. IEEE Computer
    Society (1999). https://doi.org/10.1109/SFFCS.1999.814634
22. Li, Y., Smyth, W.F.: Computing the cover array in linear time. Algorithmica 32(1),
    95–106 (2002). https://doi.org/10.1007/s00453-001-0062-2
23. Moore, D.W.G., Smyth, W.F.: An optimal algorithm to compute all the covers of
    a string. Inf. Process. Lett. 50(5), 239–246 (1994). https://doi.org/10.1016/0020-
    0190(94)00045-X
24. Moore, D.W.G., Smyth, W.F.: A correction to "An optimal algorithm to compute
    all the covers of a string". Inf. Process. Lett. 54(2), 101–103 (1995). https://doi.
    org/10.1016/0020-0190(94)00235-Q
25. Ružić, M.: Constructing efficient dictionaries in close to sorting time. In: Aceto, L.,
    Damgård, I., Goldberg, L.A., Halldórsson, M.M., Ingólfsdóttir, A., Walukiewicz,
    I. (eds.) ICALP 2008. LNCS, vol. 5125, pp. 84–95. Springer, Heidelberg (2008).
    https://doi.org/10.1007/978-3-540-70575-8_8

An Efficient Elastic-Degenerate Text Index? Not Likely

Daniel Gibney$^{(\boxtimes)}$

University of Central Florida, Orlando, FL 32816, USA
daniel.j.gibney@gmail.com
http://cs.ucf.edu/~dgibney/

Abstract. Elastic-degenerate text provides a novel and effective method for modeling collections of text that have local variations. Due to its applicability in pan-genomics, an index for an elastic-degenerate text which can efficiently report the occurrences of a given query pattern is desirable. This paper attempts to dash our hopes for such an index, one that is deterministic and has good worst-case query time. We do so by providing conditional lower bounds based on the Orthogonal Vectors Hypothesis (OVH) (and hence the Strong Exponential Time Hypothesis). We show that, even with arbitrary polynomial preprocessing time, an index for an elastic-degenerate text with n degenerate letters that can perform queries on a pattern of length m in time $O(n^\alpha m^\beta)$ for constants α and β where $\alpha < 1$ or $\beta < 1$ would violate OVH. Additionally, we provide an elastic-degenerate text index with query time $O(nm^2)$, which is independent of the size N (distinct from its length) of the elastic-degenerate text. Finally, we investigate the hardness of matching elastic-degenerate text to elastic-degenerate text.

Keywords: Elastic-degenerate text · Text indexing · Conditional lower bounds

1 Introduction

Very recently, a useful new way of modeling collections of closely related strings (or sequences) called Elastic-Degenerate Text (*ED-text*) has started receiving attention. Introduced by Iliopoulos, Kundu, and Pissis in [18], ED-text arises from the need to model collections of strings where there are regions of local variation. ED-text models these collections in a natural and convenient representation. It may be best to first look at an example. Accordingly, we have provided this Mad-libs inspired instance of an ED-text:

$$\{\text{The}\} \circ \left\{\begin{matrix} \text{dog} \\ \text{moose} \\ \text{ostrich} \end{matrix}\right\} \circ \left\{\begin{matrix} \text{jumped} \\ \text{sat} \\ \text{washed} \end{matrix}\right\} \circ \left\{\begin{matrix} \varepsilon \\ \text{but not} \end{matrix}\right\} \circ \{\text{before}\} \circ \left\{\begin{matrix} \text{dancing} \\ \text{baking} \\ \text{pontificating} \end{matrix}\right\}$$

Supported in part by the U.S. National Science Foundation (NSF) under CCF-1703489.

C. Boucher and S. V. Thankachan (Eds.): SPIRE 2020, LNCS 12303, pp. 76–88, 2020.
https://doi.org/10.1007/978-3-030-59212-7_6

One can see that within each matching pair of brackets, there is a set of ordinary strings. This set of strings models the possible variants occurring within that region. Each of these sets is referred to as a *degenerate letter*. In regions where no variation is allowed, a degenerate letter will consist of a single string. Now, having some intuition for the notion, let us formally define an ED-text.

Elastic-Degenerate Text (ED-text). A degenerate letter X is a non-empty set of strings. This can include the empty string ε. An ED-text $\tilde{T}$ of length n consists of n degenerate letters, that is $\tilde{T} = X[1] \circ X[2] \circ \ldots \circ X[n]$ where $X[i]$ refers to the ith degenerate letter in $\tilde{T}$ and $\circ$ denotes concatenation. Let $|X[i]|$ be the number of strings in $X[i]$, $X[i][j]$ the jth string in degenerate letter $X[i]$, and $|X[i][j]|$ the length of the string $X[i][j]$. The size of $\tilde{T}$ is defined as $N = \sum_{i=1}^{n} \sum_{j=1}^{|X[i]|} |X[i][j]|$. We will use the term sub-ED-text of $\tilde{T}$ to refer to a concatenation of degenerate letters that exist consecutively within $\tilde{T}$. By *alphabet*, we mean the symbols appearing within the strings. In our example, $n = 6$ and $N = 73$ (including the space within 'but not'), and the alphabet could be $\{a,b,c,...,z, , \varepsilon\}$ (not all used).

More than modeling Mad-Lib's, the ability to model collections of related strings has an important role in the field of pan-genomics [13,23,24]. One of the most fundamental problems is having to find occurrences of a pattern (consisting of just an ordinary string) within an ED-text.

Pattern Matching in ED-text. We let $\circ$ denote the concatenation of ordinary strings, as well as degenerate letters. A pattern P of length m matches a sub-ED-text $X[s] \circ \ldots \circ X[t]$ of $\tilde{T}$ if P can be written as $P = P_1 \circ P_2 \circ \ldots \circ P_{m'}$ where $P_1 \neq \varepsilon$ and is a suffix of a string in $X[s]$, P_i is equal to a string in $X[i]$ for $s < i < t$, and $P_{m'} \neq \varepsilon$ and is a prefix of a string in $X[t]$. We say that this match of P *spans* $t - s + 1$ degenerate letters and that P has a match in $\tilde{T}$ iff P has a match with some sub-ED-text of $\tilde{T}$. Returning to our example above, the pattern $P = $ "umpedbefor" is of length $m = 10$ and occurs as a match which spans $5 - 3 + 1 = 3$ degenerate letters. This work will be chiefly concerned with the indexing version of this matching problem.

Problem 1 (Elastic-Degenerate Indexing (ED-Indexing)). *Given an ED-text $\tilde{T}$ of length n and size N, construct an index that when given a query pattern P reports whether P has a match in $\tilde{T}$.*

Background and Related Work. Elastic-Degenerate String Matching (EDSM) was introduced recently in [18], however, related problems have been considered before in the form of gapped strings [20]. Within [18], the authors presented a solution using time $O(\alpha\gamma nm + N)$, where α and γ represent the maximum number of strings in any degenerate letter and γ the maximum span of any occurrence of P in $\tilde{T}$. Variants of this problem that allow for approximate matching, have appeared in [8], where the time complexity is parameterized by the number of mismatches, or the edit-distance, allowed from an exact match. Further results appearing in [7] give a randomized solution which runs in expected time $O(nm^{1.381} + N)$ by making use of fast matrix multiplication.

The work of Grossi et al. [17] addresses the online version of this problem, where preprocessing is done on the pattern, and the ED-text is provided online. This is different from the version addressed here, which is the offline variant, where the ED-text and pattern are provided as input. The authors present an algorithm requiring $O(m)$ space and $O(nm^2 + N)$ time after preprocessing, as well as a fast bit-vector algorithm requiring time $O(N\lceil\frac{m}{w}\rceil)$ with $O(m\lceil\frac{m}{w}\rceil)$ pre-processing time and space, where w is the size of the computer word. A different solution for the online problem using time $O(nm\sqrt{m\log m} + N)$ was later provided in [5]. A generalization of the online problem to a set of patterns of total length M was considered in [21]. There, an algorithm requiring time $O(N\lceil\frac{M}{w}\rceil)$ with $O(M)$ pre-processing time and space was presented.

Also in [7], a lower bound was given. It states that a solution not making use of fast matrix multiplication and running in time $O(nm^{1.5-\delta} + N)$ for $\delta > 0$ implies a strongly subcubic time algorithm for combinatorial Boolean Matrix Multiplication (BMM). The proof in [7] is distinct from the ones provided here in several ways: the reductions are from a different problem (Orthogonal vectors versus BMM), and we allow the construction of an index using arbitrary polynomial preprocessing time. Moreover, the alphabet sizes used in our reduction is four, whereas the one used in [7] is polynomial in n.

Moving forward, the following problem is the basis for the conditional lower bounds given here.

Problem 2 (Orthogonal Vectors (OV)). *Given two sets of d-dimensional binary vectors $X, Y \subseteq \{0,1\}^d$, determine whether there exist $x \in X$ and $y \in Y$ such that their inner product $x \cdot y = \sum_{i=1}^d x[i] \cdot y[i] = 0$.*

The Orthogonal Vectors Hypothesis (OVH) is a frequently used assumption in fine-grained complexity. For examples, see [1–3,6,9–12,16,22,26]. It states that for $d = \omega(\log(|X| + |Y|))$, there does not exist a solution to OV running in time $O((|X||Y|)^{1-\varepsilon})$ for any $\varepsilon > 0$. Due to a well-known reduction from k-SAT, OVH being shown to be false would prove the Strong Exponential Time Hypothesis [19] (SETH) false as well. The following extension of OV to indexing problems, along with Lemma 1 appears in [15].

Problem 3 (OV-Indexing). *Given a set of binary vectors $X \subset \{0,1\}^d$, construct an index which, when given as a query a set of binary vectors $Y \subseteq \{0,1\}^d$, reports whether there exist $x \in X$ and $y \in Y$ such that x and y are orthogonal.*

Lemma 1 [15]. *Under OVH, an index cannot support $O(|X|^\alpha|Y|^\beta)$-time queries for OV-indexing, for constant $\alpha < 1$ or constant $\beta < 1$. This is even with preprocessing of X that uses time $d^{O(1)}|X|^{O(1)}$.*

In the final section of this paper, we investigate the hardness of matching ED-text to ED-text (see Sect. 4 for details). Previous work has been done in [4] for degenerate (non-elastic) texts where each degenerate letter contains strings of equal length. In this case, the authors gave a method for determining whether the intersection of two collections of strings represented by degenerate text is empty that runs in time $O(N + M)$ for two degenerate texts of sizes N and M, respectively.

1.1   Our Contribution

Section 2 starts with a warm-up. The first reduction will help demonstrate one of the main techniques used to identify orthogonal vectors, which is then expanded upon in the next reduction.

Theorem 1. *OV-Indexing on the vector sets $X, Y \in \{0,1\}^d$ can be reduced to ED-Indexing over an ED-text $\tilde{T}$ constructed from X, of length $n = O(d|X|)$, of size $N = O(d|X|)$, and having a binary alphabet. At query time, $|Y|$ query patterns, each of length $m = O(d)$, constructed from Y need to be used.*

Based on Lemma 1 and Theorem 1, even with polynomial preprocessing of $\tilde{T}$, a significant improvement in query time over time linear in the length $\tilde{T}$ is not possible under OVH.

Corollary 1. *Over an alphabet of size two, a solution for ED-indexing that has query time $O(n^\alpha m^{O(1)} + m)$ for constant $\alpha < 1$ is not possible under OVH (and hence SETH), even with arbitrary polynomial preprocessing.*

Our second hardness result constructs larger query patterns, relating the size of the query pattern to the lower bound. The proof of Theorem 2 makes use of (as far as the authors are aware) a new technique that we call *phase-shift keying*, which may be useful in other pattern-matching-on-graph-related reductions. Combining Corollaries 1 and 2, we obtain Theorem 3.

Theorem 2. *OV-Indexing on vector sets $X, Y \in \{0,1\}^d$ can be reduced to ED-Indexing over a ED-text $\tilde{T}$, constructed from X, of length $n = O(d(|X| + |Y|))$, of size $N = O(d(|X|^2 + |Y|))$, and having an alphabet of size four. One pattern P needs to be given as a query, where P is constructed from Y and has length $m = O(d|Y||X|)$.*

Corollary 2. *Over an alphabet of size four, an ED-index that has query time $O(n^{O(1)} m^\beta + m)$ for constant $\beta < 1$ is not possible under OVH (and hence SETH), even with arbitrary polynomial preprocessing.*

Theorem 3. *Over an alphabet of size four, an ED-index that has query time $O(n^\alpha m^\beta + m)$ for constants α and β where $\alpha < 1$ or $\beta < 1$ is not possible under OVH (and hence SETH), even with arbitrary polynomial preprocessing.*

In Sect. 3 we provide an ED-index with $O(nm^2)$ query time. Lastly, in Sect. 4 we give a proof that matching two ED-texts is solvable in polynomial time when empty strings are not allowed (see Sect. 4 for exact statements of the results).

2   Lower Bounds on ED-Text Index Queries

We will use '$\bigcirc_{i=1}^{t} S_i$' to denote the concatenation of t indexed degenerate letters $S_1, ..., S_t$, or strings $S_1, ..., S_t$. For a symbol c and $\gamma \geq 1$, we use c^γ to denote c repeated γ times. We also define c^0 to be the empty string, ε. All arrays will be indexed starting at 1.

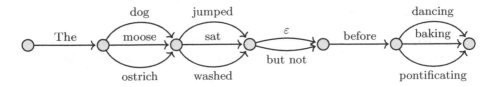

Fig. 1. A representation of the ED-text from the introduction as a directed graph.

The representation of an ED-text as a directed graph, like in Fig. 1, is helpful. Each degenerate letter can be represented as a set of edges between two vertices. The aim in finding a match of a pattern P becomes finding a path (possibly starting and ending mid-edge) whose concatenation of edge labels is the same as P. We will sometimes refer to matching a particular text within a degenerate letter with a portion of the query pattern as *taking an edge*. Note that our problem can now be viewed as the string-to-graph matching problem, whose algorithmic version as well as the indexing version were well studied from the side of lower bounds [14,15]. However, the graphs we are considering are highly restricted. Hence, these results do not apply here immediately.

For both reductions, we assume that we are given an instance of OV-Index containing sets X and Y of binary vectors of dimension d is $\omega(\log(|X|+|Y|))$ (a standard assumption due to a reduction from k-SAT) where d is $polylog(|X|+|Y|)$. The vectors in X and Y are indexed from 1 to $|X|$ and 1 to $|Y|$ (resp), but in no particular order.

2.1   Warm Up - Proof of Theorem 1 and Corollary 1.

The proof of Theorem 1 is very simple. Nonetheless, the component and vector gadgets introduced here will serve as a building blocks for the second reduction.

For a binary vector $x_i \in X$, we define the following vector and component gadgets,

$$VG(x_i) = \bigcirc_{j=1}^d CG_1(x_i[j]) \quad \text{where} \quad CG_1(b) = \begin{cases} \{00,11\} & \text{if } b=0 \\ \{11\} & \text{if } b=1 \end{cases}$$

Then, using the whole set X, we construct our ED-text as $\tilde{T} = \bigcirc_{i=1}^{|X|} (VG(x_i) \circ \{010\})$.

Next, for a binary vector $y_i \in Y$, we define the following pattern

$$P_i = \left(\bigcirc_{j=1}^d CG_2(y_i[j])\right) \circ 010 \quad \text{where} \quad CG_2(b) = \begin{cases} 11 & \text{if } b=0 \\ 00 & \text{if } b=1 \end{cases}$$

It is easy to see that a degenerate letter $\{010\}$ in $\tilde{T}$ can only match with the suffix 010 of P_j. And, that if the suffix 010 of P_j is matched with $\{010\}$ in $VG(x_i) \circ \{010\}$, then x_i must be orthogonal to y_j. From these observations we obtain the following fact, which proves Theorem 1.

Fact 1. *For all $y_j \in Y$ and $j \in [1, |Y|]$, pattern P_j has a match in $\tilde{T}$ iff $x_i \cdot y_j = 0$.*

Based on Fact 1, using $|Y|$ queries to an ED-index for $\tilde{T}$, we can determine whether there exists an orthogonal pair of vectors. Also, note that the length of $\tilde{T}$ is $n = O(d|X|)$, and the length of each pattern is $m = O(d)$. Hence, if we could answer an ED-index query in time $O(n^\alpha m^{O(1)})$ for constant $\alpha < 1$ then in time $O(|Y|(d|X|)^\alpha d^{O(1)})$ we could answer an OV-index query. Using Lemma 1, OVH would be proven false with such an ED-index. This completes the proof of Corollary 1.

2.2   Proof of Theorem 2 and Corollary 2

In this section we will provide a reduction where $m > n$. By making m this large we are able to obtain the desired results. In the previous section, since $m = O(d)$ obtaining these results was not possible.

To provide some intuition, let us start by giving the form of the final ED-text and query pattern. The ED-text $\tilde{T}$ will be of the form:

$$\tilde{T} = \{\$\} \circ \underbrace{U \circ \{\$\} \circ \ldots \circ U \circ \{\$\}}_{|Y|-1 \text{ repetitions of } U \circ \{\$\}} \circ \tilde{T}_X \circ \{\$\} \circ \underbrace{U \circ \{\$\} \circ \ldots \circ U \circ \{\$\}}_{|Y|-1 \text{ repetitions of } U \circ \{\$\}}$$

And the pattern P will be of the form:

$$P = \$ \circ VG_P(y_1) \circ \$ \circ VG_P(y_2) \circ \$ \circ \ldots \circ \$ \circ VG_P(y_{|Y|}) \circ \$$$

The gadget U is constructed as a 'universal' gadget that accepts all vector gadgets in P. The \$'s are a new symbol added to the alphabet. In $\tilde{T}$, the \$'s appear on both sides of universal vector gadgets, and nowhere within $\tilde{T}_X$. They also appear on both sides of vector gadgets within P. This causes exactly one of the vector gadgets in P to form a match in $\tilde{T}_X$ iff P has a match in $\tilde{T}$. Our main task will be to design $\tilde{T}_X$ so that this is possible iff there exists an orthogonal vector pair. For this, we will introduce the notion of *phase-shift keying*. Very loosely, our goal will be to allow for the vector gadget in P to make an initial choice but then enforce commitment. Intuitively, the vector gadget $VG_P(y_j)$ that ends up matching in $\tilde{T}_X$ gets to choose any vector gadget in $\tilde{T}_X$ upon entering $\tilde{T}_X$. However, once this choice is made, then $VG_P(y_j)$ is 'committed' and can only match degenerate letters (edges) within its choice of vector gadget.

To accomplish this, for each $i \in [1, |X|]$ we will have to define a different vector gadget. For a vector $x_i \in X$, the ED-text is

$$VG_i(x_i) = \bigcirc_{j=1}^{d} CG_1^i(x_i[j])$$

where the component gadgets are defined differently for each index i from 1 to $|X|$. Notice the 2's within the component gadgets are shifted depending on i.

$$CG_1^i(b) = \begin{cases} \{2^{|X|-(i-1)} \circ 0 \circ 2^{i-1},\ 2^{|X|-(i-1)} \circ 1 \circ 2^{i-1}, \varepsilon\} & \text{if } b = 0 \\ \{2^{|X|-(i-1)} \circ 1 \circ 2^{i-1}, \varepsilon\} & \text{if } b = 1 \end{cases}.$$

The ED-text $\tilde{T}_X$ is defined as

$$\tilde{T}_X = \{\varepsilon, 2, 22, \ldots, 2^{|X|-1}\} \circ \left(\bigcirc_{i=1}^{|X|} VG_i(x_i)\right) \circ \{\varepsilon, 2, 22, \ldots, 2^{|X|-1}\}.$$

To have our vector gadgets within P choose their 'phase-shift', we have designed our ED-text so that within the first degenerate letter of $\tilde{T}_X$ there exists several choices of phase. This choice shifts the 2's within the component gadgets in P to the left by some amount, creating a matching 'key' for some vector gadget in $\tilde{T}_X$. After this point, the pattern can only match with degenerate letters that have a matching phase-shift. See Fig. 2 for an illustration. The final degenerate letter serves the purpose of absorbing any additional unnecessary 2's.

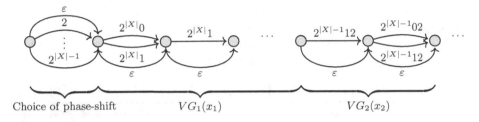

Fig. 2. Phase-shift keying: $\tilde{T}_X$ assuming $x_1 = [0, 1, \ldots]$ and $x_2 = [1, 0, \ldots]$. The edge choice in the left-most degenerate letter of $\tilde{T}_X$ determines which gadget in $\tilde{T}_X$ the gadget in P can match with.

We next need a 'universal' vector gadget U that will accept all vector gadgets in P. This can be accomplished with the following.

$$U = \underbrace{\left\{2^{|X|} \circ 0, 2^{|X|} \circ 1\right\} \circ \ldots \circ \left\{2^{|X|} \circ 0, 2^{|X|} \circ 1\right\}}_{d \text{ repetitions}} \circ \{2^{|X|}\}$$

To create the query pattern, we use the string

$$P = \$ \circ VG_P(y_1) \circ \$ \circ VG_P(y_2) \circ \$ \circ \ldots \circ \$ \circ VG_P(y_{|Y|}) \circ \$.$$

where the vector gadget VG_P is defined as

$$VG_P(y_i) = \left(\bigcirc_{j=1}^{d} \left(2^{|X|} \circ CG_2(y_i[j])\right)\right) \circ 2^{|X|} \quad \text{where} \quad CG_2(b) = \begin{cases} 1 & \text{if } b = 0 \\ 0 & \text{if } b = 1 \end{cases}$$

Lemma 2. *The pattern P occurs as a match in $\tilde{T}$ if and only if there exists an orthogonal pair of vectors x_i, y_j where $x_i \in X$ and $y_j \in Y$.*

Proof. First, suppose there exists such a pair of orthogonal vectors x_i and y_j. We will match the substring $VG_P(y_j)$ of P with the sub-ED-text $\tilde{T}_X$ as follows. Match the prefix 2^{i-1} of $VG_P(y_j)$ with the edge labeled 2^{i-1} in the first degenerate letter of $\tilde{T}_X$. Next, traverse from left-to-right until the sub-ED-text $VG_i(x_i)$ is reached in $\tilde{T}_X$ by taking only edges with label ε. Once the sub-ED-text $VG_i(x_i)$ is reached, for each component $h \in [1, d]$,

- if $y_j[h] = 0$, take edge $2^{|X|-(i-1)} \circ 1 \circ 2^{i-1}$, and match $2^{|X|-(i-1)} \circ 1 \circ 2^{i-1}$ in $VG_P(y_j)$;
- if $y_j[h] = 1$, take edge $2^{|X|-(i-1)} \circ 0 \circ 2^{i-1}$ and match $2^{|X|-(i-1)} \circ 0 \circ 2^{i-1}$ in $VG_P(y_j)$. Because $x_i[h]$ must be 0, we know this edge exists.

After matching all component gadgets in P for y_j take ε labeled edges until the last degenerate letter in $\tilde{T}_X$ is reached. On this last degenerate letter take the edge $2^{|X|-(i-1)}$. This completes matching the substring $VG_P(y_j)$ to T_X. The prefix $\$ \circ VG_P(y_1) \circ \$ \circ \ldots \circ V_P(v_{j-1})\$$ of P is matched to the sub-ED-text $\{\$\} \circ U \circ \{\$\} \ldots \{\$\} \circ U \circ \{\$\}$ which directly precedes $\tilde{T}_X$. This is possible since the universal gadget U can match any vector gadget in P, and P has at most $|X| - 1$ vector gadgets in this prefix. Similarly, we can match the suffix of P following $VG_P(y_j)$ to universal gadgets in $\tilde{T}$ directly following $\tilde{T}_X$.

Next, suppose P has a match in the ED-text $\tilde{T}$. Since there are only $|X| - 1$ universal gadgets on either side of $\tilde{T}_X$, the match has to match $\$$'s on either side of $\tilde{T}_X$. Hence, some vector gadget $VG_P(y_j)$ in P must be matched within $\tilde{T}_X$. Dependent on which ever edge 2^{i-1} is taken within the first degenerate letter of $\tilde{T}_X$, the prefix $2^{|X|-(i-1)}CG_2(y_j[1])$ of the remaining unmatched portion of $VG_P(y_j)$ can only match edges within $VG_i(x_i)$ in $\tilde{T}_X$. Moreover, once one of these edges is taken, the substring $2^{|X|-(i-1)} \circ CG_2(y_j[2]) \circ 2^{i-1}$ in $VG_P(y_j)$ is matched as well. Hence, it must again match an edge in $VG_i(x_i)$ with $\tilde{T}_X$. This happens for each of the d component gadgets in $VG_P(y_j)$, forcing $VG_P(y_j)$ to match entirely against $VG_i(x_i)$ in $\tilde{T}_X$. Because these two vector gadgets are matched, x_i and y_j are orthogonal.   □

Lemma 2 proves Theorem 2. For Corollary 2, the length of the pattern P is $m = O(d|Y||X|)$. The *lengths* of U and $VG_i(\cdot)$ are both $O(d)$ (recall the definition of length of an ED-text does not include the length of edge labels), and the length of $\tilde{T}_X$ is $O(d|X|)$. Hence, $n = O(|Y||U| + |T_x|) = O(d(|Y| + |X|))$. Suppose $|X| \geq |Y|$ so that $n = O(d|X|)$. An ED-index with query time $O(n^\alpha m^\beta)$ would imply an solution for OV-Index with query time $O((d|X|)^\alpha(d|X||Y|)^\beta) = O(d^{\alpha+\beta}|X|^{\alpha+\beta}|Y|^\beta)$. Lemma 1 implies that if β is a constant less than 1, we would prove OVH false.

3   An ED-Index

One can demonstrate that, with preprocessing, an ED-index can answer queries in time independent of N, the size of the ED-text $\tilde{T}$. Recall that n is length of $\tilde{T}$ and m the length of the query pattern.

Roughly, this is done by constructing generalized suffix trees (GST) [25] for each degenerate letter and then inserting the pattern P into each GST during query time. At each vertex in the graph-based representation of the ED-text, like in Fig. 1, a list of size $O(m)$ is kept containing all partial matches with shortest-span, one for each possible index in P. The left-most list is initially empty. We then work from left-to-right creating a new list for vertex v_{i+1} based on the list for v_i. Using the GST we can complete partial matches that end within the degenerate letter between v_i and v_{i+1}, extend partial matches that don't end, create new matches that start (and may end) during this degenerate letter, and replace longer partial matches ending the same index as shorter partial matches. Each of the $O(m)$ partial matches in the list for v_i can be processed in $O(m)$ time, resulting in $O(m^2)$ time for each of the $O(n)$ vertices.

Fact 2. *There exists an ED-index that can answer queries in time $O(nm^2)$.*

Proof. Here is a more detailed description of how to build such an index:

Each vertex v_j in the graph representation of the ED-text $\tilde{T}$ will maintain a list $list(v_j)$ of at most $m + 1$ entries. Each entry in $list(v_j)$ consists of a tuple (s, i, f) where i of an index in P such that $P[1, i]$ has been matched so far, s is a starting location of $P[1, i]$ in $\tilde{T}$ (the degenerate letter, the text within the degenerate letter, and starting index) where this match of $P[1, i]$ spans the minimum number of degenerate letters possible while also ending at v_j, and f is this minimum number of degenerate letters.

Let $F_{j,j+1}$ denote the degenerate letter between v_j and v_{j+1}. In preprocessing we construct a generalized suffix tree $ST_{j,j+1}$ for $F_{j,j+1}$. Let $size(u)$ be the size of subtree in $ST_{j,j+1}$ rooted at u. Each node of ST will maintain a bit indicating whether its subtree contains a leaf corresponding to an entire text in $F_{j,j+1}$, and if it contains such a leaf, then a pointer to the corresponding text. At query time, we can add P to $ST_{j,j+1}$ in $O(m)$ time. Suppose we have fully matched P up to vertex v_j, and wish to acquire $list(v_{j+1})$. For clarity, we will break this into three phases, *completion*, *extension*, and *creation/replacement*.

- **Completion:** In this phase, any matches that become completed within this degenerate letter will end the query by reporting a match. Iterating through $list(v_j)$, for entry (s, i, f) we first obtain the leaf ℓ corresponding to the suffix $P[i+1, m]$. If ℓ's parent node has within its subtree a leaf for an entire string within the degenerate letter we report a match and stop. After iterating through $list(v_j)$, we next obtain the leaf ℓ for $P[1, m]$. If the subtree size of ℓ's parent is more than one, we report a match. This step takes time $O(m)$ in total.

- **Extension:** In this phase we will extend any unfinished matches to $list(v_{j+1})$. Iterating through $list(v_j)$, for entry (s, i, f), we find the leaf ℓ corresponding to the suffix $P[i+1, m]$ and check for each leaf ℓ_T corresponding to an entire text $T \in F_{j,j+1}$ whether ℓ_T's parent lies on the path from root to ℓ. This can be done using lowest common ancestor (lca) queries. For $T \in F_{j,j+1}$ where this does occur, we add $(s, i + sd(lca(\ell, \ell_T)), f + 1)$ to $list(v_{j+1})$ where $sd(u)$ is the string depth of a node u in ST.

  Alternatively, we can traverse up from ℓ and check for each node on the path from ℓ to the root of ST whether it contains a leaf corresponding to the end of a string in $F_{j,j+1}$. Since, this information is precomputed, each one of these checks can be done in constant time. Combining, we can perform this in $O(\min(size(F_{j,j+1}), m))$ time per entry in $list(v_j)$ making the total time $O(m \cdot \min(size(F_{j,j+1}), m))$ for this step.

- **Creation/Replacement:** This phase ensures that any matches that start mid-degenerate letter get added to $list(v_{j+1})$ and that we only keep the partial matches for P spanning the smallest number of degenerate letters. Sort $list(v_{j+1})$ in decreasing order using the second component as the key. This can be done in linear time since the keys are bounded by m. For each node u in ST on the path from leaf ℓ for $P[1, m]$ to the root of ST, if the node u has a leaf which corresponds to the suffix of a text $T \in F_{j,j+1}$ and $sd(u)$ is not a key(a second component) in $list(v_{j+1})$, then add $(s, sd(u), 1)$ to $list(v_{j+1})$ where s is a starting location for $P[1, sd(u)]$ in T (if multiple such texts exist, we take only one). If $sd(u)$ is equal to a key i_h in $list(v_{j+1})$ then we replace the existing entry (s, i_h, f) in $list(v_{j+1})$ with $(s', i_h, 1)$ where s' is the starting location for $P[1, sd(u)]$ in T. Because $list(v_{j+1})$ is sorted by the ending index of each partial match, and the string depth of the nodes on the path is decreasing, this whole step can be done in $O(m)$ time.

The time is dominated by the second step which requires $O(m \cdot \min(size(F_{j,j+1}), m)))$ time. Summing over all degenerate letters, the total time becomes $O(m \cdot \min(N, mn))$. The construction of the generalized suffix trees in preprocessing can be done in $O(n + N)$ time.

An improved solution can likely be obtained by applying techniques based on the fast-Fourier transform such as those used [5]. Our purpose is simply to illustrate that we can decouple the size N of the degenerate text from the query time complexity. We leave further improvements for future research.

4   Matching Two Elastic-Degenerate Texts

We say an ED-text $X[1] \circ \ldots \circ X[t_1]$ matches an ED-text $Y[1] \circ \ldots \circ Y[t_2]$ if a string S exists such that S can be decomposed as $S = S_{1,1} \circ S_{1,2} \circ \ldots \circ S_{1,t_1}$ where $S_{1,1} \in X[1]$, ..., $S_{1,t_1} \in X[t_1]$ and also decomposed as $S = S_{2,1} \circ S_{2,2} \circ \ldots \circ S_{2,t_2}$ where $S_{2,1} \in Y[1]$, ..., $S_{2,t_2} \in Y[t_2]$. We say that ED-text $\tilde{T}_2$ has a match in $\tilde{T}_1$ if $\tilde{T}_2$ matches with some sub-ED-text of $\tilde{T}_1$. We note that the two ED-texts T_1 and T_2 are interchangeable in terms of this definition. Only as a matter of convention will we say that $\tilde{T}_2$ has a match within $\tilde{T}_1$.

Problem 4 (ED-text to ED-text Matching). *Given an ED-text $\tilde{T}_1$ of length n_1 and size N_1, and ED-text $\tilde{T}_2$ of length n_2 and size N_2, determine whether $\tilde{T}_2$ has a match in $\tilde{T}_1$.*

In [4] the authors provide a solution for computing whether two degenerate (non-elastic) strings match. This is for the case where within each degenerate letter all strings are of the same length (non-elastic). Their algorithm runs in time in $O(N_1 + N_2)$. It relies on the fact that the tries of these degenerate letters have a leveled structure. This gives us that the ED-text to ED-text matching problem we defined above can easily be solved in linear time for such degenerate strings.

In the case of *elastic* degenerate strings, this property no longer holds. However, we will show that when there are no empty strings within any degenerate letters, the matching problem can still be solved in polynomial time.

Theorem 4. *ED-text to ED-text Matching is solvable in polynomial time when no degenerate letters contain the empty string.*

Proof. We will first construct a DFA for $\tilde{T}_1$. For the ith degenerate letter in $\tilde{T}_1$ construct a compact trie T_i over all of its strings. Let s_i refer to the root of T_i. Next, for each trie T_i bring each leaf into correspondence with a single vertex t_i. The resulting graph we denote as G_i. By bringing t_i into correspondence with s_{i+1} for $i \in [1, n-1]$, and then making s_1 the start state and t_n the only accepting state, we obtain a DFA for $\tilde{T}_1$ of size N_1.

We perform the same steps for $\tilde{T}_2$, to form a DFA of size N_2. Then the DFA for the intersection of the two languages recognized by these DFAs can be done with the standard Cartesian product technique, which results in a DFA of size $O(N_1 N_2)$. Checking to see if the language of the resulting DFA is empty can then be done in linear time.

Repeating this process across all starting and ending degenerate letter indices in $\tilde{T}_1$ results in a total time which is $O(n_1^2 N_1 N_2)$.   □

We leave the following question open.

Open Question 1. *Is the ED-text to ED-text Matching Problem solvable in polynomial time when the empty string is allowed within degenerate letters?*

References

1. Abboud, A., Backurs, A., Hansen, T.D., Williams, V.V., Zamir, O.: Subtree isomorphism revisited. ACM Trans. Algorithms **14**(3), 27:1–27:23 (2018). https://doi.org/10.1145/3093239
2. Abboud, A., Backurs, A., Williams, V.V.: Tight hardness results for LCS and other sequence similarity measures. In: IEEE 56th Annual Symposium on Foundations of Computer Science, FOCS 2015, Berkeley, CA, USA, 17–20 October 2015, pp. 59–78 (2015). https://doi.org/10.1109/FOCS.2015.14
3. Abboud, A., Bringmann, K., Dell, H., Nederlof, J.: More consequences of falsifying SETH and the orthogonal vectors conjecture. In: Proceedings of the 50th Annual ACM SIGACT Symposium on Theory of Computing, STOC 2018, Los Angeles, CA, USA, 25–29 June 2018, pp. 253–266 (2018). https://doi.org/10.1145/3188745.3188938
4. Alzamel, M., et al.: Degenerate string comparison and applications. In: 18th International Workshop on Algorithms in Bioinformatics, WABI 2018, 20–22 August 2018, Helsinki, Finland, pp. 21:1–21:14 (2018). https://doi.org/10.4230/LIPIcs.WABI.2018.21

5. Aoyama, K., Nakashima, Y., Inenaga, S., Bannai, H., Takeda, M.: Faster online elastic degenerate string matching. In: Annual Symposium on Combinatorial Pattern Matching, CPM 2018, Qingdao, China 2–4 July 2018, pp. 9:1–9:10 (2018). https://doi.org/10.4230/LIPIcs.CPM.2018.9
6. Backurs, A., Indyk, P.: Edit distance cannot be computed in strongly subquadratic time (unless SETH is false). SIAM J. Comput. **47**(3), 1087–1097 (2018). https://doi.org/10.1137/15M1053128
7. Bernardini, G., Gawrychowski, P., Pisanti, N., Pissis, S.P., Rosone, G.: Even faster elastic-degenerate string matching via fast matrix multiplication. In: 46th International Colloquium on Automata, Languages, and Programming, ICALP 2019, Patras, Greece, 9–12 July 2019, pp. 21:1–21:15 (2019). https://doi.org/10.4230/LIPIcs.ICALP.2019.21
8. Bernardini, G., Pisanti, N., Pissis, S.P., Rosone, G.: Approximate pattern matching on elastic-degenerate text. Theor. Comput. Sci. **812**, 109–122 (2020). https://doi.org/10.1016/j.tcs.2019.08.012
9. Borassi, M., Crescenzi, P., Habib, M.: Into the square: on the complexity of some quadratic-time solvable problems. Electron. Notes Theor. Comput. Sci. **322**, 51–67 (2016). https://doi.org/10.1016/j.entcs.2016.03.005
10. Bringmann, K., Künnemann, M.: Quadratic conditional lower bounds for string problems and dynamic time warping. In: IEEE 56th Annual Symposium on Foundations of Computer Science, FOCS 2015, Berkeley, CA, USA, 17–20 October 2015, pp. 79–97 (2015). https://doi.org/10.1109/FOCS.2015.15
11. Chen, L.: On the hardness of approximate and exact (bichromatic) maximum inner product. In: 33rd Computational Complexity Conference, CCC 2018, San Diego, CA, USA, 22–24 June 2018, pp. 14:1–14:45 (2018). https://doi.org/10.4230/LIPIcs.CCC.2018.14
12. Chen, L., Williams, R.: An equivalence class for orthogonal vectors. In: Proceedings of the Thirtieth Annual ACM-SIAM Symposium on Discrete Algorithms, SODA 2019, San Diego, California, USA, 6–9 January 2019, pp. 21–40 (2019). https://doi.org/10.1137/1.9781611975482.2
13. The computational pan-genomics consortium. Computational pan-genomics: status, promises and challenges. Brief. Bioinform. **19**(1), 118–135 (2018). https://doi.org/10.1093/bib/bbw089
14. Equi, M., Grossi, R., Mäkinen, V., Tomescu, A.I.: On the complexity of string matching for graphs. In: 46th International Colloquium on Automata, Languages, and Programming, ICALP 2019, Patras, Greece, 9–12 July 2019, pp. 55:1–55:15 (2019). https://doi.org/10.4230/LIPIcs.ICALP.2019.55
15. Equi, M., Mkinen, V., Tomescu, A.I.: Graphs cannot be indexed in polynomial time for sub-quadratic time string matching, unless seth fails (2020). http://arxiv.org/abs/2002.00629
16. Gao, J., Impagliazzo, R.: Orthogonal vectors is hard for first-order properties on sparse graphs. In: Electronic Colloquium on Computational Complexity (ECCC), vol. 23, p. 53 (2016). http://eccc.hpi-web.de/report/2016/053
17. Grossi, R., et al.: On-line pattern matching on similar texts. In: 28th Annual Symposium on Combinatorial Pattern Matching, CPM 2017, Warsaw, Poland, 4–6 July 2017, pp. 9:1–9:14 (2017). https://doi.org/10.4230/LIPIcs.CPM.2017.9
18. Iliopoulos, C.S., Kundu, R., Pissis, S.P.: Efficient pattern matching in elastic-degenerate texts. In: Drewes, F., Martín-Vide, C., Truthe, B. (eds.) LATA 2017. LNCS, vol. 10168, pp. 131–142. Springer, Cham (2017). https://doi.org/10.1007/978-3-319-53733-7_9

19. Impagliazzo, R., Paturi, R., Zane, F.: Which problems have strongly exponential complexity? J. Comput. Syst. Sci. **63**(4), 512–530 (2001). https://doi.org/10.1006/jcss.2001.1774

20. Pissis, S.P.: MoTex-II: structured motif extraction from large-scale datasets. BMC Bioinform. **15**, 235 (2014). https://doi.org/10.1186/1471-2105-15-235

21. Pissis, S.P., Retha, A.: Dictionary matching in elastic-degenerate texts with applications in searching VCF files on-line. In: 17th International Symposium on Experimental Algorithms, SEA 2018, L'Aquila, Italy, 27–29 June 2018, pp. 16:1–16:14 (2018). https://doi.org/10.4230/LIPIcs.SEA.2018.16

22. Polak, A.: Why is it hard to beat $O(n^2)$ for longest common weakly increasing subsequence? Inf. Process. Lett. **132**, 1–5 (2018). https://doi.org/10.1016/j.ipl.2017.11.007

23. Sagot, M.-F., Viari, A., Pothier, J., Soldano, H.: Finding flexible patterns in a text: an application to three-dimensional molecular matching. Comput. Appl. Biosci. **11**(1), 59–70 (1995). https://doi.org/10.1093/bioinformatics/11.1.59

24. Sheikhizadeh, S., Schranz, M.E., Akdel, M., de Ridder, D., Smit, S.: Pantools: representation, storage and exploration of pan-genomic data. Bioinformatics **32**(17), 487–493 (2016). https://doi.org/10.1093/bioinformatics/btw455

25. Weiner, P.: Linear pattern matching algorithms. In: 14th Annual Symposium on Switching and Automata Theory, Iowa City, Iowa, USA, 15–17 October 1973, pp. 1–11. IEEE Computer Society (1973). https://doi.org/10.1109/SWAT.1973.13

26. Vassilevska Williams, V.: Hardness of easy problems: basing hardness on popular conjectures such as the strong exponential time hypothesis (invited talk). In: 10th International Symposium on Parameterized and Exact Computation, IPEC 2015, Patras, Greece, 16–18 September 2015, pp. 17–29 (2015). https://doi.org/10.4230/LIPIcs.IPEC.2015.17

Relative Lempel-Ziv Compression
of Suffix Arrays

Simon J. Puglisi and Bella Zhukova[✉]

Department of Computer Science, Helsinki Institute for Information Technology
(HIIT), University of Helsinki, Helsinki, Finland
{simon.puglisi,bella.zhukova}@helsinki.fi

Abstract. We show that a combination of differential encoding, random sampling, and relative Lempel-Ziv (RLZ) parsing is effective for compressing suffix arrays, while simultaneously allowing very fast decompression of arbitrary suffix array intervals, facilitating pattern matching. The resulting text index, while somewhat larger (5-10x) than the recent r-index of Gagie, Navarro, and Prezza (Proc. SODA '18)—still provides significant compression, and allows pattern location queries to be answered more than two orders of magnitude faster in practice.

1 Introduction

The suffix array [18], $\mathsf{SA}[0..n-1]$, of a text (or string, or sequence) T of length n is an array of integers containing a permutation of $(0 \ldots n-1)$, so that the suffixes of T starting at the consecutive positions indicated in SA are in lexicographical order: $\mathsf{T}[\mathsf{SA}[i]..n] < \mathsf{T}[\mathsf{SA}[i+1]..n]$. Because of the lexicographic ordering, all the suffixes starting with a given substring P of T form an interval $\mathsf{SA}[s..e]$, which can be determined by binary search in $O(|\mathsf{P}| \log n)$ time. The suffix array is thus an efficient data structure for returning all positions in T where a query pattern Q occurs; once s and e are located for $\mathsf{P} = \mathsf{Q}$, it is simple to enumerate the $occ = e - s + 1$ occurrences of Q.

An alternative to binary search is the so-called *backward search* method, which locates the interval of the SA via $2|\mathsf{P}|$ rank queries on the Burrow-Wheeler transform (BWT) of T [6,7]. Backward search is the basis for compressed text indexing, emplified by the FM-index family, which has been widely adopted in practice, for example, in Bioinformatics [17]. The BWT is easily amenable to compression (while still supporting rank queries), and so the challenge then has been to reduce the space required for the SA below its trivial $n \log n$-bit encoding, for which a handful of techniques have emerged in the past two decades. The most longstanding of these is to explicitly store the position of every bth suffix in lexicographical (i.e., SA) order. With these samples in hand, rank queries on BWT (a process called "LF mapping") allow an arbitrary $\mathsf{SA}[i]$ value can be determined in $O(b)$ time, thus allowing all occurrences of a pattern to be obtained in $O(b \cdot occ)$ time, with $O(n/b)$ extra space used for the suffix samples.

This research is supported by Academy of Finland through grant 319454.

C. Boucher and S. V. Thankachan (Eds.): SPIRE 2020, LNCS 12303, pp. 89–96, 2020.
https://doi.org/10.1007/978-3-030-59212-7_7

Very recently, Gagie, Navarro, and Prezza [8,9], exploiting an ingenious observation, showed how this can be improved to $O(occ \cdot \log \log n)$ time. They call the resulting data structure the r-index. Experiments in [8] show this improvement is not only of theoretical interest: in practice the r-index is around two orders of magnitude faster than indexes that use regular suffix sampling, and always less space consuming. Another recent alternative is the succinct compact acyclic word graph of Belazzougui, Cunial, Gagie, Prezza, and Raffinot [1], which in practice can be significantly faster than the r-index, but is much bigger (albeit much smaller than the $n \log n$ bits required by the plain SA).

Contribution. The contribution of this short paper is to show that, in practice (at least), *relative Lempel-Ziv parsing* is an effective way to compress the suffix array, and one that supports decompression of intervals especially fast. Our starting point is the differentially encoded SA, denoted SA^d, as first introduced by Gonzalez and Navarro [11]. We then derive an RLZ dictionary, R, (usually called the *reference sequence* [14]), by randomly sampling subarrays from SA^d, and parse SA^d into phrases relative to R. Supporting random access is then a matter of storing one original SA value for each phrase (to undo the differential encoding) and storing the phrase starting points in a predecessor data structure. Decompressing occ consecutive values from SA can then be performed in essentially $O(\log \log n + occ)$ time, and is very fast in practice: more than 100 times faster than the r-index [8] and the CDAWG [1], which are the fastest published methods. Depending on the dataset, our index uses 5–15 times more space than the r-index, and less than the CDAWG.

We acknowledge our approach is uncomplicated, and is essentially a new combination of known techniques: as noted above, dictionary compression of differentially encoded SAs has been explored previously by Gonzalez and Navarro [11], where they used the RePair grammar compressor [15] rather than RLZ (which was undiscovered at the time). Furthermore, RLZ is widely known to support fast random access to its underlying data, but to date has only been applied to textual data, be it natural language [3,13,16] or genomic [3,14]. However, as our experiments show, this combination turns out to be extremely effective, representing a new point on the pareto curve, and seems to simply have been overlooked to date. Another piece of related work is the relative suffix tree of Farruggia et al. [5], in which one or more suffix arrays are compressed *relative to another suffix array*, and pattern matching is supported on each individual SA. That work is different to ours in that we deal with compression of a single SA.

Our own interest in SA compression comes from our recent work developing fast indexes for gapped matching [2]. These indexes rely for their efficiency on fast scans of suffix array intervals, which is easy on an uncompressed SA, but lose significant throughput when current compressed SA implementations are used. The RLZ-compressed suffix array we describe in this paper allows us to derive compressed forms of our gapped-matching indexes that use much less space but operate at comparable speed to uncompressed ones.

Roadmap. In the following section we review the differentially encoded SA of Navarro and Gonzalez [11,12] and the way it induces sequences containing repetitions, which can then be exploited by a dictionary compressor. We also review relative Lempel-Ziv parsing [14], before describing our data structure and the way in which it supports fast subarray access. We then report on an experimental comparison of a prototype of our index, dubbed rlzsa, with the r-index and the CDAWG [1]—which represent, to our knowledge, the current state of the art. Conclusions and reflections are then offered.

2   New Locate Index

SA contains a permutation of the integers $(0 \dots n - 1)$ and so is not directly amenable to dictionary compression in the same way that, say, the text T would be—it contains no repeated elements. SA does contain repetitions of a different nature, however. In particular, because of the lexicographical order on the suffixes in SA if an interval of suffixes $SA[x, y]$ are all preceded by the same symbol c, then there must exist another interval $SA[x', x' + (y - x) + 1]$ for which $SA[x] = SA[x']+1, SA[x+1] = SA[x'+1]+1, \dots, SA[y] = SA[x'x'+(y-x)+1]+1$. Navarro and Gonzalez [11] observed that these so-called *self repetitions* can be turned into actual repetitions if one differentially encodes the suffix array as $SA^d[0] = SA[0]$ and $SA^d[i] = (SA[i] - SA[i-1] +n)$ for $i \geq 1$. Note that the "$+n$" is for technical convenience, so that all values in SA^d are positive.

Navarro and Gonzalez [11] (see also their later journal paper [12] apply a grammar compressor to SA^d, augmenting the grammar with additional pointers to facilitate random access to values in SA^d, and storing original SA values at regular intervals so that the differential encoding can be reversed.

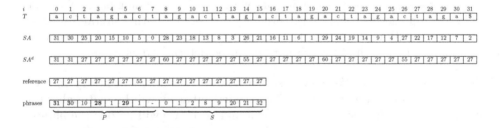

Fig. 1. An example illustrating components of our data structure.

Figure 1 shows a small example illustrating the different components of our data structure and the intermediate stages in their construction.

RLZ Parsing. A variant of the classic LZ77 parsing [21], RLZ parsing compresses a sequence X relative to a second sequence R (the reference) by encoding X as a sequence of substrings, or *phrases*, that occur in R. Our data structure is built on an atypical form of RLZ parsing that is critical to support efficient access to subarrays of the SA and which we now describe.

We derive our reference string R by randomly sampling substrings from SA^d. In Sect. 3 will return to the implementation details such as the number of samples and the size of each sample, but for the time being let us assume R is in hand. References built by random sampling have been shown to work well in practice for compressing web corpora [13] and non-trival bounds on their size have also since been proved [10].

We encode SA by parsing SA^d into phrases—represented as integer pairs—that either represent literal values from the original SA (*literal phrases*), or point to substrings that occur in the reference sequence R (*repeat phrases*). The first component of the pair is always the starting position in SA^d (equivalently SA) of the phrase. A literal phrase at position i is represented as $(i, SA[i])$. The first phrase is always the literal phrase $(0, SA[0])$. Parsing begins at position 1 in SA^d and proceeds according to the following rule. If the parsing is up to a position i in SA^d, then the next phrase is either:

- a literal phrase $(i, SA[i])$, if the previous phrase was not a literal phrase or $SA^d[i]$ does not occur in R; or
- the longest prefix of $SA^d[i, n]$ that occurs in R.

Observe that the parsing rule ensures that every repeat phrase is preceded by a literal phrase. This allows us to easily recover the portion of the SA that is covered by a repeat phrase. Let (i, p_i) be a repeat phrase of length ℓ_i and $(i-1, x)$ be the preceding literal phrase in the parsing. Then $SA[i] = SA^d[i] + x = R[p_i] + x$, $SA[i+1] = SA^d[i+1] + SA[i] = R[p_i+1] + SA[i], \ldots, SA[i+\ell_i-1] = R[p_i+\ell_i-1] + SA[i+\ell_i-2]$.

Data Structure. We store the parsing in two arrays, S and P, both of length z. S contains the starting position in SA^d of each phrase in ascending order. We build and store a predecessor data structure for S. P contains either literal SA values or positions in R as output by the parsing algorithm (the second components of each pair). The length of the ith phrase can be determined as $S[i+1] - S[i]$.

Decoding a Subarray. We now describe how to decode an arbitrary interval $SA[s, e]$ using our data structure. The decoded subarray will be materialized in an output buffer B of size $e - s + 1$. At a high level, we will decode the phrases covering $SA[s, e]$ and copy the decoded values that belong in $SA[s, e]$ (some parts of the first and last phrase may not) into B until it is full, at which point we are done. To this end, we begin by finding the index in S of the predecessor of s. Let x denote this index. If $P[x]$ is a literal phrase, we copy its value to the output buffer. Otherwise ($P[x]$ is non-literal) $P[x-1]$ is by definition literal and we set $p = P[x-1]$. The length of the phrase is $\ell = S[x+1] - S[x]$. Assuming for the moment $S[x] = s$, to decode phrase x we access $R[P[x]]$, copy $(p + R[P[x]] - n)$ to the output buffer, and then set $p = (p + R[P[x]] - n)$, continuing then to copy $(p + R[P[x]+1] - n)$ to B, and so on until either the whole phrase has been decoded, or the output buffer is full. Note that if $S[x] < s$, then we first decode (as described) and *discard* the $(s - P[x])$ symbols of phrase x that are before position s. After decoding phrase x, if the output buffer is not full, we continue to decode phrase $x + 1$, and so on, until all $e - s + 1$ values have been decoded.

Implementation Details. In our practical implementation, P is an array of 32-bit integers. We also limit the maximum phrase length to 2^{16}. For the predecessor data structure, we use the following two-layered approach. We sample every bth phrase starting position and store these in an array. In a separate array we store a differential encoding of all starting positions. Because of the aforementioned phrase length restriction, the array of differentially encoded starting positions takes 16 bits per entry. Predecessor search for a position x proceeds by first binary searching in the sampled array to find the predecessor sample at index i of that array. We then access the differentially encoded array starting at index ib and scan, summing values until the cummulative sum is greater than x, at which point we know the predecessor.

3   Experimental Evaluation

In this section we compare the practical performance of our rlzsa index to other leading compressed indexes, in particular the r-index of Gagie et al. [8] and the cdawg of Belazzougui et al. [1][1]. These indexes were selected because they are the best current approaches for locate queries according to experiments in [8][2]. We provide results for two variants of rlzsa, which are labelled rlzsa-rand and rlzsa-lz in the plots. The rlzsa-rand variant uses a reference constructed via random sampling substrings from the datasets (parameters below). The rlzsa-lz variant selects substrings for the reference based on a length-limited form of LZ77 parsing, which we describe in the full version of this paper.

Mirroring the experiments in [8], we measured memory usage and locate times per occurrence of all indexes on 1000 patterns of length 8 extracted from four repetitive datasets:

- DNA, an artificial dataset of 629145 copies of a DNA sequence of length 1000 (Human genome) where each character was mutated with probability 10^{-3};
- boost, a dataset of concatenated versions of the GitHub's boost library;
- einstein, a dataset of concatenated versions of Wikipedia's Einstein page;
- world, a collection of all pdf files of CIA World Leaders from January 2003 to December 2009 downloaded from the Pizza&Chili corpus.

The average number of occurrences per pattern was 89453 (boost), 607750 (DNA), 31788 (einstein), 29781 (world).

---

[1] The only implementation of cdawg works only for strings on {a,c,g,t}.

[2] We also tried unsuccessfully to include the Locally Compressed Suffix Array (LCSA) of Gonzalez, Navarro, and Farrada [12], which is based on differential encoding of the SA and RePair grammar compression. After expending significant effort attempting to get their code to work we discovered—in communication with the authors [4]—that our failure was due to known bugs in the (dated) LCSA codebase.

Test Machine and Environment. We used a 2.10 GHz Intel Xeon E7-4830 v3 CPU equipped with 30 MiB L3 cache and 1.5 TiB of main memory. The machine had no other significant CPU tasks running and only a single thread of execution was used. The OS was Linux (Ubuntu 16.04, 64bit) running kernel 4.10.0-38-generic. Programs were compiled using g++ version 5.4.0. All given runtimes were recorded with the C++11 high_resolution_clock time measurement facility.

Results. The results of our experiments appear in Fig. 2. On all datasets, both variants of our new rlzsa index are clearly the fastest, providing a newly relevant point on the space-time curve. We locate occurrences always at least two orders of magnitude faster than all other indexes: compared to r-index, from a minimum of 120 times on world to a maximum of 160 times on DNA. On DNA we are 100 times faster than cdawg, which is the next fastest index, and is more than twice the size of the rlzsa variants. The r-index is always the smallest index, from 5 times (world) to 14 times (DNA) smaller than rlzsa-rand.

We remark that in preliminary experiments, we observed rlzsa times to be extremely stable, and quite invariant to reference size. In the plots the rlzsa-rand variant used references size |R| of 106496 (boost), 28597248 (DNA), 6417408 einstein, 2760704 (world), with the reference sequence made up of substrings of length 4096 (boost, world) or 3072 (DNA, einstein). Finally, the rlzsa-lz index is noticeably smaller than the rlzsa-rand one on the boost dataset, but otherwise the two rlzsa indexes are very close in size.

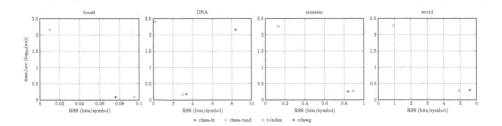

Fig. 2. Locate time per occurrence and working space (in bits per symbol) of the indexes. The vertical axis shows nanoseconds per reported occurrence and is logarithmic.

4   Concluding Remarks

We have described and tested a compressed data structure—rlzsa—that represents the suffix array and allows fast decompression of arbitrary subarrays, facilitating indexed pattern matching. The speed of interval access comes from the cache-friendly nature of RLZ decompression: after an initial predecessor query, all subarray values are obtained by a (usually small) number of cache-friendly copies from the reference sequence. Our index is also easy to construct.

There a numerous avenues for future work. Firstly, although we may never reach the impressively small size of the r-index, we believe the space usage of the rlzsa can be significantly further reduced in practice by both simple representational techniques (e.g., bit packing position values, using Elias-Fano for the predecessor structure) and by adapting improved reference construction schemes that work well for RLZ when compressing text [13,16,19,20]. Secondly, is there a way to derive a hybrid of the rlzsa and r-index approaches that is smaller than the former and faster than the latter? Finally, it may be possible to derive space bounds for the rlzsa by combining the analysis of Gagie et al. [10], which relates the size of RLZ under random sampling to grammar compression of T, with the analysis of Gonzalez and Navarro [11], which relates grammar compression of the differentially encoded SA to the kth order empirical entropy of T.

Acknowledgements. Our thanks go to Héctor Farrada, Nicola Prezza, and Daniel Valenzuela for prompt responses to our queries.

References

1. Belazzougui, D., Cunial, F., Gagie, T., Prezza, N., Raffinot, M.: Composite repetition-aware data structures. In: Cicalese, F., Porat, E., Vaccaro, U. (eds.) CPM 2015. LNCS, vol. 9133, pp. 26–39. Springer, Cham (2015). https://doi.org/10.1007/978-3-319-19929-0_3

2. Cáceres, M., Puglisi, S.J., Zhukova, B.: Fast indexes for gapped pattern matching. In: Chatzigeorgiou, A., Dondi, R., Herodotou, H., Kapoutsis, C., Manolopoulos, Y., Papadopoulos, G.A., Sikora, F. (eds.) SOFSEM 2020. LNCS, vol. 12011, pp. 493–504. Springer, Cham (2020). https://doi.org/10.1007/978-3-030-38919-2_40

3. Deorowicz, S., Grabowski, S.: Robust relative compression of genomes with random access. Bioinformatics **27**(21), 2979–2986 (2011)

4. Farrada, H.: Personal Communication

5. Farruggia, A., Gagie, T., Navarro, G., Puglisi, S.J., Sirén, J.: Relative suffix trees. Comput. J. **61**(5), 773–788 (2018)

6. Ferragina, P., Manzini, G.: Opportunistic data structures with applications. In: 41st Annual Symposium on Foundations of Computer Science, FOCS 2000, Redondo Beach, California, USA, 12–14 November 2000, pp. 390–398. IEEE Computer Society (2000)

7. Ferragina, P., Manzini, G.: Indexing compressed text. J. ACM **52**(4), 552–581 (2005)

8. Gagie, T., Navarro, G., Prezza, N.: Optimal-time text indexing in BWT-runs bounded space. In: Proceedings of SODA, pp. 1459–1477. ACM-SIAM (2018)

9. Gagie, T., Navarro, G., Prezza, N.: Fully functional suffix trees and optimal text searching in BWT-runs bounded space. J. ACM **67**(1), 2:1–2:54 (2020)

10. Gagie, T., Puglisi, S.J., Valenzuela, D.: Analyzing relative Lempel-Ziv reference construction. In: Inenaga, S., Sadakane, K., Sakai, T. (eds.) SPIRE 2016. LNCS, vol. 9954, pp. 160–165. Springer, Cham (2016). https://doi.org/10.1007/978-3-319-46049-9_16

11. González, R., Navarro, G.: Compressed text indexes with fast locate. In: Ma, B., Zhang, K. (eds.) CPM 2007. LNCS, vol. 4580, pp. 216–227. Springer, Heidelberg (2007). https://doi.org/10.1007/978-3-540-73437-6_23

12. González, R., Navarro, G., Ferrada, H.: Locally compressed suffix arrays. ACM J. Exp. Algorithmics, **19**(1), article 1 (2014)
13. Hoobin, C., Puglisi, S.J., Zobel, J.: Relative Lempel-Ziv factorization for efficient storage and retrieval of web collections. Proc. VLDB Endow. **5**(3), 265–273 (2011)
14. Kuruppu, S., Puglisi, S.J., Zobel, J.: Relative Lempel-Ziv compression of genomes for large-scale storage and retrieval. In: Chavez, E., Lonardi, S. (eds.) SPIRE 2010. LNCS, vol. 6393, pp. 201–206. Springer, Heidelberg (2010). https://doi.org/10.1007/978-3-642-16321-0_20
15. Larsson, N.J., Moffat, A.: Offline dictionary-based compression. Proc. IEEE **88**(11), 1722–1732 (2000)
16. Liao, K., Petri, M., Moffat, A., Wirth, A.: Effective construction of relative Lempel-Ziv dictionaries. In: Proceedings of 25th International Conference on the World Wide Web (WWW), pp. 807–816 (2016)
17. Mäkinen, V., Belazzougui, D., Cunial, F., Tomescu, A.I.: Genome-Scale Algorithm Design: Biological Sequence Analysis in the Era of High-Throughput Sequencing. Cambridge University Press, Cambridge (2015)
18. Manber, U., Myers, G.: Suffix arrays: a new method for on-line string searches. SIAM J. Comput. **22**(5), 935–948 (1993)
19. Tong, J., Wirth, A., Zobel, J.: Compact auxiliary dictionaries for incremental compression of large repositories. In: Proceedings of the 23rd ACM International Conference on Conference on Information and Knowledge Management, CIKM 2014, Shanghai, China, 3–7 November 2014, pp. 1629–1638. ACM (2014)
20. Tong, J., Wirth, A., Zobel, J.: Principled dictionary pruning for low-memory corpus compression. In: The 37th International ACM SIGIR Conference on Research and Development in Information Retrieval, SIGIR 2014, Gold Coast, QLD, Australia, 06–11 July 2014, pp. 283–292. ACM (2014)
21. Ziv, J., Lempel, A.: A universal algorithm for sequential data compression. IEEE Trans. Inf. Theory **23**(3), 337–343 (1977)

Algorithms

Approximating the Anticover of a String

Amihood Amir, Itai Boneh[✉], and Eitan Kondratovsky

Department of Computer Since, Bar Ilan University, Ramat Gan, Israel
amir@esc.biu.ac.il, itai.bone@live.biu.ac.il, kondrae@cs.biu.ac.il

Abstract. The k-*anticover* of a string S is a set of distinct k-length substrings such that every index in S is contained in one of these substrings. The existence of an *anticover* indicates a lack of structure in S. It was recently proven by Alzamel et al. [2] that finding whether or not a k-*anticover* exists is $\mathcal{NP}$-Hard for $k \geq 3$.

In this paper, we extend the definition to provide three optimization versions for the k-anticover problem. We provide efficient approximation algorithms for these problems.

Keywords: Anticover · NP-hardness · Approximation algorithms

1 Introduction

One of the challenges of stringology is finding regularities in a string. This task leads the theoretical interest in string combinatorics [27]. Regularities also have practical meaning. A very partial list is: palindromes play varied roles in Biology [17, 18, 26, 33], and periods and repeats are meaningful in Molecular Biology and cyber detection [9, 10, 19, 20, 28, 31, 34].

Recently there has been growing interest in the "opposite" phenomenon, i.e. strings that are *far from regular*. Fici et al. defined the concept of *antipower* [8, 15]. Since periodicity is one of the most basic string regularities then, naturally, antiperiodicity attracted attention [1]. For many phenomena, it is desirable to broaden the definition of periodicity and study wider classes of repetitive patterns in strings. One common such notion is that of a *cover*, defined as follows.

Definition 1 [Cover]. *A m-length substring C is said to be a* cover *of a n-length string T, if $n > m$ and every position of T lies within some occurrence of C.*

Note that by the definition of cover, the string C is both a prefix and a suffix of the string T. For example, consider the string $T = abaababaaba$. Clearly, T is "almost" periodic with period aba, however, as it is not completely periodic, the algorithms that exploit repetitions cannot be applied to it. On the other

---

This work was partially supported by ISF grant 1475/18 and BSF grant 2018141.
This work is part of the second author's Ph.D. dissertation.

C. Boucher and S. V. Thankachan (Eds.): SPIRE 2020, LNCS 12303, pp. 99–114, 2020.
https://doi.org/10.1007/978-3-030-59212-7_8

hand, the string $C = aba$ is a cover of T, which allows applying to T cover-based algorithms. Quasi-periodicity was introduced by Ehrenfeucht in 1990 (according to [5]). The earliest paper in which it was studied is by Apostolico, Farach and Iliopoulos [7], which defined the *quasi-period* of a string to be the length of its shortest cover and presented an algorithm for computing the quasi-period of a given string in $O(n)$ time and space. The new notion attracted immediately several groups of researchers (e.g. [11,12,25,29,30]). An overview on the first decade of the research on covers can be found in the surveys [5,21,32]. The cover concept excited later research as well. Different variants of quasi-periodicity have been introduced. These include *seeds* [23], *maximal quasi-periodic substring* [6], the notion of *k-covers* [22], λ-*cover* [35], *enhanced covers* [16], *partial cover* [24]. reconstructing a string from the cover array [14], extensions to strings in which not all letters are uniquely defined, such as *indeterminate strings* [4] or *weighted sequences* [36], and cover recovery [3]. Some of the related problems are $\mathcal{NP}$-hard (see e.g., [4,13]).

Recently, Alzamel et al. [2] defined the concept of string *k-anticover* and showed that for $k \geq 3$ finding whether a k-anticover exists is $\mathcal{NP}$-complete. In this paper we give a few definitions of the k-anticover problem as optimization problems and show approximation algorithms for them. The approximations have an appeal in that they allow very fast detection of substrings that are k-anticoverable. In reality, periodic strings are rare but the study of periodic substrings (runs) has proven quite useful. Similarly, it may be interesting to find the substrings that are very "unstructured". Our algorithms allow this to be approximated quite efficiently.

This paper is organized as follows. In Sect. 2, we define the basic notions and various approximation problems that we tackle. Section 3 shows a *linear-time* $1/2$-*approximation* of the number of indices that can be covered by a k-anticover. In Sect. 4, we show a $\log_2 n$-approximation of the *number of k-anticovers* needed for the given string, we also show that this problem is approximable but does not have a PTAS. In Sect. 5 we show that finding the smallest k for which a k-anticover exists is $\mathcal{NP}$-hard and provide a 4-approximation and a lower bound for the approximation ratio of this problem. We conclude with open problems.

2   Preliminaries

Let Σ be an alphabet. A *string* S over Σ is a finite sequence of letters from Σ. By $S[i]$, for $1 \leq i \leq |S|$, we denote the ith letter of S. The *empty string* is denoted by ϵ. By $S[i..j]$ we denote the string $S[i] \ldots S[j]$ called a *substring*, or *factor* of S (if $i > j$, then the substring is the empty string). A substring is called a *prefix* if $i = 1$ and a *suffix* if $j = |S|$. The prefix of length j is denoted by $S[..j]$. While by $S[i..]$ we denote the suffix which starts from index i in S. Denote by S^i the substring composed of i consecutive occurrences of S.

The following is the definition of k-anticover given in [2].

Definition 2 [anticover]. *Given an integer $k \geq 2$ and a string S of length $n \geq k$, let $C = \{i_1, i_2, ..., i_\ell\}$ be an ordered set of positions in S chosen from $\{1, 2, ..., n - k + 1\}$. We say that C is a k-anticover of S if*

- **Distinctness property:** *Any two substrings $S[i_j...i_j + k - 1]$ and $S[i_h...i_h + k - 1]$ are different, for $i_j, i_h \in C$ and $i_j \neq i_h$.*
- **Coverability property:** *Every position in S is covered, namely, $i_1 = 1, i_\ell = n - k + 1$, and $i_{j+1} - i_j \leq k$ for $1 \leq j \leq \ell - 1$.*

Alzamel et al. proved that for $k > 2$, deciding whether a k-anticover of S exists is $\mathcal{NP}$-hard. We consider three optimization versions of the problem.

The *MaxkAnticover* problem tries to maximize the number of indices of S that are covered by a k-anticover. In other words, we are seeking the largest number of indices of S that are covered by a distinct set of substrings (satisfying condition 2 of Definition 2). Formally.

Definition 3. *The MaxkAnticover problem has as its input a string S. We need to find a set $C = \{i_1, ..., i_\ell\}$ of indices representing distinct substrings of length k, $\{S[i_j .. i_j + k - 1] \mid j = 1, ..., \ell\}$ where the number of indices in S that are covered by C is maximized. We denote the maximal number of covered indices by $MaxkAnticover(S)$.*

Clearly the *MaxkAnticover* problem is $\mathcal{NP}$-hard, since $MaxkAnticover(S) = |S|$ if and only if S has a k-anticover.

The *MinRepkAnticover* problem insists on condition 2 of Definition 2, i.e. it requires all indices to be covered. If this is not possible, we allow some substrings starting at the indices of the k-anticover set, C, to repeat more than once. We try to minimize the number of repetitions necessary to cover all indices. This can also be viewed as using more than one k-anticover set. Each one may not cover all indices but together they do. We seek the smallest number of such sets necessary to cover S. Formally.

Definition 4. *The MinRepkAnticover problem has as its input a string S. We seek a set $C = \{i_1, ..., i_\ell\}$ of indices representing a multiset of substrings of length k, $SC = \{S[i_j .. i_j + k - 1] \mid j = 1, ..., \ell\}$ such that every index in S is covered by one of the substring of SC, and where we minimize the largest number of occurrences of any substring in SC. Denote that number by $MinRepkAnticover(S)$.*

It is clear that the *MinRepkAnticover* problem is $\mathcal{NP}$-hard, since $MinRepkAnticover(S) = 1$ if and only if S has a k-anticover.

Finally, we consider the "inverse" of the k-anticover problem. In the k-anticover problem, k given, and we are asking whether a k-anticover exists. Clearly, if $k = |S|$ there is a k-anticover for every S. Also, for $k = 1$, there is a k-anticover only if all symbols of S are different. The question is, what is the smallest k for which a k-anticover exists. Formally,

Definition 5 (*MinAnticover*).
Input: *An n-length string S.*
Output: *Smallest natural number k such that S has a k-anticover.*

It is not immediate that the *MinAnticover* problem is $\mathcal{NP}$-hard due to the lack of monotonicity. That is, a k-anticover might exist, but $(k + 1)$-anticover is not, or vice versa. For example, assume $S = abcabc$. $\{1, 3, 5\}$ is 2-anticover of S, where ab, ca, and bc are all distinct. But as for $k = 3$, S has a border of length 3, and thus there is no longer an anticover. In Sect. 5, we prove that this problem is indeed $\mathcal{NP}$-hard.

3   Approximating the Number of Covered Indices

In this section, we give a $\frac{1}{2}$-approximation for the *MaxkAnticover* problem presented in Definition 3.

3.1   The Approximation Algorithm

Algorithm's Idea: Start by a pool of all possible substrings of length k. Initially, each of these substrings, if chosen to the k-anticover, will cover k indices of S.

Iteratively, choose a substring that covers the *largest* number of uncovered indices in S and add it to the set C_A that approximates the k-anticover. Now remove from the candidate set all substrings that are equal to the chosen one. Also, update for each remaining candidate substring the number of uncovered indices that it would cover, if chosen next. Stop when there are no substrings left. Formally:

Algorithm MKA:

Initialize $Can = \{1, ..., n - k + 1\}$ as the set of candidates for the k-anticover. Each index has a list of *free indices*, which initially is $[i, ..., i + k - 1]$ for index i.

While $Can \neq \varnothing$ do:

1. Choose $j \in Can$ that has a maximal number of elements in its free list, and add j to C_A.
2. Remove from Can all indices i for which $S[i..i + k - 1] = S[j..j + k - 1]$.
3. For every index $i \in Can \cap \{j - k + 1, ..., j - 1, j + 1, ..., j + k - 1\}$, update i's free index list. This is done by deleting j's free index list from i's free index list.

Time: It is easy to implement the above algorithm in time $O(nk)$. We show an $O(n)$ implementation ($O(n \log(\min(n, \sigma)))$) for infinite alphabets, where σ is the number of different symbols in S). The while loop of algorithm MKA iterates n times. We need to show:

1. The free indices lists can be represented and maintained in space $O(n)$.
2. We can find an element with a largest free index list in constant time (Step 1. in the loop).
3. Removing indices whose k-length substring is equal to the k-length substring of candidate j amortises to $O(n)$ (Step 2. in the loop).
4. Updating all free index lists amortises to $O(n)$ (Step 3, in the loop).

The following observations solve the above four desiderata.

Observation 1. *The free list of candidate index i can be represented in constant space.*

Proof. The free list of index i is initialized as a consecutive list $[i, ..., i+k-1]$, thus it can be represented as $[i, i+k-1]$. Since all substrings in the k-anticover are of length k, it can not happen that there will be "holes" in the middle of the free index list. Therefore the free index list can only be of the form $[\ell, r]$, where ℓ is the leftmost free index and r is the rightmost free index. □

The next observation bounds the number of removals in Step 2 of Algorithm MKA.

Observation 2. *Since Can is initialized to n indices, then there are no more than n deletions from Can.*

We need to implement an efficient search for the indices i for which $S[i..i+k-1] = S[j..j+k-1]$, where j is the chosen candidate from Can. Such a search can be easily done by constructing a compact trie of all substrings of length k. This trie can be constructed in linear time for constant sized alphabets or integer alphabet and $O(n \log(\min(n, \Sigma)))$ for infinite alphabets, by pruning the suffix tree of S and having a list of all indices at the last node of each (length k) path in the trie. When an index $j \in Can$ is chosen, all equal indices are immediately accessible in the trie.

Finally, we need one last observation.

Observation 3. *Algorithm MKA attempts to change the free index list of any index i at most twice throughout its run.*

Proof. Assume index $j_1 \in Can$, $j_1 < i$ was chosen and its overlap with i caused the leftmost index in the free list to change from i to $i+d$. There can not be a j_2, $j_1 < j_2 < i$ that causes i's leftmost index to change again without completely deleting i, since i covers more free indices than j_2, and therefore i would be chosen by the algorithm before j_2. If i's and j_2's free lists are of the same size and j_2 was chosen, then, indeed i's list is changed but i is deleted, so we charge the change to the deletion. There can not be a $j_3, j_3 < j_1 < i$ that would attempt to change the leftmost index of i, because, if that were the case, j_3 would have been chosen before j_1. This is due to the fact that j_3 is to the left of j_1 therefore covers more free indices. In the initial case where both cover the same number of indices (k), j_3 gets chosen first by the algorithm because our algorithm implementation chooses the indices in the k-anticover from left to right. Therefore, the left index of the free list can only be changed once during the algorithm run.

An analogous analysis shows that the right index of the free index list can only be changed once. □

To conclude, we have the following data structures:

1. **[List L:]** A doubly linked list of all indices in Can, sorted by the index.

2. [**Trie T:**] A trie of all substrings of length k that are still in Can.
3. [**Buckets:**] Up to k buckets of indices. Bucket ℓ contains all the indices with ℓ elements in the free list. A bucket is implemented as a doubly linked list. Only the non-empty buckets are kept, and the pointers to the bucket lists are stored in a doubly linked list sorted from highest to lowest.

In addition, every index $i \in S$ appears in all three above data structures, and all its occurrences are linked.

The implementation is now clear. Initially, all indices are linked in list L, trie T is constructed, all indices have free index list (interval) of length k, and all indices are in bucket k, ordered from left to right. In addition, every index occurrence in all three lists is linked. All this takes time $O(n)$.

An index j is taken from the highest bucket, it checks in the trie and deletes all occurrences of indices equal to it (from all lists), and updates the free lists of the existing indices in proximity k to it, changing their bucket if necessary. The time is proportional to the number of indices affected, which by the above analysis is $O(n)$.

Approximation Ratio:

Lemma 1. *Let a be the number of indices of S covered by C_A constructed by Algorithm MKA, and opt the largest number of indices covered by any k-anticover C of S. Then $a \geq \frac{1}{2}opt$.*

Proof. Let b be the number of indices covered by C but not by C_A. Clearly, $opt \leq a + b$.

Consider X, an interval of indices that is covered by substring $j \in C$ but not by C_A. The reason C_A did not choose j is that it chose another copy of $S[j..j+k-1]$, say at index ℓ. But that means that ℓ covered at least $|X|$ indices, otherwise C_A would have chosen j. We conclude that $a \geq b$. But this means that $opt \leq b + a \leq 2a$, or $\frac{1}{2}opt \leq a$.   □

The approximation ratio that we proved is tight in the limit.

Lemma 2. *Algorithm MKA can not approximate MaxkAnticover to a better than $\frac{2k}{2k-1}$ ratio.*

Proof. Consider the examples: $\{a^n b^n\}$. If MKA chooses index $n - k + 1$ and index $n + 1$, then it covers $2k$ indices. The underlined symbols are in the chosen substrings of C_A: $a^{n-k} \underline{a^k} \underline{b^k} b^{n-k}$. No other indices can be covered.

However, the optimum is choosing the indices: 1, $n - k + 2$, $n - k + 3, ..., n - 1$, $2n - k + 1$. They cover $4k - 2$ indices, making the ratio: $\frac{2k}{4k-2} = \frac{k}{2k-1}$. In Fig. 1a we see an example where $n = 16$ and $k = 5$.   □

We feel that a judicious "tiebreaker", for choosing the next element of Can, from among those with a maximum size free list, can lead to a better approximation ratio. The next section shows that experiments strengthen this belief.

3.2   Simulation Results

We have run some extensive tests on actual strings and found that the approximation ratio proven above indeed shows up in practice. We also experimented with a heuristics that seems to improve the approximation, and it is a challenge to prove better bounds.

Our experiments were constructed as follows.

Our Platform: MacBook Pro, 2.7 GHz Dual-Core Intel Core i5, 8 GB RAM, using *pyspark* package with *Python* version 3.7.3. In the experiment, we fixed $k = 3, ..., 7$, binary alphabet $|\Sigma|$, and n varying from 3 to 20. For every n-length string we computed its optimal k-anticover (the number of covered indices) using a naive exponential-time algorithm. We compared it to the approximation algorithm MKA and computed the approximation ratio for each length.

We then introduced an additional heuristic. If two substrings cover the same amount of free indices, we choose the one with the smaller lexicographic order. The experiments show that a better approximation ratio is achieved using the "infrequent" tie breaker. For the MKA algorithm, our experiments indeed achieved a $\frac{1}{2}$-approximation, but the infrequent heuristic converged to a $\frac{2}{3}$-approximation ratio. The exact numbers appear in the figures. In Fig. 1b we have the results of running MKA. The y-axis is the simulation ratio, The x-axis is the length of string S. The results for $k = 3 - 7$ are plotted. As can be seen, the ratio converges to 0.5. In fact, the graph matches $\frac{k}{2k-1}$.

In Fig. 1c we have the results of running MKA with the added heuristic of choosing the least frequent substring that covers most uncovered indices. Again, the y-axis is the simulation ratio, The x-axis is the length of string S. The results for $k = 3 - 7$ are plotted. As can be seen, the ratio converges to 0.66. Here the graph matches $\frac{2k}{3k-1}$.

All experiments were also run over alphabets of sizes 3, 4, and 5, with the same results.

4   Approximating the Number of k-covers

We start with an observation on the lower bound for a polynomial approximation for this problem

Observation 4. *There is no polynomial approximation algorithm for MinRep-kAnticover with an approximation ratio $\alpha < 2$, unless $\mathcal{P} = \mathcal{NP}$.*

Proof. Assume, to the contrary, that such an algorithm A exists. For strings that have a k-Anticover, A must output a cover where the number of times that a k-anticover substring is used is bounded by $1 * \alpha < 2$. Since the maximal number of repetitions is an integer, that number is 1, which is a proper k-Anticover. It follows that A can be used to recognize strings that have a k-Anticover in polynomial time.   □

We define a generalization of Algorithm MKA denoted as *SubsetMKA* (S, AC). SubsetMKA has an additional input AC, that is a set of substrings of length k in S. Let I be the set of indices that are **not** covered by the substrings in AC. $SubsetMKA$ will attempt to yield a k-anticover that maximizes the amount of covered indices in I.

$SubsetMKA$ runs exactly as MKA with one exception: for every $i \in Can$, the free index list is initialized as $[i..i + k] \setminus AC$.

Algorithm MRA:

Initialize $UC = \{1, ..., n\}$ as the set of uncovered indices.
Initialize the covering set $C = \varnothing$.

While $UC \neq \varnothing$ do:

1. Use Algorithm $SubsetMKA$ to obtain a k-anticover c of the remaining indices UC.
2. Remove from UC all the indices that are covered by c.
3. Set $C = C \bigcup c$

Return C

Approximation Ratio: Let C^i be the set of new indices covered by the cover obtained from Subset MKA in the ith iteration of Algorithm MRA. Let C_{opt} be a k cover of S that is optimal for **MinRepkAnticover**. Denote the maximal amount of repetitions for a single k-length substring in C_{opt} as m. Consider a partition of C_{opt} to m sets of distinct k-length substrings S^i_{opt} for $1 \leq i \leq m$. Let the set of indices covered by S^i_{opt} be C^i_{opt} The following lemma is the key to proving the $\log_2(n)$ approximation ratio achieved by algorithm MRA:

Lemma 3. *The first m sets obtained from MRA (namely: C^i for $1 \leq i \leq m$) cover at least $\frac{n}{2}$ indices.*

Proof. Denote as UC_i the indices that remain uncovered by C in step i of the algorithm **after** C^i is added (so $UC_0 = [1..n]$, $UC_1 = UC_0 \setminus C_1$). For every $1 \leq i \leq m$, partition C^i_{opt} into two distinct sets : $New^i = C^i_{opt} \bigcap UC_i$ and $Old^i = C^i_{opt} \setminus UC_i$. Since every Old^i is contained within $I[1..n] \setminus UC_m$, it holds that $\bigcup_{i=1}^m Old^i \subseteq \bigcup_{i=1}^m C^i$ and $|\bigcup_{i=1}^m Old^i| \leq |\bigcup_{i=1}^m C^i|$. Similar arguments from the proof of Lemma 1 can be made to show that for every $1 \leq i \leq m$, $|C^i| \geq |New^i|$. Since C^i are distinct, it follows that $|\bigcup_{i=1}^m C^i| \geq |\bigcup_{i=1}^m New^i|$. Putting the two inequalities together, we have $2|\bigcup_{i=1}^m C^i| \geq |\bigcup_{i=1}^m New^i| + |\bigcup_{i=1}^m Old^i|$. Since New^i and Old^i are distinct and $New^i \bigcup Old^i = C^i_{ops}$ we have $2|\bigcup_{i=1}^m C^i| \geq |\bigcup_{i=1}^m C^i_{opt}|$. Finally, since C^i_{ops} covers all the indices of S we have $|\bigcup_{i=1}^m C^i| \geq \frac{n}{2}$. $\square$

Observation 5. *Lemma 3 can be generalized to make the following claim: for every integer $x \geq 1$ it holds that $|UC_{xm}| < \frac{n}{2^x}$.*

Proof. By induction. $x = 1$ is simply Lemma 3. For $x > 1$, we assume that the claim holds for x. Consider the set $C_x = \{C^i | x \cdot m + 1 \leq i \leq (x + 1) \cdot m\}$.

Consider the following partition of C^i_{opt} into two distinct sets: $New^i_x = C^i_{opt} \cap UC_{x \cdot m+i}$ and $Old^i_x = (C^i_{opt} \setminus UC_{x \cdot m+i}) \cap UC_{x \cdot m}$. Notice that unlike the partition in the proof of Lemma 3 where $Old^i \cup New^i = C^i_{opt}$, With this partition we have $Old^i_x \cup New^i_x = C^i_{opt} \cap UC_{x \cdot m}$. The same arguments as in the proof of Lemma 3 can be made to show that $2|\bigcup_{i=1}^m C^{x \cdot m+i}| \geq |\bigcup_{i=1}^m New^i_x| + |\bigcup_{i=1}^m Old^i_x|$. It follows that $2|\bigcup_{i=1}^m C^{x \cdot m+i}| \geq |\bigcup_{i=1}^m (C^i opt \cap UC_{x \cdot m})|$ Since $\bigcup_{i=1}^m C^i_{opt}$ covers every index in S, we are left with $2|\bigcup_{i=1}^m C^{x \cdot m+i}| \geq |UC_{x \cdot m}|$ which indicates that C_x covers at least half of the remaining uncovered indices. The induction hypothesis suggests that $|UC_{x \cdot n}| < \frac{n}{2^x}$. The next m covers cover at least half of $UC_{x \cdot nm}$, so $UC_{(x+1) \cdot m} \leq \frac{n}{2^{x+1}}$.   □

The approximation ratio is derived from Observation 5. After $m \cdot \log_2(n)$ iterations, there is at most 1 uncovered index. The next iteration will surely cover it. Every iteration of Algorithm MRA increases the maximal repetition of a single substring by at most 1, so the maximal repetition in the output is at most $m \cdot \log_2(n) + 1$. It follows directly from the approximation ratio that the running time of Algorithm MRA is bounded by $O(nm \log(n))$.

Definition 6. *SubsetkAnticover(S, I) is the optimization problem of finding a set A of distinct substrings in S of size k that maximizes the amount of indices in $I \subseteq [1..n]$ covered by any of A's substrings.*

Observation 6. *Algorithm SubsetMKA is a $\frac{1}{2}$ approximation for Subsetk-Anticover(S, I).*

Observation 6 can be proven via similar reasoning as in the proof of Observation 1. Let $SAlg$ be an approximation algorithm for $SubsetkAnticover(S, I)$ with approximation ratio α.

Lemma 4. *Replacing Algorithm subsetMKA with SAlg in Algorithm MRA will yield a $\frac{m}{\alpha} \ln n + 1$ Approximation for MinRepkAnticover.*

Proof. Let m be the optimal value for *MinRepkAnticover*. Let $C_{opt} = C^1_{opt}, C^2_{opt}..C^m_{opt}$ be a set of m distinct sets of strings that collectively cover every index in S. The existence of such C_{opt} suggests that for **every** subset of indices $I \subseteq [1..n]$ there exists a subset C^i_{opt} that covers at least $\frac{|I|}{m}$ of I's indices. Denote the remaining indices in step i of Algorithm MRA as $\tilde{U}C_i$. The existence of a distinct set that covers $\frac{|UC_i|}{m}$ indices of UC_i suggests that $SAlg$ will cover at least $\frac{\alpha}{m}|UC_i|$ indices. Therefore, in every iteration UC_i decreases by a multiplicative factor of $1 - \frac{\alpha}{m}$. It follows that UC_i is bounded by $n(1 - \frac{\alpha}{m})^i$. For $i = \frac{m}{\alpha} \ln n$, This expression is bounded by 1. The following iteration must cover the single remaining index.   □

In our specific construction, we managed to prove an approximation ratio of $m \log_2(n) + 1$, which is tighter than the $2m \ln(n) + 1$ approximation ratio derived from Lemma 4. This may not be the case if we implement this construction with another, possibly better, approximation algorithm.

5   Approximating the Smallest k for Which a k-anticover Exists

We start by proving a lower bound for the approximation ratio achieved by a polynomial algorithm.

We present a construction that outputs a string S' from an input string S, where S' has no 2-anticover, and it has a 3-anticover if and only if the original S has a 3-anticover.

Jolly Character: We use the notation of a jolly character "$\star$" as presented in [2]. Each instance of the symbol "$\star$" in S represents a distinct unique character that does not occur anywhere else in S. The jolly character has the useful property that every substring containing it is unique, and therefore can be added to the anticover without causing any substring repetition. This property makes every symbol within a distance of at most $k-1$ from the jolly symbol **trivially coverable** by a k-anticover.

The Construction: Given a string $S[1..n]$ of length at least 3. Let its last 2 symbols be $S[n-1..n] = \sigma_2, \sigma_3$. Let x and y be two distinct symbols such that $x, y \notin \Sigma$. We construct S' as follows: $S' = S \cdot xx \star yyyyy \star (x\sigma_2\sigma_3xx\star)^3$

Lemma 5. S' does not have a 2-anticover.

Proof. Any string containing the substring $yyyyy$ can not have a 2-anticover. This is due to the fact that the y's in the second and in the fourth places must be covered by distinct instances of the substring yy.   □

Lemma 6. S' has a 3-anticover if and only if S has a 3-Anticover.

Proof. If S has a 3-anticover then it can be extended to a 3-anticover of S' as follows: The 2 instances of x concatenated at the end of the original S are trivially covered. All the y's are trivially covered apart from the middle one. In order to cover it, add the middle instance of yyy to the 3-anticover. In the three instances of $x\sigma_2\sigma_3xx$ only the middle σ_3 is not trivially covered. There are three distinct substrings that can cover it: $x\sigma_2\sigma_3, \sigma_2\sigma_3x$ and σ_3xx. Add a different one of these substrings for each instance of $x\sigma_2\sigma_3xx$ to cover its respective σ_3. Every substring we added to the cover contained either x, y or a jolly character, so it can neither appear in the original S nor in its 3-anticover. It follows that the distinctness property remains in our extended 3-anticover.

If S' has a 3-anticover, all the symbols of S are covered by distinct substrings. We proceed to show that every symbols of S is covered by substrings of S, rather than by a new substring that was added by our construction. The only two substrings that we added and may cover symbols in S are the instances of $\sigma_2\sigma_3x$ and σ_3xx appended immediately after S. But if any of these substrings is in the 3-anticover of S', then the three instances of $x\sigma_2\sigma_3xx$ can not be covered. This is due to the fact that the σ_3 in one of them must be covered by $\sigma_2\sigma_3x$, and in one of the others by σ_3xx.   □

Finally, we show that a $\alpha < \frac{4}{3}$ approximation for $MinAnticover$ can not exist. Assume, to the contrary, the existence of a polynomial algorithm A that approximates $MinAnticover$ within a ratio of $\alpha < \frac{4}{3}$. Given a string S, we can construct S' and use A to decide whether or not S' has a 3-anticover as in Observation 4. It is clear that S' can be constructed in polynomial time.

We proceed to provide a 4-approximation for $MinAnticover$.

Definition 7. *For a word w with $|w| \geq k$, we define the k-covering set of w as the set of distinct length k subwords of w. We denote the k-covering set of w as $C_k(w)$.*

The following observation is the key for bounding the minimal possible size of k such that a k-anticover exists.

Observation 7. *Let S be a string. If there is a word w with size $|w| \geq 2k - 1$, where there is a set of more than $|C_k(w)|$ occurrences of w in S, and every two occurrences in the set start within a distance of at least k from each other, then S does not have a k-anticover.*

Proof. Consider the middle index of every occurrence of w in the set. If a k-anticover exists, every one of these indices needs to be covered by a substring from the set $C_k(w)$. Since there is a distance of k between two occurrences - every center must be covered by a **different** element of $C_k(w)$. Since there are more than $|C_k(w)|$ instances, a substring of size k must be selected at least twice to cover all of the centers. □

Given S, it is easy to find, in polynomial time, the minimum k such that every subword w with size $|w| = 2k-1$ has at most $|C_k(w)|$ occurrences with a distance of at least k from each other. In our approximation algorithm, we start by finding this value, denoted as k from now on. Observation 7 guarantees that the value k^* that we approximate is at least k. We proceed to find a $(4k - 1)$-anticover. For simplicity, we start with the assumption that $S[1..4k-1] \neq S[n-(4k-1)+1..n]$.

Algorithm MAC:

Initialize $C = \{0, n - (4k - 1) + 1\}$ as the currently selected indices for the $4k - 1$-anticover. Initialize $i = 4k$ as a pointer for the next index that needs to be covered.
While $i < n - (4k - 1) + 1$:
1. Choose an index $j \in [i - (2k - 1)..i - (k - 1)]$ such that there does not exist an index ℓ in C such that $S[\ell..\ell + (4k - 1)] = S[j..j + (4k - 1)]$.   2. Add j to C.   3. Set i to $j + 4k$.

Lemma 7. *Algorithm MAC always terminates with a $(4k - 1)$-anticover.*

Proof. We need to show that Step 1 never fails to choose a word starting in $[i - (2k - 1)..i - (k - 1)]$. Assume that i is an index for which every word starting in this interval has already been selected. Consider the word $w = S[i - (k - 1)..i +$

$(k-1)]$ with size $|w| = 2k-1$. There are at least $|C_k(w)|$ different words of size $4k-1$ starting in $[i-(2k-1)..i-(k-1)]$. According to our assumption, every one of these words was selected before we reached i. Every one of these words contain an instance of w starting within its first k indices. After we pick a word in index j, the next index to cover will be $j+4k-1$, and the minimal starting index of a word selected to cover this index will be $j+4k-1-(2k-1) = j+2k$. It follows that the words we pick start within a distance of at least $2k$ from each other. So the selected instances of the words from $[i-(2k-1)..i-(k-1)]$ corresponds to at least $|C_k(w)|$ instances of w that are at least k indices apart from each other. This is making $S[i-(k-1)..i+(k-1)]$ the $|C_k(w)|+1$ instance of w (also within a distance of at least k from the previous one), in contradiction to our selection of k. It's also possible that the suffix $S[n-(4k-1)+1..n]$ is one of the words that could cover i, but have been already selected. In this case, w will be the $|C_k(w)|$ occurrence and the suffix will be the $|C_k(w)|+1$. It can be easily verified that the suffix must also start within a distance of at least k from its predecessor in the Anticover generated by MAC.   □

It is straightforward to implement Algorithm MAC in polynomial time.

As for the case in which $S[1..4k-1] = S[n-(4k-1)+1..n]$, we run Algorithm MAC with $4k$ rather than with $4k-1$ to find a $4k$-anticover. If we have $S[1..4k] = S[n-4k+1..n]$ too, it means than S must have a prefix and a suffix of the form σ^t with $t \geq 4k$, making any k-anticover impossible for $k \leq t$. In this case, we run Algorithm MAC with $t+1$ instead of $4k-1$ and it is guaranteed to find a $(t+1)$-anticover, which is optimal.

6   Conclusion and Open Problems

We defined three natural optimization versions for the k-Anticover problem and provided upper and lower bounds for the approximation ratio achieved by a polynomial time algorithm. There are a myriad open problems left. We don't have a lower bound for the approximation ratio of $MaxkAnticover$. In particular, it would be interesting to prove the $\frac{2}{3}$ ratio achieved by the experiments on the change to algorithm MKA that involves choosing the most infrequent of the substrings that cover the most indices. It is also not clear whether the problem has a PTAS. In the $MinRepAnticover$ problem, we showed a $O(\log n)$-approximation algorithm, and proved that no approximation algorithm can achieve a better than $\frac{1}{2}$ ratio. That gap needs to be closed. Finally, for the $MinAnticover$ problem, the gap between the $\frac{4}{3}$ lower bound on the approximation ratio and the 4-approximation algorithm still needs to be closed.

7   Appendix

7.1   Figures

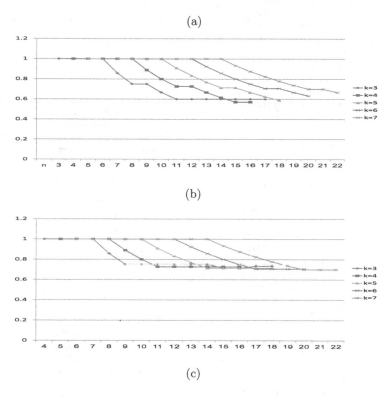

$n=16, k=5$

aaaaaaaaaaaaaaaaabbbbbbbbbbbbbbbbb

C_A chosen by algorithm MKA.

aaaaaaaaaaaaaaaaabbbbbbbbbbbbbbbbb

Optimal k-anticover.

(a)

(b)

(c)

Fig. 1. (a) An example for MKA's approximation ratio lower bound. (b) Simulation of MKA. (c) Simulation of MKA with least frequent heuristic

7.2   The Experiment Results

See Table 1.

Table 1. Minimal ratio algorithm MKA (α) and minimal ratio algorithm infrequent tiebreaker (β).

k	n	α	β
3	3	1.0	0.1
	4	1.0	1.0
	5	1.0	1.0
	6	1.0	1.0
	7	0.8571428	0.857143
	8	0.75	0.75
	9	0.75	0.75
	10	0.666666	0.75
	11	0.6	0.75
	12	0.6	0.75
	13	0.6	0.75
	14	0.6	0.714285
	15	0.6	0.714285
	16	0.6	0.733333
	17	0.6	0.733333
4	4	1.0	1.0
	5	1.0	1.0
	6	1.0	1.0
	7	1.0	1.0
	8	1.0	1.0
	9	0.888889	0.888889
	10	0.8	0.8
	11	0.727273	0.727273
	12	0.727273	0.727273
	13	0.666667	0.727273
	14	0.615384	0.727273
	15	0.571428	0.727273
	16	0.571428	0.727273
5	5	1.0	1.0
	6	1.0	1.0
	7	1.0	1.0
	8	1.0	1.0
	9	1.0	1.0
	10	1.0	1.0
	11	0.909091	0.909091
	12	0.833333	0.833333
	13	0.769230	0.769230
	14	0.714285	0.714285
	15	0.714285	0.714285
	16	0.666667	0.714285
	17	0.625	0.714285
	18	0.588235	0.714285

k	n	α	β
6	6	1.0	1.0
	7	1.0	1.0
	8	1.0	1.0
	9	1.0	1.0
	10	1.0	1.0
	11	1.0	1.0
	12	1.0	1.0
	13	0.923077	0.923077
	14	0.857143	0.857143
	15	0.8	0.8
	16	0.75	0.75
	17	0.705882	0.705882
	18	0.705882	0.705882
	19	0.666667	0.705882
	20	0.631579	0.705882
7	7	1.0	1.0
	8	1.0	1.0
	9	1.0	1.0
	10	1.0	1.0
	11	1.0	1.0
	12	1.0	1.0
	13	1.0	1.0
	14	1.0	1.0
	15	0.933333	0.933333
	16	0.875	0.875
	17	0.823529	0.823529
	18	0.777778	0.777778
	19	0.736842	0.736842
	20	0.7	0.7
	21	0.7	0.7
	22	0.666667	0.7

References

1. Alamro, H., Badkobeh, G., Belazzougui, D., Iliopoulos, C.S., Puglisi, S.J.: Computing the antiperiod(s) of a string. In: Pisanti, N., Pissis, S.P. (eds.) Proceedings of 30th Combinatorial Pattern Matching, (CPM), LIPIcs, vol. 128, pp. 32:1–32:11. Schloss Dagstuhl - Leibniz-Zentrum für Informatik (2019)
2. Alzamel, A., et al.: Finding the anticover of a string. In: Proceedings of 31st Combinatorial Pattern Matching (CPM), LIPIcs (2020, to appear)
3. Amir, A., Levy, A., Lewenstein, M., Lubin, R., Porat, B.: Can we recover the cover? In: Proceedings of 28st Annual Symposium on Combinatorial Pattern Matching (CPM), LIPICS (2017)
4. Antoniou, P., Crochemore, M., Iliopoulos, C.S., Jayasekera, I., Landau, G.M.: Conservative string covering of indeterminate strings. In: Proceedings of Prague Stringology Conference, pp. 108–115 (2008)
5. Apostolico, A., Breslauer, D.: Of periods, quasiperiods, repetitions and covers. In: Mycielski, J., Rozenberg, G., Salomaa, A. (eds.) Structures in Logic and Computer Science. LNCS, vol. 1261, pp. 236–248. Springer, Heidelberg (1997). https://doi.org/10.1007/3-540-63246-8_14
6. Apostolico, A., Ehrenfeucht, A.: Efficient detection of quasiperiodicities in strings. Theoret. Comput. Sci. **119**(2), 247–265 (1993)
7. Apostolico, A., Iliopoulos, C., Farach, M.: Optimal superprimitivity testing for strings. Inf. Process. Lett. **39**, 17–20 (1991)
8. Badkobeh, G., Fici, G., Puglisi, S.J.: Algorithms for anti-powers in strings. Inf. Process. Lett. **137**, 57–60 (2018)
9. Bar-Noy, A., Nisgav, A., Patt-Shamir, B.: Nearly optimal perfectly periodic schedules. Distrib. Comput. **15**(4), 207–220 (2002). https://doi.org/10.1007/s00446-002-0085-1
10. Benson, G.: Tandem repeats finder: a program to analyze DNA sequence. Nucleic Acids Res. **27**(2), 573–580 (1999)
11. Breslauer, D.: An on-line string superprimitivity test. Inf. Process. Lett. **44**, 345–347 (1992)
12. Breslauer, D.: Testing string superprimitivity in parallel. Inf. Process. Lett. **49**(5), 235–241 (1994)
13. Christodoulakis, M., Iliopoulos, C.S., Park, K., Sim, J.S.: Approximate seeds of strings. J. Automata, Lang. Comb. **10**, 609–626 (2005)
14. Crochemore, M., Iliopoulos, C.S., Pissis, S.P., Tischler, G.: Cover array string reconstruction. In: Amir, A., Parida, L. (eds.) CPM 2010. LNCS, vol. 6129, pp. 251–259. Springer, Heidelberg (2010). https://doi.org/10.1007/978-3-642-13509-5_23
15. Fici, G., Restivo, A., Silva, M., Zamboni, L.Q.: Anti-powers in infinite words. J. Comb. Theory Ser. A **157**, 109–119 (2018)
16. Flouri, T., et al.: Enhanced string covering. Theoret. Comput. Sci. **506**, 102–114 (2013)
17. Fuglsang, A.: Distribution of potential type ii restriction sites (palindromes) in prokaryotes. Biochem. Biophys. Res. Commun. **310**(2), 280–285 (2003)
18. Gelfand, M.S., Koonin, E.V.: Avoidance of palindromic words in bacterial and archaeal genomes: a close connection with restriction enzymes. Nucleic Acids Res. **25**, 2430–2439 (1997)
19. Gfeller, B.: Finding longest approximate periodic patterns. In: Dehne, F., Iacono, J., Sack, J.-R. (eds.) WADS 2011. LNCS, vol. 6844, pp. 463–474. Springer, Heidelberg (2011). https://doi.org/10.1007/978-3-642-22300-6_39

20. Han, J., Dong, G., Yin, Y.: Efficient mining of partial periodic patterns in time series database. In: Proceedings of 15th International Conference on Data Engineering (ICDE), pp. 106–115 (1999)
21. Iliopoulos, C.S., Mouchard, L.: Quasiperiodicity and string covering. Theoret. Comput. Sci. **218**(1), 205–216 (1999)
22. Iliopoulos, C.S., Smyth, W.F.: An on-line algorithm of computing a minimum set of k-covers of a string. In: Proceedings of 9th Australian Workshop on Combinatorial Algorithms (AWOCA), pp. 97–106 (1998)
23. Iliopoulus, C.S., Moore, D.W.G., Park, K.: Covering a string. Algorithmica **16**(3), 288–297 (1996). https://doi.org/10.1007/BF01955677
24. Kociumaka, T., Pissis, S.P., Radoszewski, J., Rytter, W., Waleń, T.: Fast algorithm for partial covers in words. In: Fischer, J., Sanders, P. (eds.) CPM 2013. LNCS, vol. 7922, pp. 177–188. Springer, Heidelberg (2013). https://doi.org/10.1007/978-3-642-38905-4_18
25. Li, Y., Smyth, W.F.: Computing the cover array in linear time. Algorithmica **32**(1), 95–106 (2002). https://doi.org/10.1007/s00453-001-0062-2
26. Lisnic, B., Svetec, I.K., Saric, H., Nikolic, I., Zgaga, Z.: Palindrome content of the yeast Saccharomyces cerevisiae genome. Curr. Genet. **47**, 289–297 (2005). https://doi.org/10.1007/s00294-005-0573-5
27. Lothaire, M. (ed.): Combinatorics on Words, 2nd edn. Cambridge University Press, Cambridge (1997)
28. Loving, J., Scaduto, J.P., Benson, G.: An SIMD algorithm for wraparound tandem alignment. In: Cai, Z., Daescu, O., Li, M. (eds.) ISBRA 2017. LNCS, vol. 10330, pp. 140–149. Springer, Cham (2017). https://doi.org/10.1007/978-3-319-59575-7_13
29. Moore, D., Smyth, W.F.: An optimal algorithm to compute all the covers of a string. Inf. Process. Lett. **50**(5), 239–246 (1994)
30. Moore, D., Smyth, W.F.: A correction to: an optimal algorithm to compute all the covers of a string. Inf. Process. Lett. **54**, 101–103 (1995)
31. Pellegrini, M., Renda, M.E., Vecchio, A.: TRStalker: an efficient heuristic for finding fuzzy tandem repeats. Bioinformatics [ISMB] **26**(12), 358–366 (2010)
32. Smyth, W.F.: Repetitive perhaps, but certainly not boring. Theoret. Comput. Sci. **249**(2), 343–355 (2000)
33. Srivastava, S.K., Robins, H.S.: Palindromic nucleotide analysis in human T cell receptor rearrangements. PLoS ONE **7**(12), e52250 (2012)
34. Wexler, Y., Yakhini, Z., Kashi, Y., Geiger, D.: Finding approximate tandem repeats in genomic sequences. In: RECOMB, pp. 223–232 (2004)
35. Zhang, H., Guo, Q., Iliopoulos, C.S.: Algorithms for computing the lambda-regularities in strings. Fundamenta Informaticae **84**(1), 33–49 (2008)
36. Zhang, H., Guo, Q., Iliopoulos, C.S.: Varieties of regularities in weighted sequences. In: Chen, B. (ed.) AAIM 2010. LNCS, vol. 6124, pp. 271–280. Springer, Heidelberg (2010). https://doi.org/10.1007/978-3-642-14355-7_28

Multidimensional Period Recovery

Amihood Amir[1], Ayelet Butman[2], Eitan Kondratovsky[1], Avivit Levy[3(✉)], and Dina Sokol[4]

[1] Department of Computer Science, Bar-Ilan University, Ramat Gan, Israel
[2] Department of Computer Science, Holon Institute of Technology, Holon, Israel
[3] Department of Software Engineering, Shenkar College, 52526 Ramat-Gan, Israel
avivitlevy@gmail.com
[4] Department of Computer and Information Science,
Brooklyn College of the City University of New York, New York City, USA

Abstract. Multidimensional data are widely used in real-life applications. Intel's new brand of SSDs, called 3D XPoint, is an example of three-dimensional data. Motivated by a structural analysis of multidimensional data, we introduce the multidimensional period recovery problem, defined as follows. The input is a d-dimensional text array, with dimensions $n_1 \times n_2 \times \cdots \times n_d$, that contains corruptions, while the original text without the corruptions is periodic. The goal is then to report the period of the original text. We show that, if the number of corruptions is at most $\lfloor \frac{1}{2+\epsilon} \lfloor \frac{n_1}{p_1} \rfloor \cdots \lfloor \frac{n_d}{p_d} \rfloor \rfloor$, where $\epsilon > 0$ and $p_1 \times \cdots \times p_d$ are the period's dimensions, then the amount of possible period candidates is $O(\log N)$, where $N = \Pi_{i=1}^{d} n_i$. The independency of this bound of the number of dimensions is a surprising key contribution of this paper. We present an $O(\Pi_{i=1}^{d} n_i \Pi_{i=1}^{d} \log n_i)$ algorithm, for any constant dimension d, (linear time up to logarithmic factor) to report these candidates. The tightness of the bound on the number of errors enabling a small size candidate set is demonstrated by showing that if the number of errors is equal to $\lfloor \frac{1}{2} \lfloor \frac{n_1}{p_1} \rfloor \cdots \lfloor \frac{n_d}{p_d} \rfloor \rfloor$, a family of texts with $\Theta(N)$ period candidates can be constructed for any dimension $d \geq 2$.

1 Introduction

Periodicity is an important feature of strings suggesting a clean mathematical formalization to describe cyclic phenomena in many fields such as astronomy, geology, earth science, oceanography, meteorology, biological systems, genomics, economics, and more.

Periodicity has been extensively studied over the years and linear time algorithms for exploring the periodic nature of a string were presented (e.g. [15, 22, 25]). Multidimensional periodicity [2,20,29] and periodicity in parameterized strings [12] were also explored. In addition, periodicity has played a role in efficient parallel [3,4,14,18] and dynamic string algorithms [5,6].

A. Amir—Partly supported by ISF grant 1475/18 and BSF grant 2018141.
D. Sokol—Partly supported by BSF grant 2018141.

C. Boucher and S. V. Thankachan (Eds.): SPIRE 2020, LNCS 12303, pp. 115–130, 2020.
https://doi.org/10.1007/978-3-030-59212-7_9

However, realistic data may contain errors. These errors can happen because of real data sampling or they may result from the representation of data as a string. Hence, *approximate periodicity* has been explored from different angles [10,26,27,30]. Recently, Amir et al. [7] presented an algorithm to construct an approximate period in $O(nk \log \log n)$ time, where n is the string length and k is the minimal Hamming distance to the closest periodic string. For swap distance, where the errors are the exchange of two adjacent symbols (with no symbol participating in more than one exchange), their algorithm runs in $O(n^2)$ time.

The goal of this paper is to discover the original period of a periodic multi-dimensional text. This problem has been termed *period recovery*. However, even in strings, and even under a single error, there can be a few indistinguishable candidates. For example, *aaaabaaaaa* has two indistinguishable periodic sources *aaaabaaaab* and *aaaaaaaaaa*. Previous work on one-dimensional period recovery [8] shows that for a reasonable bound on the number of errors, although the original periodic behavior cannot be recovered by a single solution, there are at most $O(\log n)$ possible solutions, where n is the input text length.

In this paper, we take a step forward in the area of period recovery studying the problem for multi-dimensional text. For simplicity of exposition, we first thoroughly discuss the two-dimensional definitions and techniques. We then show that our techniques can be generalized multidimensional texts.

We begin by defining the two-dimensional (2D) period recovery problem for a two-dimensional text with dimensions $n \times m$. We then prove that for a reasonable number of errors, there are $O(\log(nm))$ possible candidate 2D periods of a 2D text. Proving this bound on the number of candidates is already a non-trivial key contribution of this paper. Note that, the reduction of the two-dimensional text to a one-dimensional text, which is the technique we use, results in a logarithmic factor in the candidates bound for the additional dimension. For the d-dimensional case, it yields an $O(\log^d N)$ bound. However, we are able to reduce this bound to $O(\log N)$, *which is independent of the number of dimensions*.

Furthermore, for the two-dimensional case, we present an efficient KMR algorithm for renaming rectangles. Our algorithm works in $O(nm \log n \log m)$ time and space and answers renaming queries in $O(1)$ time. Moreover, we also provide an example such that, if the number of errors is equal to $\lfloor \frac{1}{2} \lfloor \frac{n}{p} \rfloor \lfloor \frac{m}{q} \rfloor \rfloor$, then there are $\Theta(nm)$ false positive candidates, where $p \times q$ are the period's dimensions. That is, having a bigger number of errors makes it inefficient to search for the original period within such a huge set of candidates. We finally generalize the results on two-dimensional text to higher dimensions.

The Paper Contribution. The main contribution of this paper is:

- Proving that the logarithmic bound on the possible period candidates of [8] for 1-dimensional text holds for multidimensional texts as well, and is independent of the number of dimensions.
- Presenting an efficient period recovery algorithm for multidimensional texts.

2   Background and Problem Definition

2.1   1D Periodicity

A string r is *periodic* if its longest proper prefix that is also a suffix is at least half the length of r. A string s is *primitive* if it cannot be expressed in the form $s = u^j$, for some integer $j > 1$ and some prefix u of s. This notion of primitivity[1] is derived from the study of *squares*, where xx is defined as a square if x is primitive, otherwise it would be a *run*. A periodic string r can be expressed as $u^j u'$, where $j > 1$ and u' is a prefix of u for one unique primitive u, which is called *the period* of r. The *exponent* e of r is defined as the rational number that satisfies $e = \frac{|r|}{|u|}$, that is $u^e = r$.

Every non-primitive string is periodic but not every periodic string is non-primitive. For example, abc, abcab are both primitive and non-periodic, abcabc is non-primitive (and hence periodic), while abcabca is primitive and periodic with period abc.

2.2   2D Periodicity

We say that U is a horizontal prefix (resp. suffix) in rectangular array M if U is an initial (resp. ending) sequence of contiguous columns in M. A *horizontal border* of M is a proper horizontal prefix that is also a horizontal suffix of M. We say that B is the *longest horizontal border* of M if it is the horizontal border of M that spans the largest number of columns among the horizontal borders of M. The *horizontal period*, or *h-period*, of a rectangular array M is $n - b$ where b is the number of columns contained in the longest horizontal border of M.

Definition 1 [16,28]. *An $n \times m$ array M with h-period q is horizontally periodic, or h-periodic, if $q \leq \lfloor \frac{m}{2} \rfloor$.*

The h-period of an h-periodic array M is the least common multiple of the periods that occur in the rows of M. Vertical periodicity and the vertical period of a rectangular v-periodic array are defined analogously.

Definition 2 [11]. *An $n \times m$ array M is a 2D repetition if M is h-periodic and v-periodic.*

Consider an $n \times m$ array M and rational numbers $x > 0$, $y > 0$. $M^{x,y}$ is the array constructed by repeating M x times vertically and y times horizontally, yielding an $\lfloor xn \rfloor \times \lfloor ym \rfloor$ array. For example, $M = \begin{bmatrix} a & b & c \\ e & f & g \end{bmatrix}$

---

[1] This notion should not be confused with other notions of primitivity in stringology, such as in *covers*. The difference in the definition of primitivity for covers stems from the fact that the string must end with a complete occurrence of a cover, which is not the case for a period.

$$M^{\frac{5}{2},\frac{7}{3}} = \begin{bmatrix} a\ b\ c\ a\ b\ c\ a \\ e\ f\ g\ e\ f\ g\ e \\ a\ b\ c\ a\ b\ c\ a \\ e\ f\ g\ e\ f\ g\ e \\ a\ b\ c\ a\ b\ c\ a \end{bmatrix} \qquad M^{2,\frac{8}{3}} = \begin{bmatrix} a\ b\ c\ a\ b\ c\ a\ b \\ e\ f\ g\ e\ f\ g\ e\ f \\ a\ b\ c\ a\ b\ c\ a\ b \\ e\ f\ g\ e\ f\ g\ e\ f \end{bmatrix}$$

Definition 3 [11,21]. *An $n \times m$ array M is primitive if it cannot be partitioned into more than one non-overlapping complete occurrence of some block P. M is non-primitive if M can be expressed as $M = P^{r,s}$ for real numbers r, s such that $r > 1$ is an integer or $s > 1$ is an integer.*

The different basic configurations of a non-primitive rectangular array are:

P
P

P	P

P	P
P	P

As in the string terminology, a 2D repetition can be either primitive or non-primitive. In the above example, $M^{\frac{5}{2},\frac{7}{3}}$ is both a 2D repetition and primitive, while $M^{2,\frac{8}{3}}$ is a 2D repetition and non-primitive.

Definition 4 [9]. *A primitive root P of 2D repetition M is a primitive sub-array such that $M = P^{r,s}$ for rational numbers r, s. M begins with P at its upper left corner and can be partitioned into non-overlapping replicas of P, possibly including partial occurrences of P at its right and/or lower ends.*

Lemma 1 [9]. *Every 2D repetition M has a unique primitive root P such that $M = P^{r,s}$ for rational numbers r, s.*

Definition 5 [9]. *Let M be a 2D repetition of dimensions $n \times m$ with primitive root P of dimensions $p \times q$. The exponent of M is a tuple (e_1, e_2), in which e_1 and e_2 are rational numbers, that satisfy $e_1 = \frac{n}{p}$ and $e_2 = \frac{m}{q}$.*

Note that by the definition, in a 2D repetition $M =$

P	...	P	P'
...	...	...	...
P	...	P	P'
P''	...	P''	P'''

there are at least two P-blocks horizontally and vertically. That is, the primitive root P repeats both to the right and underneath its initial occurrence in M.

2.3   Problem Definition

The goal is to recover the original period of a periodic text that may be corrupted by replacement errors, where their number is measured by the *Hamming distance*. Clearly, if too many errors are introduced to the text, there is no hope of recovering the original period. Intuitively, if 50% or more of the copies of the original period are corrupted, a recovering process would not have a sufficient information for candidate elimination. We prove that if the number of errors is smaller, the number of possible candidates can be significantly reduced, regardless of the error distribution. The problem is formally defined below.

Period Recovery over the Hamming Distance: Given a $n \times m$ array T defined over alphabet Σ and a real constant ϵ ($\epsilon > 0$), find all primitive roots P, with dimensions $p \times q$, such that the Hamming distance between $P^{n/p,m/q}$ and T is at most $\lfloor \frac{1}{2+\epsilon} \lfloor \frac{n}{p} \rfloor \lfloor \frac{m}{q} \rfloor \rfloor$.

3   The Bound on the Candidates Set Size

In this section we prove the bound on the number of errors that enables feasible amount of approximate period candidates for any error distribution. For the rest of the paper we choose the bound on the number of errors to be at most $\lfloor \frac{1}{2+\epsilon} \lfloor \frac{n}{p} \rfloor \lfloor \frac{m}{q} \rfloor \rfloor$. Subsection 3.1 explains the intuition behind this bound.

3.1   The Intuition Behind the Bound on the Number of Errors

The recovery problem is to restore the source text, a text before any substitution error has been applied, under the assumption that this source text is periodic. There are cases in which it is impossible to distinguish between different source texts without additional information. Even with only two substitution errors there can be few indistinguishable source texts, as the following example shows.

Example 1. Let $T = \begin{bmatrix} abaa \\ aaaa \\ abaa \\ aaaa \end{bmatrix}$ be the input to the recovery problem. Assume that T is the resulted text after a single substitution error occurred in some unknown source text. The source text candidates (or candidates for short) are the following.

$$C = \left\{ \begin{bmatrix} aaaa \\ aaaa \\ aaaa \\ aaaa \end{bmatrix}, \begin{bmatrix} abab \\ aaaa \\ abab \\ aaaa \end{bmatrix}, \right\} = \left\{ [a]^{4,4}, \begin{bmatrix} ab \\ aa \end{bmatrix}^{2,2} \right\}$$

That is, it is not always possible to put the finger on a single source text based on the periodicity property. Therefore, our recovery algorithm reports a set of indistinguishable source text candidates. We reduce the candidates set to a sublinear size by assuming a bound on the number of corruptions to the input text. When the number of errors is unbounded, the number of candidates is equal to $\Theta(|\Sigma|^{\lfloor n/2 \rfloor \lfloor m/2 \rfloor})$. The lower bound comes from taking all matrices of size $\lfloor n/2 \rfloor \times \lfloor m/2 \rfloor$ as a 2D-repetition of a period. The upper bound comes from counting $\sum_{i=1}^{x} \sum_{j=1}^{y} |\Sigma|^{ij} <= 4|\Sigma|^{xy}$ by using sum of geometric progression formula twice.

Error and Approximation Models. Before introducing the chosen bound, we describe the error model and the approximation model. In this paper we consider substitution errors, which preserve the array dimensions. We focus on the worst case model of errors, where there is no assumption about the errors distribution. Thus, the recovery algorithm needs to handle any error distribution.

Even when restricting to cases where the period occurs in a constant number of positions without any error, there can still be examples with a huge amount of source text candidates. For example, if the text has a period occurring at least c times with no corruptions, for some constant c, then it is possible to construct $O(\frac{n}{c})$ source text candidates by concatenating $\frac{n}{c}$ different periods each c times. This amount is already huge to enable detection of the correct source text, as it would be necessary to iterate over all these candidates.

When setting the error bound, we take into account the amount of uncorrupted period occurrences as a percentage of the overall occurrences. Observing the case when half of the period occurrences are uncorrupted, it is possible to provide an example with a huge candidates set size.

Example 2. The 1D string $a^{2k}ba^{4k+1}$ with one error has $\frac{n}{6}$ periodic candidates, where $n = 6k + 2$. The 2D text T is a replication of this string in m rows:

$$T = \begin{bmatrix} a^{2k}ba^{4k+1} \\ \vdots \\ a^{2k}ba^{4k+1} \end{bmatrix}$$

The source text candidates are the following:

$$C = \left\{ \begin{bmatrix} a^{2k}b\ a^{2k}b\ a^{2k} \\ \vdots \\ a^{2k}b\ a^{2k}b\ a^{2k} \end{bmatrix}, \begin{bmatrix} a^{2k}ba\ a^{2k}ba\ a^{2k-2} \\ \vdots \\ a^{2k}ba\ a^{2k}ba\ a^{2k-2} \end{bmatrix}, \begin{bmatrix} a^{2k}ba^2\ a^{2k}ba^2\ a^{2k-4} \\ \vdots \\ a^{2k}ba^2\ a^{2k}ba^2\ a^{2k-4} \end{bmatrix}, \ldots, \begin{bmatrix} a^{2k}ba^k\ a^{2k}ba^k \\ \vdots \\ a^{2k}ba^k\ a^{2k}ba^k \end{bmatrix} \right\}$$

$$= \left\{ [a^{2k}b]^{2+\frac{2k}{2k+1},m}, [a^{2k}ba]^{2+\frac{2k-2}{2k+2},m}, [a^{2k}ba^2]^{2+\frac{2k-4}{2k+3},m}, \ldots, [a^{2k}ba^k]^{2,m} \right\}$$

Each such candidate text has different period's dimensions and for each the corruptions introduced are exactly in half of the period's occurrences.

3.2 The Tightness of Our Bound

Example 2 (in Subsect. 3.1) presents a family of arrays that causes linear amount of candidates. In this example, there are at most $\lfloor \frac{1}{2} \lfloor \frac{n_1}{p_1} \rfloor \cdots \lfloor \frac{n_d}{p_d} \rfloor \rfloor$ errors, where $n_1 \times \cdots \times n_d$ are the array's dimensions, and $p_1 \times \cdots \times p_d$ are the period's dimensions. On the other hand, below we prove that if the number of errors is at most $\lfloor \frac{1}{2+\epsilon} \lfloor \frac{n_1}{p_1} \rfloor \cdots \lfloor \frac{n_d}{p_d} \rfloor \rfloor$, then there is a logarithmic number of candidates. We show that, for the right choice of ϵ, the bounds are tight.

Lemma 2. *For* $\epsilon < \frac{4}{\Pi_{i=1}^d n_i}$,

$$\left\lfloor \frac{1}{2} \left\lfloor \frac{n_1}{p_1} \right\rfloor \cdots \left\lfloor \frac{n_d}{p_d} \right\rfloor \right\rfloor \geq \left\lfloor \frac{1}{2+\epsilon} \left\lfloor \frac{n_1}{p_1} \right\rfloor \cdots \left\lfloor \frac{n_d}{p_d} \right\rfloor \right\rfloor \geq \left\lfloor \frac{1}{2} \left\lfloor \frac{n_1}{p_1} \right\rfloor \cdots \left\lfloor \frac{n_d}{p_d} \right\rfloor \right\rfloor - 1$$

Proof. Let $X = \lfloor \frac{n_1}{p_1} \rfloor \cdots \lfloor \frac{n_d}{p_d} \rfloor$. We wish to prove that, $\lfloor \frac{X}{2} \rfloor \geq \lfloor \frac{X}{2+\epsilon} \rfloor \geq \lfloor \frac{X}{2} \rfloor - 1$. The left inequality is always true for $\epsilon > 0$. Therefore, the remaining part is to show that, $\frac{X}{2+\epsilon} \geq \frac{X}{2} - 1$. Thus, $1 \geq \frac{\epsilon X}{4+2\epsilon}$, which means that $\frac{4}{\epsilon} \geq X - 2$. We have that: $X \leq \Pi_{i=1}^d n_i$, thus, by choosing $\epsilon < \frac{4}{\Pi_{i=1}^d n_i}$ the condition holds. □

3.3 The Candidate Set Size Bound

In this subsection, we present the lower bound on $Ham(T_1, T_2)$ where T_1 and T_2 are periodic arrays of same dimensions, and explain its novelty. As a first step, we intend to generalize the following lemma stated for one-dimensional (1D) texts.

Lemma 3 [Amir et al. [8]]. *Let T_1 and T_2 be two 1D n-length texts, with P_1 of length p_1 and P_2 of length p_2 their 1D primitive periods, respectively, where, without loss of generality, assume that $p_1 \geq p_2$. Then, $Ham(T_1, T_2) \geq \lfloor \frac{n}{p_1} \rfloor$.*

Using the same terms of Lemma 3 we deal with 2D arrays. Let T_1 and T_2 be 2D texts. We denote P_1 and P_2 to be the 2D periods of T_1 and T_2, respectively, and assume, without loss of generality, that P_1's area is bigger than the area of P_2. The main difficulty in this type of proofs is handling the arrays' dimensions. That is, even if an array has a bigger area it does not mean that both dimensions of the array are bigger. Let M and M' be a $n \times m$-array and $n' \times m'$-array, respectively. Without loss of generality, assume $n \cdot m \geq n' \cdot m'$. Then there are several cases each of which requires different handling during proofs on two-dimensional arrays of different dimensions sizes.

To overcome the above problem a pseudo renaming technique is used. Pseudo renaming relies on an existence of a renaming process such that equal elements are transformed into the same letter. We stress that the renaming process time might not be bounded. For the sake of the bound theorem's proof, only the existence of such renaming needs to be assumed.

For proving our bound, a pseudo renaming on vertical sub-strings is applied to reduce to the 1D-recovery problem. However, the heights of P_1 and P_2 might be of different size not knowing which is smaller. Assume without loss of generality, that the area of P_1 is bigger than the area of P_2. Thus, at least one of P_1's dimensions is bigger than the corresponding dimension of P_2. Assume without loss of generality that P_1 is wider than P_2. Otherwise, if P_1 is higher than P_2, applying transpose on the input and the arrays P_1, P_2 results in an analogous problem, in which P_1^T is wider than P_2^T. That is, a transpose operation preserves all other properties.

Lemma 4. *Let T_1 and T_2 be two $n \times m$ 2D texts, with P_1 and P_2 their 2D primitive periods, respectively. P_1 is a $p_1 \times q_1$ array and P_2 is a $p_2 \times q_2$ array.*

1. *If $q_1 > q_2$, i.e., P_1 is wider than P_2 then, $Ham(T_1, T_2) \geq \lfloor \frac{n}{p_1} \rfloor \cdot \lfloor \frac{m}{q_1} \rfloor$.*
2. *If $p_1 > p_2$, i.e., P_1 is higher than P_2 then, $Ham(T_1, T_2) \geq \lceil \frac{n}{p_1} \rceil \cdot \lceil \frac{m}{q_1} \rfloor$.*
3. *If P_2 is wider or higher than P_1, then, $Ham(T_1, T_2) \geq \lfloor \frac{n}{p_2} \rfloor \cdot \lfloor \frac{m}{q_2} \rfloor$.*

Lemma 5. *Let $\epsilon > 0$, T an $n \times m$ 2D text, Let S be the set of candidate approximate 2D primitive periods of T, i.e., $S = \{P_i | Ham(T, T_i) \leq \lfloor \frac{1}{2+\epsilon} \lfloor \frac{n}{p_i} \rfloor \lfloor \frac{m}{q_i} \rfloor \rfloor \}$, where P_i has dimensions $p_i \times q_i$, and T_i is the $n \times m$ array that is a 2D repetition of the primitive period P_i. Then, S has a nested rectangles structure, i.e., sorting the candidates P_i by non-decreasing order of their areas results in the same order of sorting them by non-decreasing order of any of their dimensions.*

Proof. Let P_1 and P_2 be approximate 2D primitive periods of T in S. Without loss of generality, assume that $\lfloor \frac{n}{p_2} \rfloor \lfloor \frac{m}{q_2} \rfloor \geq \lfloor \frac{n}{p_1} \rfloor \lfloor \frac{m}{q_1} \rfloor$. For simplicity, denote $b = \lfloor \frac{n}{p_2} \rfloor \lfloor \frac{m}{q_2} \rfloor \geq \lfloor \frac{n}{p_1} \rfloor \lfloor \frac{m}{q_1} \rfloor = a$.

Assume to the contrary that $p_1 > p_2$ and $q_1 \leq q_2$. The symmetric case, where $p_1 \leq p_2$ and $q_1 > q_2$, results in the same expressions. Lemma 4 requires that either P_1 is wider or higher than P_2 or vise versa, thus, $Ham(T_1, T_2) \geq \max\{\lfloor \frac{n}{p_1} \rfloor \lfloor \frac{m}{q_1} \rfloor, \lfloor \frac{n}{p_2} \rfloor \lfloor \frac{m}{q_2} \rfloor \} = \max\{a, b\} = b$. By the triangle inequality, we have: $Ham(T_1, T) + Ham(T_2, T) \geq Ham(T_1, T_2)$. Therefore, we have that:

$$\frac{1}{2+\epsilon} \lfloor \frac{n}{p_1} \rfloor \lfloor \frac{m}{q_1} \rfloor + \frac{1}{2+\epsilon} \lfloor \frac{n}{p_2} \rfloor \lfloor \frac{m}{q_2} \rfloor \geq \lfloor \frac{1}{2+\epsilon} \lfloor \frac{n}{p_1} \rfloor \lfloor \frac{m}{q_1} \rfloor \rfloor + \lfloor \frac{1}{2+\epsilon} \lfloor \frac{n}{p_2} \rfloor \lfloor \frac{m}{q_2} \rfloor \rfloor \geq$$

$$\max \left\{ \lfloor \frac{n}{p_1} \rfloor \lfloor \frac{m}{q_1} \rfloor, \lfloor \frac{n}{p_2} \rfloor \lfloor \frac{m}{q_2} \rfloor \right\}$$

However, then we get:

$$\frac{1}{2+\epsilon} a + \frac{1}{2+\epsilon} b \geq b \implies \frac{1}{2+\epsilon} a \geq \frac{1+\epsilon}{2+\epsilon} b \implies a \geq (1+\epsilon)b,$$

which contradicts the fact that $b = \max\{a, b\}$. Therefore, $p_1 \geq p_2$ and $q_1 \geq q_2$.
□

Lemma 6. *Let $\epsilon > 0$, T an $n \times m$ 2D text, P_1 and P_2 approximate 2D primitive periods of T. Let T_1 and T_2 be the $n \times m$ arrays that are 2D repetitions of the primitive periods P_1 and P_2, respectively. Assume that the number of mismatches between T_1 and T and between T_2 and T is less than $\lfloor \frac{1}{2+\epsilon} \lfloor \frac{n}{p_1} \rfloor \lfloor \frac{m}{q_1} \rfloor \rfloor$, $\lfloor \frac{1}{2+\epsilon} \lfloor \frac{n}{p_2} \rfloor \lfloor \frac{m}{q_2} \rfloor \rfloor$, respectively. Without loss of generality, assume that $\lfloor \frac{n}{p_2} \rfloor \lfloor \frac{m}{q_2} \rfloor \geq \lfloor \frac{n}{p_1} \rfloor \lfloor \frac{m}{q_1} \rfloor$. Then, $\lfloor \frac{n}{p_2} \rfloor \lfloor \frac{m}{q_2} \rfloor \geq (1+\epsilon) \lfloor \frac{n}{p_1} \rfloor \lfloor \frac{m}{q_1} \rfloor$.*

Proof. For simplicity, denote $b = \lfloor \frac{n}{p_2} \rfloor \lfloor \frac{m}{q_2} \rfloor \geq \lfloor \frac{n}{p_1} \rfloor \lfloor \frac{m}{q_1} \rfloor = a$. By Lemma 5, we have, without loss of generality, that $p_1 \geq p_2$ and $q_1 \geq q_2$. By the triangle inequality, we have: $Ham(T_1, T) + Ham(T_2, T) \geq Ham(T_1, T_2)$. Therefore, we get using Lemma 4:

$$\frac{1}{2+\epsilon} a + \frac{1}{2+\epsilon} b \geq a \implies \frac{1}{2+\epsilon} b \geq \frac{1+\epsilon}{2+\epsilon} a \implies b \geq (1+\epsilon)a$$

□

Corollary 1. *Let T be an $n \times m$ text. Then, there are at most $\log_{1+\epsilon}(nm) + 1$ different approximate periods P of T with at most $\lfloor \frac{1}{2+\epsilon} \lfloor \frac{n}{p} \rfloor \lfloor \frac{m}{q} \rfloor \rfloor$ errors.*

4   Two-Dimensional KMR

In 1972, Karp, Miller and Rosenberg (KMR) [23] presented the one-dimensional renaming problem. In this problem one aims to preprocess the text to answer the renaming query, i.e., provide a name for a substring of the text, such that equal substrings receive the same name. Assuming the RAM model in which words, single cells of memory, are of size $\log n$ bits, the key idea is to give every substring a name that is stored in a constant amount of words. The one-dimensional KMR algorithm preprocess n-length text in $O(n \log n)$ time to answer the renaming query in constant time.

In 1991, Crochemore and Rytter [17] extended this method to two-dimensional squared sub-arrays, i.e., sub-arrays with equal amount of rows and columns. Their algorithm works in $O(N \log N)$ time, where N is the input size, i.e., a multiplication of its two dimensions. In this section, we describe our generalization of the KMR renaming to any two-dimensional sub-array.

Definition 6. [Two-Dimensional KMR Renaming]
Input: *Two-dimensional text T with dimensions $n \times m$.*
Output: *A data structure that answers the query $Name_{r,q}[i, j]$, the name of the sub-array with dimensions $r \times q$ that starts at position i, j inside T, such that two sub-arrays have the same name if and only if they are equal to each other.*

Theorem 1. *Let T be an array with dimensions $n \times m$. After preprocessing T in $O(nm \log n \log m)$ time, we can answer $Name_{r,q}[i, j]$ queries in $O(1)$ time.*

In the one-dimensional KMR algorithm, the main observation is that it is enough to only answer the queries of substrings whose length is a power of two. Indeed, the name of every substring that is not a power of two can be built by concatenating the names of its largest prefix and suffix, both of lengths that are equal to the largest power of two that is smaller than the substring's length.

The one-dimensional KMR algorithm has $\lceil \log n \rceil$ stages, where n is the text length. The stages are numbered by index k, $0 \leq k \leq \lceil \log n \rceil - 1$. Such that at the kth stage, we compute an array $Name_{2^k}$ of size n, where the element at index i is the name of the substring $T[i..i + 2^k]$. Notice that, the text should be padded with $2^{\lceil \log n \rceil - 1} - 1$ special characters $\$ \notin \Sigma$, where Σ is the alphabet of the text. Each stage of the renaming algorithm relies on the previous stage. The use of BucketSort allows the algorithm to create names efficiently without additional logarithmic factors. At the $(k + 1)$th stage the names $Name_{2^k}[i]$ and $Name_{2^k}[i + 2^k]$ are combined to create the name $Name_{2^{k+1}}[i]$ for the substring $T[i..i + 2^{k+1}]$.

The same observation holds also for two-dimensional sub-arrays. In this case, it is enough to support renaming queries only for sub-arrays with dimensions $2^i \times 2^j$, where $0 \leq i \leq \lceil \log n \rceil - 1$ and $0 \leq j \leq \lceil \log m \rceil - 1$. If the sub-array has a dimension that is not a power of 2, then the algorithm considers the KMR name to be the clockwise (starting from the top-left corner) concatenation of the largest four named sub-arrays that cover it, starting from each of the four corners.

Let M be an $n \times m$ array. The generalization of KMR algorithm for the two-dimensional case is done in the natural manner by first renaming the rows using the one-dimensional renaming. That is, the rows are concatenated into nm-length string. The output is $\lceil \log n \rceil$ arrays of names each of length nm. Note that, we stop the KMR construction algorithm after $\lceil \log n \rceil$ iterations. These arrays are then rearranged back to 2D arrays of dimensions $n \times m$. Symmetrically, each of these $\lceil \log n \rceil$ arrays is processed by using the one-dimensional renaming on columns. The overall process results in $\lceil \log n \rceil \lceil \log m \rceil$ arrays each containing the names of sub-arrays of dimensions $2^i \times 2^j$. If nm is not a power of 2, then the renaming for the nm-length requires ℓ-length padded extension at its end, where $\ell \geq 0$ is chosen such that $\ell + nm$ is the smallest power of 2, larger than nm.

5   The Recovery Algorithm

In the 2D period recovery problem, the input is a $n \times m$ 2D text T and a number $\epsilon > 0$. The output is the set of all primitive periods P with dimensions $p \times q$, such that the Hamming distance between T and $P^{\frac{n}{p}, \frac{m}{q}}$ is at most $\lfloor \frac{1}{2+\epsilon} \lfloor \frac{n}{p} \rfloor \lfloor \frac{m}{q} \rfloor \rfloor$.

This section is organized as follows. We begin with some claims that reduce the number of candidates for being an approximate period of the 2D text T. We then show how to verify these candidates by two procedures: primitivity check and Hamming distance calculation query to check whether the Hamming distance is $\leq \lfloor \frac{1}{2+\epsilon} \lfloor \frac{n}{p} \rfloor \lfloor \frac{m}{q} \rfloor \rfloor$.

Definition 7. *Let T be a $n \times m$ 2D text, and P an approximate period candidate for T with dimensions $p \times q$. We call a position ℓ in T an exact position with respect to P if the occurrence of P in T at position ℓ is not a partial occurrence, i.e., where $\ell \in \{0, p, 2p, \ldots, (\lfloor \frac{n}{p} \rfloor - 1)p\}$ or $\ell \in \{0, q, 2q, \ldots, (\lfloor \frac{m}{q} \rfloor - 1)q\}$.*

Lemma 7. *Let T be a $n \times m$ 2D text, and P an approximate period candidate for T with dimensions $p \times q$. Then, there are more than $\frac{1+\epsilon}{2+\epsilon} \lfloor \frac{n}{p} \rfloor \lfloor \frac{m}{q} \rfloor > \frac{1}{2} \lfloor \frac{n}{p} \rfloor \lfloor \frac{m}{q} \rfloor$ exact positions i, j in T with respect to P.*

The above lemma follows immediately from the fact that for a solution P with dimensions $p \times q$, there are at most $\lfloor \frac{1}{2+\epsilon} \lfloor \frac{n}{p} \rfloor \lfloor \frac{m}{q} \rfloor \rfloor$ mismatches between T and $P^{\frac{n}{p}, \frac{m}{q}}$. In addition, since there are strictly more than $\frac{1}{2} \lfloor \frac{n}{p} \rfloor \lfloor \frac{m}{q} \rfloor$ exact copies of sub-array P in T, at most one sub-array P can fulfill this requirement per each period dimensions $p \times q$. This leads to the following corollary.

Corollary 2. *For each period's dimensions $p \times q$, there can be at most one candidate for being an approximate primitive root of T.*

Notice that, a sub-array P with dimensions $p \times q$ that has more than $\frac{1}{2} \lfloor \frac{n}{p} \rfloor \lfloor \frac{m}{q} \rfloor$ exact copies in T is only a *candidate* sub-array for being an approximate primitive root of T. The algorithm iterates over all possible dimensions $p \times q$ and looks for the sub-array that occurs the majority of times. This sub-array is then to be examined by the verification stage. The novelty of this idea is that by giving

input : 2D array T of dimensions $n \times m$
output: All approximate periods of T

for $p \leftarrow 1$ to $\frac{n}{2}$ do
    for $q \leftarrow 1$ to $\frac{m}{2}$ do
        /* Step 1: find candidate sub-array with dimensions $p \times q$ */
        $names \leftarrow$ EmptyList($[]$)
        for $i \leftarrow 0$ to $\lfloor \frac{n}{p} \rfloor - 1$ do
            for $j \leftarrow 0$ to $\lfloor \frac{m}{q} \rfloor - 1$ do
                $names$.Append($Name_{p \times q}[i,j]$)
            end
        end
        $P \leftarrow$ FindMajority ($names$)
        if P occurs $\leq \lfloor \frac{1}{2} \lfloor \frac{n}{p} \rfloor \lfloor \frac{m}{q} \rfloor \rfloor$ in $names$ then continue to next p, q

        /* Step 2: check primitivity of the candidate P */
        if ! PrimitivityCheck (P) then continue to the next p, q

        /* Step 3: compute Hamming distance from T */
        $d \leftarrow$ HammingDistance ($T, P^{\frac{n}{p}, \frac{m}{q}}$)
        if $d \leq \lfloor \frac{1}{2+\epsilon} \lfloor \frac{n}{p} \rfloor \lfloor \frac{m}{q} \rfloor \rfloor$ then Report P as an approximate period of T
    end
end

Algorithm 1: Period recovery over the Hamming distance

a single candidate per each possible dimensions of P, it is possible to save time by applying a more expensive verification on a smaller number of candidates. Similar idea was shown in [1].

Let $p \times q$ be some dimensions. Each sub-array with dimensions $p \times q$ that is at exact positions is encoded by its KMR name. The *majority* algorithm [13] then applied on these names. This algorithm uses $O(1)$ space and requires two passes over the names. The second pass verifies that we indeed find the majority between the elements.

5.1 Computing the Hamming Distance

The Hamming distance between $P^{\frac{n}{p}, \frac{m}{q}}$ and T must be at most $\lfloor \frac{1}{2+\epsilon} \lfloor \frac{n}{p} \rfloor \lfloor \frac{m}{q} \rfloor \rfloor$ for P to be a valid solution. We compute the Hamming distance between all the possible positions for P in the text and the candidate array P. If the number of mismatches is at most $\lfloor \frac{1}{2+\epsilon} \lfloor \frac{n}{p} \rfloor \lfloor \frac{m}{q} \rfloor \rfloor$, then P is an approximate period of T. Note that, P can be represented in constant space by the position at which it occurs in T as an uncorrupted occurrence.

Definition 8 [Hamming Distance between Sub-arrays]. *Let M_1, M_2 be two-dimensional arrays with dimensions $n \times m$. In a sub-arrays Hamming distance data structure the arrays are preprocessed to answer $Ham(S_1, S_2)$ queries, where S_1 and S_2 are sub-arrays of M_1 and M_2, respectively, having the same dimensions $p \times q$.*

Theorem 2. *A sub-arrays Hamming distance data structure can be constructed in $O(nm \min\{\log n, \log m\})$ time and space with $O(\Delta)$ query time, where $\Delta = Ham(S_1, S_2)$. Moreover, queries of the form "does $Ham(S_1, S_2) \leq \delta$?" can be answered in time $O(\min\{\delta, Ham(S_1, S_2)\})$.*

Assume without loss of generality, that $n \leq m$. The idea is using the "Kangaroo Jumps" [19] to detect horizontal mismatches between the columns. Then repeat the "Kangaroo Jumps" vertically to count the amount of mismatches. There are at most Δ unequal columns, and the overall amount of mismatches among these columns is Δ, where $\Delta = Ham(S_1, S_2)$. "Kangaroo Jumps" is a one-dimensional technique that is based on *lcp* queries, which is a query that inputs two suffixes i, j of 1D text T and outputs the length of their longest common prefix. By repeating the *lcp* query, we can calculate the Hamming distance between any two 1D substrings in $O(\Delta)$, where Δ is the Hamming distance between these substrings. It is also possible to stop if the Hamming distance exceeds some threshold δ after spending $O(\delta)$ time. We apply the "Kangaroo Jumps" structure on the KMR internal arrays for names of the form $2^i \times 1$.

5.2   Primitivity Check

We only consider a sub-array P to be the approximate primitive root of T if P is primitive. In order to decide whether a candidate sub-array is primitive, we combine the *2-Period Queries* algorithm presented by Kociumaka et al. [24] with the 2D KMR renaming algorithm. The *2-Period Queries* algorithm preprocesses a one-dimensional text to answer the query of the form whether a substring A is periodic and, if so, compute its shortest period. Once the period p of a substring A is computed, the substring is non-primitive if the algorithm returns $p < |A|$ such that $p \mid |A|$, otherwise it is primitive. The algorithm preprocessing time is linear with constant time query.

We process each of the KMR 2D arrays for *2-Period Queries* by concatenating the rows one after the other. Assume that we have a sub-array of dimensions $p \times q$ and we wish to check its primitivity, horizontally and vertically. Without loss of generality, assume that we want to check horizontal primitivity. If p is a power of 2 ($p = 2^k$), then directly from the KMR array that is related to names of dimensions $2^k \times 1$ we can check if a substring is primitive using the *2-Period Queries*. Otherwise, let k be the maximal integer such that $2^k < p$. That is, by two *2-Period Queries* we can check if the upper sub-array of P of dimensions $2^k \times q$ is primitive and similarly about the bottom sub-array of dimensions $2^k \times q$. If one of them is primitive, then we finished and P is also horizontally primitive. However, the case where S_1 and S_2 are both non-primitive requires special verification. Let p_1 and p_2 be the periods of S_1 and S_2, respectively. Then, the length of the period of P equals to the *least common multiple (lcm)* between the p_1 and p_2. That is, if $lcm(p_1, p_2) < q$ and $lcm(p_1, p_2) \mid q$, then P is non-primitive. Otherwise, P is horizontally primitive.

5.3   Time and Space Complexity

The KMR renaming is performed once in $O(nm \log n \log m)$ time and space by Theorem 1. The sub-array primitivity check is also performed once in $O(nm \log n)$ time and space. In addition, $O(\log n)$ suffix trees are constructed on the KMR arrays to support "Kangaroo Jumps" in $O(nm \log n)$ time and space by Theorem 2.

In Step 1, for each period's dimensions $p \times q$, the procedure of finding a candidate substring with dimensions $p \times q$ is done in $O(\frac{nm}{pq})$ time, as follows. First, for pair of positions $i \in \{0, p, \ldots, (\lfloor \frac{n}{p} \rfloor - 1)p\}$ and $j \in \{0, q, \ldots, (\lfloor \frac{m}{q} \rfloor - 1)q\}$, the name of the sub-array at position (i, j) is found and inserted to a list in constant time. Then, the algorithm of [13] for finding the majority sub-array, P, is performed in linear time of the list size, which is equal to $\lfloor \frac{n}{p} \rfloor \lfloor \frac{m}{q} \rfloor$.

In Step 2, primitivity check costs constant time per candidate using the *2-Period Queries* of [24].

Finally, in Step 3, we compute the total number of mismatches between P and the sub-array at positions i, j, where $i = 0_{mod \, p}$ and $j = 0_{mod \, q}$ (some P occurrences might be partial). Note that this procedure runs at most in $O(\frac{nm}{pq})$ time, since having more than $\lfloor \frac{1}{2+\epsilon} \lfloor \frac{n}{p} \rfloor \lfloor \frac{m}{q} \rfloor \rfloor$ mismatches means that P is not a valid solution.

This gives a total of $\sum_{p=1}^{n} \sum_{q=1}^{m} \frac{nm}{pq} = nm \cdot \sum_{p=1}^{n} \frac{1}{p} \sum_{q=1}^{m} \frac{1}{q} = O(nm \log n \log m)$ time for finding all valid solutions in T. Thus, the total time and space complexity of the algorithm is bounded by $O(nm \log n \log m)$.

6   Multidimensional Generalization

In this section, we present the generalization of all the techniques that were presented in this paper to the multidimensional case.

In Subsect. 3.3, we prove that for the 2D texts, with dimensions $n \times m$, there are at most $O(\log(nm))$ approximate periods candidates. Here, we generalize the techniques to work for the multidimensional case.

Lemma 8. *Let T_1 and T_2 be two d-dimensional texts with dimensions $n_1 \times n_2 \times \cdots \times n_d$. Let P_1 and P_2 be their d-dimensional primitive periods, respectively. P_1 is a $p_1 \times p_2 \times \cdots \times p_d$ array and P_2 is a $q_1 \times q_2 \times \cdots \times q_d$ array.*

1. *If there exists some index i such that $p_i > q_i$, then, $Ham(T_1, T_2) \geq \lfloor \frac{n_1}{p_1} \rfloor \cdot \lfloor \frac{n_2}{p_2} \rfloor \cdots \lfloor \frac{n_d}{p_d} \rfloor$.*
2. *If there exists some index i such that $p_i < q_i$, then, $Ham(T_1, T_2) \geq \lfloor \frac{n_1}{q_1} \rfloor \cdot \lfloor \frac{n_2}{q_2} \rfloor \cdot \lfloor \frac{n_d}{q_d} \rfloor$.*

Proving the d-dimensional variations of Lemma 5 and Lemma 6 is done by the redefinition of a and b as $\lfloor \frac{n_1}{p_1} \rfloor \cdots \lfloor \frac{n_d}{p_d} \rfloor$ and $\lfloor \frac{n_1}{q_1} \rfloor \cdots \lfloor \frac{n_d}{q_d} \rfloor$, respectively. The following corollary holds.

Corollary 3. *Let T be an $n_1 \times n_2 \times \cdots \times n_d$ text. Then, there are at most $\log_{1+\epsilon}(\Pi_{i=1}^{d} n_i) + 1$ different approximate periods P of T with at most $\lfloor \frac{1}{2+\epsilon} \lfloor \frac{n_1}{p_1} \rfloor \cdot \lfloor \frac{n_2}{p_2} \rfloor \cdots \lfloor \frac{n_d}{p_d} \rfloor \rfloor$ errors.*

Definition 9. [Multidimensional KMR Renaming]
Input: *d-dimensional text T with dimensions $n_1 \times n_2 \times \cdots \times n_d$.*
Output: *A data structure that answers the query $Name_{p_1, p_2, \ldots, p_d}[i_1, i_2, \ldots, i_d]$, the name of the sub-array with dimensions $p_1 \times p_2 \times \cdots \times p_d$ that starts at position $i_1, i_2, \ldots, i_d$ inside T, such that two d-dimensional sub-arrays have the same name if and only if they are equal.*

Theorem 3. *Let T be an array with dimensions $n_1 \times n_2 \times \cdots \times n_d$. After preprocessing T in $O(\Pi_{i=1}^{d} n_i \Pi_{i=1}^{d} \log n_i)$ time, we can answer in $O(2^d)$ time $Name_{p_1, p_2, \ldots, p_d}[i_1, i_2, \ldots, i_d]$ queries.*

Definition 10 [d-Dimensional Sub-arrays Hamming Distance Data Structure].
Let M_1, M_2 be d-dimensional arrays with dimensions $n_1 \times \cdots \times n_d$. Preprocess them to answer $Ham(S_1, S_2)$ queries, where S_1 and S_2 are sub-arrays of M_1 and M_2, respectively, having the same dimensions $p_1 \times \cdots \times p_d$.

Theorem 4. *Without loss of generality, assume that $n_1 \leq n_2 \leq \ldots \leq n_d$. A d-dimensional sub-arrays Hamming distance data structure can be constructed in $O(\Pi_{i=1}^{d} n_i \Pi_{i=1}^{d-1} \log n_i)$ time and space to support queries in $O(2^d \cdot \Delta)$ time, where $\Delta = Ham(S_1, S_2)$. Moreover, queries of the form "does $Ham(S_1, S_2) \leq \delta$?" are answered in $O(2^d \cdot \min\{\delta, Ham(S_1, S_2)\})$ time.*

The multidimensional primitivity check problem is the following.

Definition 11. [Multidimensional Primitivity Check]
Input: *d-dimensional text T with dimensions $n_1 \times n_2 \times \cdots \times n_d$.*
Output: *A data structure for answering query if a sub-array P of T is primitive.*

Theorem 5. *Multidimensional primitivity check data structure can be constructed in $O(\Pi_{i=1}^{d} n_i \Pi_{i=1}^{d} \log n_i)$ time and space to support query in $O(d \cdot 2^{d-1})$ time.*

7   Conclusion and Open Problems

We presented a recovery algorithm for 2D periodicity. In fact, our technique generalizes to multidimensional, with $d \cdot 2^d$ factor being a constant multiple in our time complexity, where d is the number of dimensions. This overhead is caused by the representation of the KMR names that is consuming $O(2^d)$ words. It would be interesting to get rid of this overhead. As in 1D, it is of interest to see whether a smaller number of allowed errors may lead to a smaller recovery set. In addition, it is of interest to explore the concept of recovery in a 2D cover.

References

1. Amir, A., Amit, M., Landau, G.M., Sokol, D.: Period recovery of strings over the hamming and edit distances. Theor. Comput. Sci. **710**, 2–18 (2018)
2. Amir, A., Benson, G.: Two-dimensional periodicity in rectangular arrays. SIAM J. Comput. **27**(1), 90–106 (1998)
3. Amir, A., Benson, G., Farach, M.: Optimal parallel two dimensional pattern matching. In: Snyder, L. (ed.) Proceedings of the 5th Annual ACM Symposium on Parallel Algorithms and Architectures, SPAA 1993, Velen, Germany, 30 June–2 July 1993, pp. 79–85. ACM (1993)
4. Amir, A., Benson, G., Farach, M.: Optimal parallel two dimensional text searching on a CREW PRAM. Inf. Comput. **144**(1), 1–17 (1998)
5. Amir, A., Boneh, I.: Dynamic palindrome detection. CoRR, abs/1906.09732 (2019)
6. Amir, A., Boneh, I., Charalampopoulos, P., Kondratovsky, E.: Repetition detection in a dynamic string. In: ESA, LIPIcs, vol. 144, pp. 5:1–5:18. Schloss Dagstuhl - Leibniz-Zentrum für Informatik (2019)
7. Amir, A., Eisenberg, E., Levy, A.: Approximate periodicity. Inf. Comput. **241**, 215–226 (2015)
8. Amir, A., Eisenberg, E., Levy, A., Porat, E., Shapira, N.: Cycle detection and correction. ACM Trans. Algorithms **9**(1), 13:1–13:20 (2012)
9. Amir, A., Landau, G.M., Marcus, S., Sokol, D.: Two-dimensional maximal repetitions. Theoret. Comput. Sci. **812**, 49–61 (2019)
10. Amit, M., Crochemore, M., Landau, G.M.: Locating all maximal approximate runs in a string. In: Fischer, J., Sanders, P. (eds.) CPM 2013. LNCS, vol. 7922, pp. 13–27. Springer, Heidelberg (2013). https://doi.org/10.1007/978-3-642-38905-4_4
11. Apostolico, A., Brimkov, V.E.: Fibonacci arrays and their two-dimensional repetitions. Theor. Comput. Sci. **237**(1–2), 263–273 (2000)
12. Apostolico, A., Giancarlo, R.: Periodicity and repetitions in parameterized strings. Discret. Appl. Math. **156**(9), 1389–1398 (2008). General Theory of Information Transfer and Combinatorics
13. Boyer, R.S., Moore, J.S.: MJRTY: a fast majority vote algorithm. In: Boyer, R.S. (ed.) Automated Reasoning: Essays in Honor of Woody Bledsoe. Automated Reasoning Series, pp. 105–118. Kluwer Academic Publishers (1991)
14. Cole, R., et al.: Optimally fast parallel algorithms for preprocessing and pattern matching in one and two dimensions. In: 34th Annual Symposium on Foundations of Computer Science, Palo Alto, California, USA, 3–5 November 1993, pp. 248–258. IEEE Computer Society (1993)
15. Crochemore, M.: An optimal algorithm for computing the repetitions in a word. Inf. Process. Lett. **12**(5), 244–250 (1981)
16. Crochemore, M., Gasieniec, L., Hariharan, R., Muthukrishnan, S., Rytter, W.: A constant time optimal parallel algorithm for two-dimensional pattern matching. SIAM J. Comput. **27**(3), 668–681 (1998)
17. Crochemore, M., Rytter, W.: Usefulness of the Karp-Miller-Rosenberg algorithm in parallel computations on strings and arrays. Theoret. Comput. Sci. **88**(1), 59–82 (1991)
18. Galil, Z.: Optimal parallel algorithms for string matching. Inf. Control **67**(1–3), 144–157 (1985)
19. Galil, Z., Giancarlo, R.: Improved string matching with k mismatches. SIGACT News **17**(4), 52–54 (1986)

20. Galil, Z., Park, K.: Alphabet-independent two-dimensional witness computation. SIAM J. Comput. **25**(5), 907–935 (1996)
21. Gamard, G., Richomme, G., Shallit, J., Smith, T.J.: Periodicity in rectangular arrays. Inf. Process. Lett. **118**, 58–63 (2017)
22. Gusfield, D., Stoye, J.: Linear time algorithms for finding and representing all the tandem repeats in a string. J. Comput. Syst. Sci. **69**(4), 525–546 (2004)
23. Karp, R.M., Miller, R.E., Rosenberg, A.L.: Rapid identification of repeated patterns in strings, trees and arrays. In: Fischer, P.C., Zeiger, H.P., Ullman, J.D., Rosenberg, A.L. (ed.) Proceedings of the 4th Annual ACM Symposium on Theory of Computing, Denver, Colorado, USA, 1–3 May 1972, pp. 125–136. ACM (1972)
24. Kociumaka, T., Radoszewski, J., Rytter, W., Walen, T.: Internal pattern matching queries in a text and applications. In: Indyk, P. (ed.) Proceedings of the Twenty-Sixth Annual ACM-SIAM Symposium on Discrete Algorithms, SODA 2015, San Diego, CA, USA, 4–6 January 2015, pp. 532–551. SIAM (2015)
25. Kolpakov, R.M., Kucherov, G.: Finding maximal repetitions in a word in linear time. In: 40th Annual Symposium on Foundations of Computer Science, FOCS 1999, New York, NY, USA, 17–18 October 1999, pp. 596–604. IEEE Computer Society (1999)
26. Kolpakov, R.M., Kucherov, G.: Finding approximate repetitions under hamming distance. Theor. Comput. Sci. **303**(1), 135–156 (2003)
27. Landau, G.M., Schmidt, J.P., Sokol, D.: An algorithm for approximate tandem repeats. J. Comput. Biol. **8**(1), 1–18 (2001)
28. Marcus, S., Sokol, D.: 2d Lyndon words and applications. Algorithmica **77**(1), 116–133 (2017). https://doi.org/10.1007/s00453-015-0065-z
29. Régnier, M., Rostami, L.: A unifying look at d-dimensional periodicities and space coverings. In: Apostolico, A., Crochemore, M., Galil, Z., Manber, U. (eds.) CPM 1993. LNCS, vol. 684, pp. 215–227. Springer, Heidelberg (1993). https://doi.org/10.1007/BFb0029807
30. Sim, J.S., Iliopoulos, C.S., Park, K., Smyth, W.F.: Approximate periods of strings. Theoret. Comput. Sci. **262**(1), 557–568 (2001)

Computing Covers Under Substring Consistent Equivalence Relations

Natsumi Kikuchi$^{(\boxtimes)}$, Diptarama Hendrian⬤, Ryo Yoshinaka⬤,
and Ayumi Shinohara⬤

Graduate School of Information Sciences, Tohoku University, Sendai, Japan
natsumi_kikuchi@shino.ecei.tohoku.ac.jp,
{diptarama,ryoshinaka,ayumis}@tohoku.ac.jp

Abstract. Covers are a kind of quasiperiodicity in strings. A string C is a cover of another string T if any position of T is inside some occurrence of C in T. The shortest and longest cover arrays of T have the lengths of the shortest and longest covers of each prefix of T, respectively. The literature has proposed linear-time algorithms computing longest and shortest cover arrays taking border arrays as input. An equivalence relation $\approx$ over strings is called a *substring consistent equivalence relation (SCER)* iff $X \approx Y$ implies (1) $|X| = |Y|$ and (2) $X[i:j] \approx Y[i:j]$ for all $1 \le i \le j \le |X|$. In this paper, we generalize the notion of covers for SCERs and prove that existing algorithms to compute the shortest cover array and the longest cover array of a string T under the identity relation will work for any SCERs taking the accordingly generalized border arrays.

Keywords: String covers · Substring consistent equivalence relations · String regularities

1 Introduction

Finding regularities in strings is an important task in string processing due to its applications such as pattern matching and string compression. Many variants of regularities in strings have been studied including periods, covers, and seeds [6,7,20]. One of the most studied regularities is periods due to their mathematical combinatoric properties and their applications to string processing algorithms [14]. The notion of periods has been generalized concerning various kinds of equivalence relations. Apostolico and Giancarlo [8] studied periods on parameterized strings. Gourdel et al. [17] studied string periods on the order-preserving model.

Covers are another kind of regularities that have extensively been studied. For two strings T and C, C is a *cover* of T if any position of T is inside some occurrences of C in T. For example, aba is a cover of $T = \underline{aba}ab\underline{aba}ab\underline{aba}ab\underline{aba}$ because all positions in T are inside occurrences of aba. The other covers of T are abaaba, abaababaaba and T itself. Apostolico and Ehrenfeucht [6] called

© Springer Nature Switzerland AG 2020
C. Boucher and S. V. Thankachan (Eds.): SPIRE 2020, LNCS 12303, pp. 131–146, 2020.
https://doi.org/10.1007/978-3-030-59212-7_10

Table 1. The time complexity of computing border (*Border*), shortest cover (*SCover*) and longest cover (*LCover*) arrays under SCERs, where n is the input length, Π is the parameter set in parameterized equivalence, and k is the number of input strings in permuted equivalence.

Equivalence relation	*Border*	*SCover*	*LCover*
Identity equivalence	$O(n)$ [23]	$O(n)$ [10]	$O(n)$ [25]
Parameterized equivalence	$O(n \log |\Pi|)$ [4]	$O(n \log |\Pi|)$	$O(n \log |\Pi|)$
Order-isomorphism	$O(n \log n)$ [22, 24]	$O(n \log n)$	$O(n \log n)$
Permuted equivalence	$O(nk)$ [15, 18]	$O(nk)$	$O(nk)$

a string having a cover besides itself *quasiperiodic* and proposed an algorithm that computes all maximal quasiperiodic substrings of a string. Later, Iliopoulos and Mouchard [19] and Brodal and Pedersen [11] proposed $O(n \log n)$ time algorithm for this task. Apostolico et al. [7] presented a linear-time algorithm to test whether a string is quasiperiodic. Breslauer [10] proposed an online linear-time algorithm that computes the shortest covers of all prefixes as *the shortest cover array* of a string. Moore and Smyth [27,28] proposed a linear-time algorithm to compute all covers of a string. Later, Li and Smyth [25] proposed an online linear-time algorithm to compute the longest proper covers of all prefixes of a string as *the longest cover array*. Amir et al. [2] defined the approximate cover problem and showed its NP-hardness.

Recently, Matsuoka et al. [26] introduced the notion of *substring consistent equivalence relations (SCERs)*, which are equivalence relations $\approx$ on strings such that $X \approx Y$ implies (1) $|X| = |Y|$ and (2) $X[i : j] \approx Y[i : j]$ for all $1 \leq i \leq j \leq |X|$, where $X[i : j]$ denotes the substring of X starting at i and ending at j. Clearly the identity relation is an SCER. Moreover, many variants of equivalence relations used in pattern matching are SCERs, such as parameterized pattern matching [9], order-preserving pattern matching [22,24], permuted pattern matching [21], and Cartesian tree matching [29]. Matsuoka et al. [26] proposed an algorithm to compute the border array of an input string T under an SCER, which can be used for pattern matching under SCERs.

In this paper, we generalize the notion of covers, which used to be defined based on the identity relation, to be based on SCERs, and prove that both of the algorithms for the shortest and longest cover arrays by Breslauer [10] and Li and Smyth [25], respectively, work under SCERs with no changes: just by replacing the input of those algorithms from the border array under the identity relation to the one under a concerned SCER, their algorithms compute the shortest and longest cover arrays under the SCER. As a minor contribution, we present a slightly simplified version of Li and Smyth's algorithm, with a correctness proof. Table 1 summarizes implications of our results. The time complexities for computing shortest and longest cover arrays based on various SCERs are the same as those for border arrays. Moreover, if border arrays under an equivalence relation can be computed online, e.g., parameterized equivalence and order-isomorphism, these cover arrays can be computed online by computing border arrays with existing online algorithms at the same time.

2   Preliminaries

For an alphabet Σ, Σ^* denotes the set of all strings over Σ, including the empty string ε. The length of a string $T \in \Sigma^*$ is denoted as $|T|$. For $1 \le i \le j \le |T|$, $T[i:j]$ denotes the substring of T that starts at i and ends at j. By $T[:j] = T[1:j]$ we denote the *prefix* of T that ends at j and by $T[i:] = T[i:|T|]$ the *suffix* of T that starts at i.

Matsuoka et al. [26] introduced the notion of substring consistent equivalence relations, generalizing several equivalence relations proposed so far in pattern matching.

Definition 1 (Substring Consistent Equivalence Relation (SCER) $\approx$). *An equivalence relation $\approx \subseteq \Sigma^* \times \Sigma^*$ is an SCER if for two strings X and Y, $X \approx Y$ implies (1) $|X| = |Y|$ and (2) $X[i:j] \approx Y[i:j]$ for all $1 \le i \le j \le |X|$. By $[X]_\approx$ we denote the $\approx$-equivalence class of X.*

For instance, matching relations in parameterized pattern matching [9], order-preserving pattern matching [22,24], and permuted pattern matching [21] are SCERs, while matching relations in abelian pattern matching [16], indeterminate string pattern matching [5] and function matching [3] are not.

Definition 2 (Parameterized equivalence [9]). *Two strings X and Y of the same length are a parameterized match, denoted as $X \overset{\mathrm{pr}}{\approx} Y$, if X can be transformed into Y by applying a renaming bijection g from the characters of X to the characters of Y.*

Definition 3 (Order-isomorphism [22,24]). *Two strings X and Y of the same length over an alphabet with a linear order $\prec$ are order isomorphic, denoted as $X \overset{\mathrm{op}}{\approx} Y$, if $X[i] \prec X[j] \Leftrightarrow Y[i] \prec Y[j]$ for all $1 \le i, j \le |X|$.*

Definition 4 ($\approx$-occurrence [26]). *For two strings T and P, a position $1 \le i \le |T| - |P| + 1$ is an $\approx$-occurrence of P in T if $P \approx T[i : i + |P| - 1]$. The set of $\approx$-occurrence positions of P in T is denoted by $\mathrm{Occ}_{P,T}$.*

Definition 5 ($\approx$-border [26]). *A string B is a $\approx$-border of T if $B \approx T[: |B|] \approx T[|T| - |B| + 1 :]$. We denote by $\mathrm{Bord}_\approx(T)$ the set of all $\approx$-borders of T. A $\approx$-border B of T is called proper if $|B| < |T|$, and called trivial if $B = \varepsilon$.*

Lemma 1 ([26]). *(1) $B \in \mathrm{Bord}_\approx(S)$ and $B' \in \mathrm{Bord}_\approx(B)$ implies $B' \in \mathrm{Bord}_\approx(S)$. (2) $B, B' \in \mathrm{Bord}_\approx(S)$ and $|B'| \le |B|$ implies $B' \in \mathrm{Bord}_\approx(B)$.*

Based on Lemma 1, Matsuoka et al. [26] proposed an algorithm to compute border arrays under SCERs, which are defined as follows.

Definition 6 ($\approx$-border array). *The $\approx$-border array $Border_T$ of T is an array of length $|T|$ such that $Border_T[i] = \max\{|B| \mid B$ is a proper $\approx$-border of $T[: i]\}$ for $1 \le i \le |T|$.*

Table 2. The $=$-border array, the shortest $=$-cover array, and the longest $=$-cover array of $T = $ **abaababaabaababa**.

T	1	2	3	4	5	6	7	8	9	10	11	12	13	14	15	16
	a	b	a	a	b	a	b	a	a	b	a	a	b	a	b	a
$Border_T$	0	0	1	1	2	3	2	3	4	5	6	4	5	6	7	8
$SCover_T$	1	2	3	4	5	3	7	3	9	5	3	12	5	3	15	3
$LCover_T$	0	0	0	0	0	3	0	3	0	5	6	0	5	6	0	8

Table 3. The $\overset{pr}{\approx}$-border array, the shortest $\overset{pr}{\approx}$-cover array, and the longest $\overset{pr}{\approx}$-cover array of $T = $ **abaababaabaababa**. Notice that $SCover_T[i] = 1$ for all i, for a $\overset{pr}{\approx}$ b.

T	1	2	3	4	5	6	7	8	9	10	11	12	13	14	15	16
	a	b	a	a	b	a	b	a	a	b	a	a	b	a	b	a
$Border_T$	0	1	2	1	2	3	3	3	4	5	6	4	5	6	7	8
$SCover_T$	1	1	1	1	1	1	1	1	1	1	1	1	1	1	1	1
$LCover_T$	0	1	2	1	2	3	3	3	1	5	6	1	5	6	3	8

Tables 2 and 3 show examples of $\approx$-border arrays. We use the identity relation in Table 2 and the parameterized equivalence (Definition 2) in Table 3.

The well-known property on $=$-borders (e.g., [1]) holds for $\approx$-borders, too.

Lemma 2. *For any $1 < i \le n$, $Border_T[i-1] + 1 \ge Border_T[i]$.*

3   Covers Under SCERs

In this section, we define covers under SCERs ($\approx$-covers) and present some properties of $\approx$-covers, which prepares for the succeeding sections. Section 4 shows that the algorithm to compute shortest cover arrays by Breslauer [10] will work under SCERs with no change. Section 5 presents a slight variant of the algorithm by Li and Smyth [25] for computing the longest cover arrays and proves its correctness.

Definition 7 ($\approx$-cover). *We say that a string C of length c is an $\approx$-cover of a string T of length n if there are $x_1, x_2, \ldots, x_m \in \mathsf{Occ}_{C,T}$ such that $x_1 = 1$, $x_m = n - c + 1$ and $x_{i-1} < x_i \le x_{i-1} + c$ for all $1 < i \le m$. Moreover, we say that an $\approx$-cover C of T is proper if $c < n$. The set of all $\approx$-covers of T is denoted by $\mathsf{Cov}_\approx(T)$. A string T is primitive[1] if T has no proper $\approx$-cover.*

By definition, $\mathsf{Cov}_\approx(T) \subseteq \mathsf{Bord}_\approx(T)$. Below we observe that basic lemmas in [10] on $=$-covers and $=$-borders hold for $\approx$-covers and $\approx$-borders.

---

[1] In some references it is called *superprimitive*, reserving the term "primitive" for strings that cannot be represented as S^k for some string S and integer $k \ge 2$.

Algorithm 1: Algorithm computing the shortest $\approx$-cover array

---

1   let *Border* be the $\approx$-border array of T;

2   $Reach[i] \leftarrow 0$ for $1 \le i \le n$;

3   **for** $1 \le i \le n$ **do**

4   **if** $Border[i] > 0$ **and** $Reach[SCover[Border[i]]] \ge i - SCover[Border[i]]$ **then**

5          $SCover[i] \leftarrow SCover[Border[i]]$;

6          $Reach[SCover[i]] \leftarrow i$;

7   **else**

8          $SCover[i] \leftarrow i$;

9          $Reach[i] \leftarrow i$;

Lemma 3. *If $C \in \mathsf{Cov}_\approx(T)$, $B \in \mathsf{Bord}_\approx(T)$, and $|C| \le |B|$, then $C \in \mathsf{Cov}_\approx(B)$.*

Lemma 4. *For any $C, C' \in \mathsf{Cov}_\approx(T)$ such that $|C| \le |C'|$, $C \in \mathsf{Cov}_\approx(C')$.*

Lemma 5. *If $C \in \mathsf{Cov}_\approx(T)$ and $C' \in \mathsf{Cov}_\approx(C)$, then $C' \in \mathsf{Cov}_\approx(T)$.*

Lemma 6. *An $\approx$-cover C of T is primitive iff it is a shortest $\approx$-cover of T.*

Lemma 7. *For $0 \le i - 1 \le j \le |T|$, $\mathsf{Cov}_\approx(T[:j]) \cap \mathsf{Cov}_\approx(T[i:]) \subseteq \mathsf{Cov}_\approx(T)$.*

Lemma 8. *A string C of length c is a proper $\approx$-cover of T of length n iff $C \in \mathsf{Bord}_\approx(T)$ and $C \in \mathsf{Cov}_\approx(T[:n-i])$ for some $1 \le i \le c$.*

In the seaquel of this paper, we fix an input string T of length n.

4   Shortest $\approx$-cover array

In this section we prove that Algorithm 1 by Breslauer [10] computes the shortest $\approx$-cover array for an input string T based on the $\approx$-border array.

Definition 8 (Shortest $\approx$-cover array). *The shortest $\approx$-cover array $SCover_T$ of T is an array of length n such that $SCover_T[i] = \min\{\,|C| \mid C \in \mathsf{Cov}_\approx(T[:i])\,\}$ for $1 \le i \le n$.*

Tables 2 and 3 show examples of shortest $\approx$-cover arrays. Note that $SCover_T[i]$ is the length of the unique (modulo $\approx$-equivalence) primitive cover of $T[:i]$ by Lemma 6.

    Algorithm 1 uses an additional array *Reach* to compute *SCover*. The algorithm updates *Reach* and *SCover* incrementally so that $Reach[j]$ shall be the length of the longest prefix of T of which $T[:j]$ is a $\approx$-cover and *SCover* shall be the shortest $\approx$-cover array. More precisely, in each iteration i, the algorithm updates *Reach* and *SCover* so that they satisfy the following properties at the end of the i-th iteration.

R(i) $Reach[j] = 0$ if $j > i$ or $T[:j]$ is not primitive. Otherwise, $Reach[j] = \max\{\,p \mid T[:j]$ is a $\approx -$cover of $T[:p]$ and $p \le i\,\}$.

S(i) For $1 \leq j \leq i$, $SCover[j] = \min\{ |C| \mid C \in \mathsf{Cov}_{\approx}(T[:j]) \}$.

If **S**(n) holds, we have $SCover = SCover_T$.

Theorem 1. *Given the $\approx$-border array of text T of length n, Algorithm 1 computes the shortest $\approx$-cover array $SCover_T$ of T in $O(n)$ time.*

Proof. The linear time complexity is obvious.

We show the above invariants **R**(i) and **S**(i) by induction on i. Clearly the invariant holds for $i = 0$, i.e., the initial values of $Reach[j] = 0$ for all $j > 0$ satisfy the invariant **R**(0). Vacuously **S**(0) is true.

Assume that **R**(i − 1) and **S**(i − 1) hold at the beginning of the i-th iteration. Let $b = Border[i]$ and $c = SCover[b]$.

Suppose the **if**-condition of Line 4 is satisfied in the i-th iteration. By the induction hypothesis on $Reach[c]$, which is at least as large as $i - c \geq 1$ at the beginning of the i-th iteration, $T[:c]$ is a primitive $\approx$-cover of $T[:i-l]$ for some $1 \leq l \leq c$. Then the algorithm updates the value of $Reach[c]$ to $i \geq 1$, which is still positive at the end of the i-th iteration. By $T[:b] \in \mathsf{Bord}_{\approx}(T[:i])$ and $T[:c] \in \mathsf{Cov}_{\approx}(T[:b]) \subseteq \mathsf{Bord}_{\approx}(T[:b])$ (by **S**(i − 1)), Lemma 1 (1) implies $T[:c] \in \mathsf{Bord}_{\approx}(T[:i])$. Therefore, $T[:c]$ is a proper $\approx$-cover of $T[:i]$ by Lemma 8. Thus, $Reach[c] = i$ satisfies the invariant. On the other hand, the value $Reach[i]$ is not changed from its initial value 0, while we get $SCover[i] = c$. Indeed $T[:i]$ is not primitive as it has a $\approx$-cover $T[:c]$. That is, $Reach[i]$ and $SCover[i]$ satisfy the invariants. Since $T[:c]$ is the unique primitive $\approx$-cover prefix of $T[:i]$, for other j, $Reach[j]$ need not be updated.

Suppose the **if**-condition is not satisfied in the i-th iteration, where both $Reach[i]$ and $SCover[i]$ are set to be i. If $b = 0$, $T[:i]$ has no proper $\approx$-cover. Thus $T[:i]$ is primitive and the lemma holds. Next, consider the case where $b \neq 0$ and $Reach[c] < i - c$. To show by contradiction that $T[:i]$ is primitive, assume that $T[:i]$ has a primitive proper $\approx$-cover $T[:k]$. By $T[:k] \in \mathsf{Cov}_{\approx}(T[:i]) \subseteq \mathsf{Bord}_{\approx}(T[:i])$ and Lemma 3, we have $T[:k] \in \mathsf{Cov}_{\approx}(T[:b])$. Since $T[:b]$ has only one (up to $\approx$-equivalence) primitive $\approx$-cover by Lemma 6, we have $k = c$, i.e., $T[:c] \in \mathsf{Cov}_{\approx}(T[:i])$. By Lemma 8, $T[:c] \in \mathsf{Cov}_{\approx}(T[:i-j])$ for some $1 \leq j \leq c$, which contradicts the fact $Reach[c] < i - c$ with the induction hypothesis. Therefore, $T[:i]$ has no primitive proper $\approx$-cover and thus $T[:i]$ is primitive by Lemma 6. We conclude that $Reach[i] = SCover[i] = i$ satisfies **R**(i) and **S**(i) and $Reach[j]$ need not be updated for other j. □

Corollary 1. *If $Border_T$ can be computed in $\beta(n)$ time, $SCover_T$ can be computed in $O(\beta(n) + n)$ time.*

5 Longest $\approx$-cover array

This section discusses computing the longest $\approx$-cover array of a text. Tables 2 and 3 show examples of longest $\approx$-cover arrays.

Definition 9 (Longest $\approx$-cover array). *The* longest $\approx$-cover array *$LCover_T$ of T is an array of length n such that for $1 \leq i \leq n$, $LCover_T[i] = \max(\{\, |C| \mid C$ is a proper $\approx -cover$ of $T[: i] \,\} \cup \{0\})$.*

Let $LCover_T^0[i] = i$ and $LCover_T^q[i] = LCover_T[LCover_T^{q-1}[i]]$ for $q \geq 1$. The following lemma is a corollary to Lemmas 4 and 5.

Lemma 9. *For any $1 \leq j \leq i$, $T[: j] \in \mathsf{Cov}_\approx(T[: i])$ iff $j = LCover_T^q[i]$ for some $q \geq 0$.*

Therefore, using the longest $\approx$-cover array, one can easily obtain all the $\approx$-covers up to $\approx$-equivalence.

Li and Smyth [25] presented an online linear-time algorithm to compute the longest $=$-cover array from the $=$-border array of a text T. We will present a slight variant of theirs for computing the longest $\approx$-cover array. Our modification is not due to the generalization. In fact their algorithm works for computing $\approx$-covers as it is. We changed their algorithm just for simplicity. We will briefly discuss the difference of their and our algorithms later.

Li and Smyth showed some properties of longest $=$-cover arrays, but not all of them hold under SCERs. For instance, the longest $\approx$-cover array in Table 3 is a counterexample to Theorems 2.2 and 2.3 in [25]. So it is not trivial that their algorithm and our variant work under SCERs and we need to carefully check the correctness of the algorithms.

Their algorithm involves an auxiliary array of length n based on the notion of "live" prefixes. A prefix S of T is said to be *live* if T can be extended so that S will be a cover of TU for some $U \in \Sigma^*$. This notion is also known as "left seeds" [12,13]. We generalize the notion for SCERs as follows.

Definition 10 (left $\approx$-seed). *For strings T of length n and S of length m, S is said to be a* left $\approx$-seed *of T if there exist k and l such that $k \leq l < m$, $S \in \mathsf{Cov}_\approx(T[: n - k])$ and $S[: l] \approx T[n - l + 1 :]$. We denote by $\mathsf{LSeed}_\approx(T)$ the set of all left $\approx$-seeds of T.*

We remark that it is not necessarily true that $\mathsf{LSeed}_\approx(T) = \{\, S \mid S \in \mathsf{Cov}_\approx(TU)$ for some $U \,\}$ according to the above definition, contrarily to the case of the identity relation. Consider the order-isomorphism $\overset{\mathrm{op}}{\approx}$ (Definition 3) on $\Sigma = \{\mathsf{a, b, c, d}\}$ with $\mathsf{a} \prec \mathsf{b} \prec \mathsf{c} \prec \mathsf{d}$. Then $S = \mathsf{acb}$ is a left $\overset{\mathrm{op}}{\approx}$-seed of $T = \mathsf{adcbc}$, since $S \overset{\mathrm{op}}{\approx} T[: 3]$ and $S[: 2] \overset{\mathrm{op}}{\approx} T[4 :]$. However, for no character $U \in \Sigma$, we have $S \overset{\mathrm{op}}{\approx} (TU)[4 : 6]$, since U needs to be a character bigger than b and smaller than c.

Clearly $\mathsf{Cov}_\approx(T) \subseteq \mathsf{LSeed}_\approx(T)$. Moreover, $S \in \mathsf{LSeed}_\approx(T)$ implies $S \in \mathsf{LSeed}_\approx(T')$ for any prefix T' of T unless $|S| > |T'|$. Being a left $\approx$-seed is a weaker property than being an $\approx$-cover, but it is easier to handle in an online algorithm, due to the monotonicity that $T[: j] \notin \mathsf{LSeed}_\approx(T[: i - 1])$ implies $T[: j] \notin \mathsf{LSeed}_\approx(T[: i])$ for every $j < i$. The following series of lemmas investigate the relation among left $\approx$-seeds and $\approx$-covers.

Lemma 10. *If $k \leq l$, then $\mathsf{Cov}_{\approx}(T[: n-k]) \cap \mathsf{LSeed}_{\approx}(T[n-l+1 :]) \subseteq \mathsf{LSeed}_{\approx}(T)$.*

Proof. Suppose $S \in \mathsf{Cov}_{\approx}(T[: n - k]) \cap \mathsf{LSeed}_{\approx}(T[n - l + 1 :])$. By $S \in \mathsf{LSeed}_{\approx}(T[n - l + 1 :])$, there are k', l' such that $k' \leq l' < |S| \leq l$, $S \in \mathsf{Cov}_{\approx}(T[n-l+1 : n-k'])$ and $S[: l'] \approx T[n-l'+1 :]$. We have $S \in \mathsf{Cov}_{\approx}(T[: n-k'])$ by $S \in \mathsf{Cov}_{\approx}(T[: n - k])$, $n - k \geq n - l$, and Lemma 7. Hence $S \in \mathsf{LSeed}_{\approx}(T[: i])$ by Definition 10. □

Lemma 11 says somewhat long prefixes are all left $\approx$-seeds, which we call *primary*. Lemma 12 says shorter left $\approx$-seeds are $\approx$-covers of long left $\approx$-seeds. As a corollary, we obtain Lemma 13, which corresponds to Lemma 2.5 in [25].

Lemma 11 (Primary left $\approx$-seeds). *For any $1 \leq i \leq n$ and $i - Border_T[i] \leq j \leq i$, we have $T[: j] \in \mathsf{LSeed}_{\approx}(T[: i])$.*

Proof. Let $b = Border_T[i]$, $m = \lfloor (i - j)/(i - b) \rfloor$, $l = i - (m + 1)(i - b)$ and $x_k = k(i - b) + 1$ for $k \geq 0$. It is enough to show that (a) $\{x_0, \ldots, x_m\}$ witnesses $T[: j] \in \mathsf{Cov}_{\approx}(T[: x_m + j - 1])$, (b) $T[: l] \approx T[i - l + 1 : i]$, and (c) $i - (x_m + j - 1) \leq l < j$. The equation (c) can be verified by simple calculation.

(a) Since $x_{k+1} - x_k = i - b \leq j$, it is enough to show $x_k \in \mathsf{Occ}_{T[:j],T[:i]}$ for all $k \leq m$. Since $T[: b] \approx T[i - b + 1 : i]$, any "corresponding" substrings of $T[1 : b]$ and $T[i-b+1 : i]$ are $\approx$-equivalent. In particular, $T[x_k : x_k + j - 1] \approx T[x_k + i - b : x_k + i - b + j - 1] = T[x_{k+1} : x_{k+1} + j - 1]$ for all $0 \leq k < m$. That is, $T[: j] \approx T[x_k : x_k + j - 1]$ and thus $x_k \in \mathsf{Occ}_{T[:j],T[:i]}$ for all $0 \leq k \leq m$.
(b) The same argument for corresponding substrings of $T[1 : b]$ and $T[i-b+1 : i]$ of length l establishes $T[: l] \approx T[x_m : x_m + l - 1] \approx T[x_{m+1} : x_{m+1} + l - 1] = T[i - l + 1 : i]$. □

Lemma 12. *For any $1 \leq i \leq n$, $T[: j]$ for $1 \leq j < i - Border_T[i]$ is a left $\approx$-seed of $T[: i]$ iff $T[: j]$ is the longest proper $\approx$-cover of a left $\approx$-seed of $T[: i]$.*

Proof. Let $b = Border_T[i]$. ($\Longrightarrow$) Assume that for $1 \leq j < i - b$, $T[: j] \in \mathsf{LSeed}_{\approx}(T[: i])$, namely, there exist k and l such that $k \leq l < j$, $T[: j] \in \mathsf{Cov}_{\approx}(T[: i - k])$ and $T[: l] \in \mathsf{Bord}_{\approx}(T[: i])$. Since $T[: b]$ is the longest proper $\approx$-border of $T[: i]$, $k \leq l \leq b$ and $j < i - b \leq i - k$. By Lemma 9, there exists $T[: m] \in \mathsf{Cov}_{\approx}(T[: i - k])$ such that $j = LCover_T[m]$. Moreover, since $j < m \leq i - k$ and $k \leq l < m$, we have $T[: m] \in \mathsf{LSeed}_{\approx}(T[: i])$. Therefore $T[: j]$ is the longest proper $\approx$-cover of $T[: m]$, which is a left $\approx$-seed of $T[: i]$.

($\Longleftarrow$) Assume there is a left $\approx$-seed prefix $T[: m]$ of $T[: i]$ that is properly covered by $T[: j]$. By Definition 10, there exist k and l such that $k \leq l < m$, $T[: m] \in \mathsf{Cov}_{\approx}(T[: i-k])$ and $T[: l] \in \mathsf{Bord}_{\approx}(T[: i])$. Thus we have $T[: j] \in \mathsf{Cov}_{\approx}(T[: i - k])$ by Lemma 5. If $j \geq l$, $T[: j] \in \mathsf{Cov}_{\approx}(T[: i - k])$ and $T[: l] \in \mathsf{Bord}_{\approx}(T[: i])$, which implies $T[: j] \in \mathsf{LSeed}_{\approx}(T[: i])$ by Definition 10. If $j < l < m$, $T[: j] \in \mathsf{Cov}_{\approx}(T[: m]) \subseteq \mathsf{LSeed}_{\approx}(T[: m])$ implies $T[: j] \in \mathsf{LSeed}_{\approx}(T[: l])$. By Lemma 10, $T[: j] \in \mathsf{LSeed}_{\approx}(T[: i])$. □

Lemma 13. *For any $1 \leq i \leq n$ and $1 \leq j \leq i$, $T[: j] \in \mathsf{LSeed}_{\approx}(T[: i])$ iff there exists k such that $i - Border_T[i] \leq k \leq i$ and $j = LCover_T^q[k]$ for some $q \geq 0$.*

Proof. By Lemmas 9, 11 and 12. □

Our algorithm involves an auxiliary array based on the following function LongestLSeedCov$_T$, which is updated by Lemma 15. The significance of this function is shown as Lemma 14.

Definition 11 (LongestLSeedCov$_T(i,j)$). *For a string T, define*

$$\text{LongestLSeedCov}_T(i,j) = \max(\{\, l \mid T[: l] \in \text{LSeed}_\approx(T[: i]) \cap \text{Cov}_\approx(T[: j]) \,\} \cup \{0\}).$$

Lemma 14. *For any $1 \leq i \leq n$, $LCover_T[i] = \text{LongestLSeedCov}_T(i, Border_T[i])$.*

Proof. It suffices to show $\text{Cov}_\approx(T[: i]) \setminus [T[: i]]_\approx = \text{LSeed}_\approx(T[: i]) \cap \text{Cov}_\approx(T[: b])$ for $b = Border_T[i]$. If $C \in \text{Cov}_\approx(T[: i])$ with $|C| \neq i$, then obviously $C \in \text{Bord}_\approx(T[: i]) \cap \text{LSeed}_\approx(T[: i])$. By Lemma 1, $C \in \text{Bord}_\approx(T[: b])$. Suppose $S \in \text{LSeed}_\approx(T[: i]) \cap \text{Cov}_\approx(T[: b])$. There is $k < |S|$ such that $S \in \text{Cov}_\approx(T[: i - k])$. By $k < |S| \leq b$ and Lemma 7, we have $S \in \text{Cov}_\approx(T[: i])$. □

Lemma 15. $\text{LongestLSeedCov}_T(i,j) = \text{LongestLSeedCov}_T(i - 1, j)$ *for $1 \leq j \leq Border_T[i]$. Moreover, for $j = Border_T[i]$, if $T[: j] \notin \text{LSeed}_\approx(T[: i - 1])$, then* $\text{LongestLSeedCov}_T(i,j) = \text{LongestLSeedCov}_T(i - 1, LCover[j])$.

Proof. Let $l = \text{LongestLSeedCov}_T(i-1,j)$ and $l' = \text{LongestLSeedCov}_T(i,j)$. Since $j \leq Border_T[i] < i$, we have $l' < i$, which implies $l' \leq l$.

Suppose $l = 0$. This implies $l' = 0$ and thus $l' = l$ holds. Suppose in addition that $j = Border_T[i]$ and $T[: j] \notin \text{LSeed}_\approx(T[: i - 1])$. The fact $l = 0$ means $\text{LSeed}_\approx(T[: i - 1]) \cap \text{Cov}_\approx(T[: j]) = \emptyset$, which implies $\text{LSeed}_\approx(T[: i - 1]) \cap \text{Cov}_\approx(T[: LCover_T[j]]) = \emptyset$ by Lemmas 4 and 5. Therefore, $\text{LongestLSeedCov}_T(i - 1, [2]LCover_T[j]) = 0$. So the lemma holds.

Hereafter we assume $l \geq 1$. Let $b_i = Border_T[i] \geq 1$. By $T[: l] \in \text{LSeed}_\approx(T[: i - 1])$, there exists $k < l$ such that $T[: l] \in \text{Cov}_\approx(T[: i - 1 - k])$. On the other hand, by $b_i \leq i - 1$, $T[: l] \in \text{LSeed}_\approx(T[: b_i]) = \text{LSeed}_\approx(T[i - b_i + 1 : i])$. Since $k < l \leq j \leq b_i$, by Lemma 10, $T[: l] \in \text{Cov}_\approx(T[: i-1-k]) \cap \text{LSeed}_\approx(T[i-b_i+1 : i])$ implies $l \in \text{LSeed}_\approx(T[: i])$. Thus $l' = l$.

Suppose $j = b_i$ and $T[: j] \notin \text{LSeed}_\approx(T[: i - 1])$. Since $\text{Cov}_\approx(T[: j]) = \text{Cov}_\approx(T[: LCover_T[j]]) \cup [T[: j]]_\approx$ by Lemmas 4 and 5, $T[: j] \notin \text{LSeed}_\approx(T[: i - 1])$ implies $\text{LongestLSeedCov}_T(i - 1, LCover_T[j]) = l = l'$. □

Algorithm 2 computes the longest $\approx$-cover array $LCover_T$ of T as $LCover$ taking the $\approx$-border array $Border_T$ as input. Following Li and Smyth [25], we explain the algorithm using a tree formed by $LCover_T$, called the $\approx$-*cover tree*. The $\approx$-cover tree consists of nodes $0, \ldots, n$. The root is 0 and the parent of $j \neq 0$ is $LCover_T[j]$. By Lemma 9, $T[: k] \in \text{Cov}_\approx(T[: j])$ if and only if $k \neq 0$ and k is an ancestor of j (including the case where $k = j$) in the $\approx$-cover tree. Hereafter, we casually use the index j to mean (any string $\approx$-equivalent to) the prefix $T[: j]$ of T, if no confusion arises. We use two additional arrays $LSChildren$ and $LongestLSAnc$, which have zero-based indices in accordance with the $\approx$-cover tree's nodes. $LSChildren[j]$ counts the number of children of j that are left $\approx$-seeds of T. $LongestLSAnc[j]$ points at the lowest ancestor of j that is a left

---

Algorithm 2: Algorithm computing the longest $\approx$-cover array

---

1 let $Border$ be the $\approx$-border array of T;
2 $LSChildren[i] \leftarrow 0$, $LongestLSAnc[i] \leftarrow i$ for $0 \le i \le n$;
3 **for** $1 \le i \le n$ **do**
4   **if** $LSChildren[Border[i]] = 0$ **and** $0 < 2 \cdot Border[i] < i$ **then**
5        $LongestLSAnc[Border[i]] \leftarrow LongestLSAnc[LCover[Border[i]]]$;
6     $LCover[i] \leftarrow LongestLSAnc[Border[i]]$;
7     $LSChildren[LCover[i]] \leftarrow LSChildren[LCover[i]] + 1$;
8   **if** $i > 1$ **then**
9        $c_1 \leftarrow i - Border[i]$;
10       $c_2 \leftarrow (i-1) - Border[i-1]$;
11   **for** j **from** c_2 **to** $c_1 - 1$ **do**
12   **while** $LSChildren[j] = 0$ **do**
13          $LSChildren[LCover[j]] \leftarrow LSChildren[LCover[j]] - 1$;
14          $j \leftarrow LCover[j]$;

$\approx$-seed of T. More precisely, the algorithm maintains them so that they satisfy the following invariants at the end of the i-th iteration of the outer **for** loop.

1. $LongestLSAnc[j] = j$ if $LSChildren[Border_T[j]] > 0$ or $Border_T[j] \ge i - Border_T[i]$ or $Border_T[j] = 0$ for $0 \le j \le n$.
2. $LongestLSAnc[j] = \mathsf{LongestLSeedCov}_T(i, j)$ for $0 \le j \le Border_T[i]$.
3. $LCover[j] = LCover_T[j]$ for $1 \le j \le i$.
4. $LSChildren[j] = |\mathsf{LSChildren}(i, j)|$, where

$$\mathsf{LSChildren}(i, j) = \{\, k \mid T[: k] \in \mathsf{LSeed}_{\approx}(T[: i]) \text{ and } j = LCover_T[k] \,\},$$

for $0 \le j \le n$. Note that $\mathsf{LSChildren}(i, j) = \emptyset$ for $j \ge i$.

Suppose we already have the $\approx$-cover tree for $T[: i-1]$. To update it for $T[: i]$ by adding a node i, we must determine the parent $LCover_T[i]$ of i. By Lemma 14 and the invariant, we know that $LCover_T[i] = LongestLSAnc[Border_T[i]]$. The array $LongestLSAnc$ can be maintained by Lemma 15, where we must update $LongestLSAnc[j]$ when $T[: j] \notin \mathsf{LSeed}_{\approx}(T[: i-1])$ for $j = Border_T[i] > Border_T[i-1]$. By Lemma 13, $T[: j] \in \mathsf{LSeed}_{\approx}(T[: i-1])$ iff $i-1-Border_T[i-1] \le j \le i-1$ or $LSChildren[j] > 0$ assuming that $LSChildren$ satisfies the invariant for $i-1$. Therefore, constructing the $\approx$-cover tree is reduced to maintaining the array $LSChildren$. By Lemma 13, $LSChildren[j]$ counts the number of children of j that are ancestors of an element of the set $P_i = \{\, k \mid i - Border_T[i] \le k \le i \,\}$, which is the index range of primary left $\approx$-seeds. At the beginning of the i-th iteration, $LSChildren$ is based on P_{i-1}, and we must update $LSChildren$ to be based on P_i by the end of the i-th iteration. $LSChildren[j]$ needs to be updated only when j is an ancestor of some k in the difference of P_{i-1} and P_i. So, we first increment the value $LSChildren[LCover[i]]$ by one as $LCover[i]$ has got a new child $i \in P_i \setminus P_{i-1}$. Since $LCover[i]$ is a left $\approx$-seed of $T[: i-1]$,

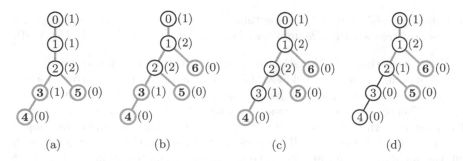

Fig. 1. Updating the $\overset{\text{pr}}{\approx}$-cover tree of $T[:5] = \text{abcac}$ (a) for that of $T[:6] = \text{abcacc}$ (d). *LSChildren* counts the numbers of children which are ancestors of some nodes drawn as thick red circles. Those highlighted nodes represent primary left $\overset{\text{pr}}{\approx}$-seeds $\{3,4,5\}$ of $T[:5]$ in (a) and those $\{5,6\}$ of $T[:6]$ in (d). Paths from highlighted nodes to the root are highlighted, so that *LSChildren*$[j]$ is the number of highlighted edges from j.

we need not increment *LSChildren*$[j]$ for further ancestors j of *LCover*$[i]$. For those $k \in P_{i-1} \setminus P_i$, we decrement *LSChildren*$[LCover[k]]$ unless k is an ancestor of P_i. If this results in *LSChildren*$[LCover[k]] = 0$, we recursively decrement *LSChildren*$[LCover^2[k]]$, and so on.

Example 1. We consider the parameterized-equivalence $\overset{\text{pr}}{\approx}$ (Definition 2) as an SCER. Suppose we have computed the $\overset{\text{pr}}{\approx}$-cover tree for $T[:5] = \text{abcac}$ as shown in Fig. 1 (a). Our goal is to obtain the one for $T[:6] = \text{abcacc}$ shown in Fig. 1 (d). Since *LongestLSAnc*$[j] = j$ for all j throughout this example, we focus on updating *LCover* and *LSChildren*. In the figures, *LSChildren* is shown in parentheses beside each node. We have $Border_T[5] = 2$ and $Border_T[6] = 1$, so the index sets of the primary left $\overset{\text{pr}}{\approx}$-seeds of $T[:5]$ and $T[:6]$ are $P_5 = \{3,4,5\}$ and $P_6 = \{5,6\}$, respectively. Since $Border_T[6] = 1$, Algorithm 2 first lets *LCover*$[6] = LongestLSAnc[Border[6]] = 1$. In other words, a new node 6 is added as a child of 1. It remains to update *LSChildren*, which is now based on $P_5 = \{3,4,5\}$ but shall be based on $P_6 = \{5,6\}$. First we increment *LSChildren*$[LCover[6]] = LSChildren[1]$ by one, as illustrated in Fig. 1 (b). At this moment, *LSChildren*$[j]$ counts the number of children of j which are ancestors of some of $P_5 \cup P_6 = \{3,4,5,6\}$. The inner **for**-loop of Line 11 modifies *LSChildren* so that it shall be based on $\{4,5,6\}$ first and then on $\{5,6\}$. Since the node 3 is the parent of 4, the *LSChildren* arrays based on $\{3,4,5,6\}$ and $\{4,5,6\}$ are identical, as shown in Figs. 1 (b) and (c), respectively. To modify *LSChildren* to be based on $\{5,6\}$, we decrement *LSChildren*$[j]$ if j has a child which is an ancestor of 4 but not that of 5 or 6. Since the node 4 is such a child of *LCover*$[4] = 3$ (4 is an ancestor of 4, and *LSChildren*$[4] = 0$ means that 4 is not an ancestor of 5 or 6), so we decrement *LSChildren*$[3]$ by one. This results in *LSChildren*$[3] = 0$, by which we know that the node 3 is an ancestor of 4 but not that of 5 or 6. Hence we decrement *LSChildren*$[LCover[3]] = LSChildren[2]$. This results in *LSChildren*$[2] = 1$, which means that the node 2 is an ancestor of 4 and that of 5 or 6 at the same time. So, we stop the recursion and obtain the $\approx$-cover tree with $LSChildren$ based on $P_6 = \{5, 6\}$, as shown in Fig. 1 (d).

We remark that Li & Smyth's original algorithm maintains an array $Dead$ that represents whether $j \notin \mathsf{LSeed}_{\approx}(T[:i])$ in addition to the arrays used in our algorithm. Our algorithm judges the property using two arrays $Border$ and $LSChildren$ based on Lemmas 11 and 12. The reason why their algorithm requires the additional array is that it performs the inner **for** loop of Line 11 in the reverse order. If we perform the loop in the reverse order without the auxiliary array, in the above example, in the iteration on $j = 4$, we obtain the tree in Fig. 1 (d), and then in the iteration on $j = 3$, the value of $LSChildren[LCover[3]] = LSChildren[2]$ is decremented to 0 and further more $LSChildren[LCover[2]] = LSChildren[1]$ is decremented to 1. Their algorithm stops iteration of the **while** loop at Line 12 if $Dead[j] = $ **True**, to restrain excessive decrement of $LSChildren[j]$.

Theorem 2. *Given the $\approx$-border array $Border_T$ of T, Algorithm 2 computes the longest $\approx$-cover array $LCover_T$ of T in $O(n)$ time.*

Proof. We prove the above invariants by induction on i. In the first iteration, neither of the **if** antecedents are satisfied. At the end of the iteration, we have $LCover[1] = LongestLSAnc[Border[1]] = LongestLSAnc[0] = 0$ and $LSChildren[0] = 1$. Together with the initialization, all the arrays satisfy the above invariants. By Lemmas 16 and 17, finally the algorithm computes $LCover_T$. The linear-time complexity is shown in Lemma 18.   □

Corollary 2. *If $Border_T$ can be computed in $\beta(n)$ time, $LCover_T$ can be computed in $O(\beta(n) + n)$ time.*

Lemma 16. *Suppose that all the invariants hold at the beginning of the i-th iteration of the outer **for** loop. Then, at the end of the i-th loop, the invariants on $LongestLSAnc$ and $LCover$ are satisfied.*

Proof. Assume that $LSChildren$, $LongestLSAnc$, and $LCover$ hold the above properties at the end of the $(i-1)$-th iteration. Let $b_i = Border[i]$ and $b_{i-1} = Border[i-1]$.

We first show that the invariant on $LongestLSAnc$ is satisfied. Concerning the first claim on $LongestLSAnc$, the value of $LongestLSAnc[j]$ can be altered from its initial value j only when $LSChildren[j] = 0$, $0 < 2j < i$ and $j = b_i$, in which case, the invariant does not necessitate $LongestLSAnc[j] = j$. On the other hand, by Lemma 2, if $Border_T[j] < i - 1 - Border_T[i - 1]$, then $Border_T[j] < i - Border_T[i]$. Therefore, once the value of $LongestLSAnc[j]$ has been altered from j, the invariant will never necessitate $LongestLSAnc[j] = j$.

Concerning the second claim on $LongestLSAnc$, suppose $j \le b_i$. If $j < b_i$, then $j \le b_{i-1}$ by Lemma 2. By the induction hypothesis on $LongestLSAnc[j]$ and Lemma 15, $LongestLSAnc[j] = \mathsf{LongestLSeedCov}(i - 1, j) = \mathsf{LongestLSeedCov}(i, j)$. It remains to show $LongestLSAnc[b_i] = \mathsf{LongestLSeedCov}(i, b_i)$.

If $b_i = 0$, $\mathsf{LongestLSeedCov}_T(i, b_i) = \mathsf{LongestLSeedCov}_T(i - 1, b_i) = 0$. Suppose $b_i > 0$ and $T[: b_i] \notin \mathsf{LSeed}_{\approx}(T[: i - 1])$. Let $m = LCover[b_i]$, for which $m < b_i \leq b_{i-1} + 1$. By Lemma 15 and the induction hypothesis on $LongestLSAnc[m]$, we have $\mathsf{LongestLSeedCov}_T(i, b_i) = \mathsf{LongestLSeedCov}_T(i - 1, m) = LongestLSAnc[m]$. By Lemmas 11 and 12 and the induction hypothesis, $b_i < i - 1 - b_{i-1}$ and $LSChildren[b_i] = 0$. Thus, since $2b_i \leq b_i + b_{i-1} + 1 < i - 1 + 1 = i$, the algorithm lets $LongestLSAnc[b_i] = LongestLSAnc[m]$ in Line 5, which fulfills the invariant on $LongestLSAnc$.

Suppose $T[: b_i] \in \mathsf{LSeed}_{\approx}(T[: i - 1])$. In this case, there is $k < b_i$ such that $T[: b_i] \in \mathsf{Cov}_{\approx}(T[: i-1-k])$. By Lemma 7, $T[: b_i] \in \mathsf{Cov}_{\approx}(T[: i]) \subseteq \mathsf{LSeed}_{\approx}(T[: i])$ and thus $\mathsf{LongestLSeedCov}_T(i, b_i) = b_i$. By Lemma 2, $b_i = b_{i-1} + 1$ holds. By Lemmas 11 and 12, either $b_i \geq i - 1 - b_{i-1}$ or $LSChildren[b_i] > 0$. The former case implies $2b_i \geq i$ and thus in either case the algorithm does not execute Line 5. By the induction hypothesis, $LongestLSAnc[b_i] = b_i$, which fulfills the invariant on $LongestLSAnc[b_i]$.

The invariant on $LCover$ is fulfilled in Line 6, which makes $LCover[i] = LongestLSAnc[b_i]$ in accordance with Lemma 14. $\qquad\square$

Lemma 17. *If the invariants hold at the beginning of the i-th iteration of the outer **for** loop, the invariant on $LSChildren$ holds at the end of the i-th loop.*

Proof. Assume that at the end of the $(i - 1)$-th iteration, the invariants hold. Let $b_i = Border[i]$, $b_{i-1} = Border[i - 1]$, $c_1 = i - b_i$, and $c_2 = (i - 1) - b_{i-1}$. Note that $c_1 \geq c_2$ by Lemma 2.

First we discuss $LSChildren[j]$ for $j \geq c_1$. For any k with $c_2 \leq c_1 \leq k < i$, by Lemma 11, $T[: k] \in \mathsf{LSeed}_{\approx}(T[: i - 1]) \cap \mathsf{LSeed}_{\approx}(T[: i])$. This means that for any j with $c_2 \leq c_1 \leq j \leq i$,

$$\mathsf{LSChildren}(i, j) = \mathsf{LSChildren}(i - 1, j) \cup I_j$$

where $I_j = \{i\}$ for $j = LCover_T[i]$ and $I_j = \emptyset$ for $j \neq LCover_T[i]$. Accordingly, for those $j \geq c_1$, the algorithm realizes $LSChildren[j] = |\mathsf{LSChildren}(i - 1, j)| + |I_j| = |\mathsf{LSChildren}(i, j)|$.

It remains to show the invariants on $LSChildren[j]$ for $j < c_1$. By Lemma 13, $\mathsf{LSChildren}(i, j)$ can be rewritten as $\mathsf{LSChildren}(i, j) = \mathsf{rangeChildren}(c_1, i, j)$ for

$$\mathsf{rangeChildren}(k, l, j) = LCover^{-1}[j] \cap \{ LCover^q[h] \mid k \leq h \leq l \text{ and } q \geq 0 \}$$

where $LCover^{-1}[j] = \{ h \mid j = LCover[h] \}$. In terms of the $\approx$-cover tree, $LCover^{-1}[j]$ is the set of children of j and $\mathsf{rangeChildren}(k, l, j)$ is the set of children which have an element between k and l as a descendant (a node is thought to be a descendant of itself). Note that $0 \notin LCover^{-1}[j]$ for any $j \geq 0$. After executing Line 7 of Algorithm 2, together with the induction hypothesis, we have $LSChildren[j] = |\mathsf{rangeChildren}(c_2, i, j)|$. If $c_1 = c_2$, then the algorithm does not go into the inner **for** loop of Line 11 and we have done the proof. If $c_1 > c_2$, it is enough to show that at the end of each iteration of the inner **for** loop of Line 11,

$$LSChildren[l] = |\mathsf{rangeChildren}(j + 1, i, l)| \tag{1}$$

for all $l < c_1$. For $j = c_1 - 1$, we have $LSChildren[l] = |\text{rangeChildren}(c_1, i, l)| = |\text{LSChildren}(i, l)|$ for all $l < c_1$. For this purpose, we show by induction on r that at the end of the r-th iteration of the **while** loop (Line 12), we have

$$LSChildren[l] = |\text{rangeChildren}(j + 1, i, l) \cup (LCover^{-1}[l] \cap \{LCover^q[j] \mid q \geq r\})| \tag{2}$$

for all $l < c_1$. Note that there always exists r_j such that $LCover^{r_j}[j] = 0$, for which $LCover^{-1}[l] \cap \{LCover^q[j] \mid q \geq r_j\} = \emptyset$, i.e., Eq. (2) is equivalent to (1).

For $r = 0$, i.e., at the beginning of the first iteration of the **while** loop, Eq. (1) for $j - 1$ holds, i.e., $LSChildren[l] = |\text{rangeChildren}(j, i, l)|$, which is equivalent to (2) with $r = 0$.

Assuming the induction hypothesis (2) for r holds, we show that it is the case for $r + 1$. Increasing r by one never expands the set on the right hand of (2). The set will lose an element h iff $h = LCover^r[j]$, $l = LCover^{r+1}[j]$ and

$$LCover^r[j] \notin \{LCover^q[k] \mid j < k \leq i,\ q \geq 0\}. \tag{3}$$

If $LSChildren[LCover^r[j]] \neq 0$, the loop is not repeated. It is enough to show that for any $l < c_1$

$$LCover^{-1}[l] \cap \{LCover^q[j] \mid q \geq r\} \subseteq \text{rangeChildren}(j + 1, i, l), \tag{4}$$

so that we establish (1). If $LCover^r[j] = 0$, $LCover^{-1}[l] \cap \{LCover^q[j] \mid q \geq r\} = \emptyset$. Clearly (4) holds. Suppose $LCover^r[j] \neq 0$. The assumption that $LSChildren[LCover^r[j]] \neq 0$ means, by induction hypothesis (2), there is

$$k \in \text{rangeChildren}(j + 1, i, LCover^r[j])$$
$$\cup (LCover^{-1}[LCover^r[j]] \cap \{LCover^q[j] \mid q \geq r\}).$$

By $LCover^{-1}[LCover^r[j]] \cap \{LCover^q[j] \mid q \geq r\} = \emptyset$, $k \in \text{rangeChildren}(j + 1, i, LCover^r[j])$, which means $k = LCover^s[h] \in LCover^{-1}[LCover^r[j]]$ for some $j < h \leq i$ and $s \geq 0$, i.e., $LCover^{s+1}[h] = LCover^r[j]$. For $1 \leq l \leq c_1$, if $LCover^q[j] \in LCover^{-1}[l]$ for some $q \geq r$, then

$$LCover^{q-r+s+1}[h] = LCover^q[j] \in LCover^{-1}[l].$$

That is, $LCover^q[j] \in \text{rangeChildren}(j + 1, i, l)$, which shows (4) and thus (1).

Suppose $LSChildren[LCover^r[j]] = 0$. We show that (3) holds. By the induction hypothesis (2) for r, $LSChildren[LCover^r[j]] = 0$ means

$$\text{rangeChildren}(j + 1, i, LCover^r[j])$$
$$\cup (LCover^{-1}[LCover^r[j]] \cap \{LCover^q[j] \mid q \geq r\}) = \emptyset.$$

If (3) did not hold, there were j' and q such that $LCover^r[j] = LCover^q[j']$ and $j < j' \leq i$, where $q \geq 1$ by $LCover^r[j] \leq j < j'$. Then $LCover^{q-1}[j'] \in LCover^{-1}[LCover^r[j]]$, which is a contradiction. So, the condition (3) holds. □

Lemma 18. *Algorithm 2 runs in $O(n)$ time.*

Proof. Let $t(j)$ and $f(j)$ be the numbers of times that the **while** condition on j (Line 12) is judged true and false, respectively. Since $\sum_{j=0}^{n} f(j) \leq n + \sum_{j=1}^{n} t(j)$, it is enough to show $t(j) \leq 1$ for every j to establish the linear-time complexity. Suppose that the algorithm finds $LSChildren[j] = 0$ at the **while** loop in the i-th iteration of the outer **for** loop. We show that it happens for the least $i > j$ such that $T[: j] \notin \mathsf{LSeed}_{\approx}(T[: i])$. Note that the condition is checked only for $j < c_1$, where $c_1 = i - Border[i]$. Therefore, $LSChildren[j] = 0$ implies $T[: j] \notin \mathsf{LSeed}_{\approx}(T[: i])$ by Lemma 12. Since $T[: j] \notin \mathsf{LSeed}_{\approx}(T[: i])$ implies $T[: j] \notin \mathsf{LSeed}_{\approx}(T[: i'])$ for any $i' > i$, it is enough to show $T[: j] \in \mathsf{LSeed}_{\approx}(T[: i - 1])$. For $c_2 = i - 1 - Border[i - 1]$, by the algorithm, $j = LCover^q[k]$ for some $c_2 \leq k < c_1$ and $q \geq 0$. If $q = 0$, i.e., $c_2 \leq j = k < c_1$, by Lemma 11, $T[: j] \in \mathsf{LSeed}_{\approx}(T[: i - 1])$. If $q \geq 1$, the value $LSChildren[j]$ is decremented in the q-th iteration of the **while** loop, just before deciding $LSChildren[j] = 0$. Moreover, $T[: j] \notin \mathsf{LSeed}_{\approx}(T[: i])$ implies $j \neq LCover[i]$, and hence $LSChildren[j]$ was strictly positive at the end of the $(i - 1)$-th iteration of the outer **for** loop. By the invariant, $\mathsf{LSChildren}(i - 1, j) \neq \emptyset$, which means $T[: j] \in \mathsf{LSeed}_{\approx}(T[: i - 1])$. $\square$

References

1. Aho, A.V., Hopcroft, J.E.: The design and analysis of computer algorithms. Pearson Education India (1974)
2. Amir, A., Levy, A., Lubin, R., Porat, E.: Approximate cover of strings. Theor. Comput. Sci. **793**, 59–69 (2019). https://doi.org/10.1016/j.tcs.2019.05.020
3. Amir, A., Aumann, Y., Lewenstein, M., Porat, E.: Function matching. SIAM J. Comput. **35**(5), 1007–1022 (2006)
4. Amir, A., Farach, M., Muthukrishnan, S.: Alphabet dependence in parameterized matching. Inf. Process. Lett. **49**(3), 111–115 (1994). https://doi.org/10.1016/0020-0190(94)90086-8
5. Antoniou, P., Crochemore, M., Iliopoulos, C., Jayasekera, I., Landau, G.: Conservative string covering of indeterminate strings. In: Prague Stringology Conference 2008, pp. 108–115 (2008)
6. Apostolico, A., Ehrenfeucht, A.: Efficient detection of quasiperiodicities in strings. Theor. Comput. Sci. **119**(2), 247–265 (1993). https://doi.org/10.1016/0304-3975(93)90159-Q
7. Apostolico, A., Farach, M., Iliopoulos, C.S.: Optimal superprimitivity testing for strings. Inf. Process. Lett. **39**(1), 17–20 (1991). https://doi.org/10.1016/0020-0190(91)90056-N
8. Apostolico, A., Giancarlo, R.: Periodicity and repetitions in parameterized strings. Discrete Appl. Math. **156**(9), 1389–1398 (2008). https://doi.org/10.1016/j.dam.2006.11.017
9. Baker, B.S.: Parameterized pattern matching: algorithms and applications. J. Comput. Syst. Sci. **52**(1), 28–42 (1996). https://doi.org/10.1006/jcss.1996.0003
10. Breslauer, D.: An on-line string superprimitivity test. Inf. Process. Lett. **44**(6), 345–347 (1992). https://doi.org/10.1016/0020-0190(92)90111-8

11. Brodal, G.S., Pedersen, C.N.S.: Finding maximal quasiperiodicities in strings. In: Giancarlo, R., Sankoff, D. (eds.) CPM 2000. LNCS, vol. 1848, pp. 397–411. Springer, Heidelberg (2000). https://doi.org/10.1007/3-540-45123-4_33
12. Christou, M., Crochemore, M., Guth, O., Iliopoulos, C.S., Pissis, S.P.: On left and right seeds of a string. J. Discrete Algorithms 17, 31–44 (2012)
13. Christou, M., et al.: Efficient seed computation revisited. Theor. Comput. Sci. 483, 171–181 (2013). https://doi.org/10.1016/j.tcs.2011.12.078
14. Crochemore, M., Rytter, W.: Jewels of Stringology. World Scientific Publishing Co., Pte. Ltd. (2002). https://doi.org/10.1142/9789812778222
15. Diptarama, Ueki, Y., Narisawa, K., Shinohara, A.: KMP based pattern matching algorithms for multi-track strings. In: Proceedings of Student Research Forum Papers and Posters at SOFSEM2016, pp. 100–107 (2016)
16. Ehlers, T., Manea, F., Mercaş, R., Nowotka, D.: k-abelian pattern matching. J. Discrete Algorithms 34, 37–48 (2015)
17. Gourdel, G., Kociumaka, T., Radoszewski, J., Rytter, W., Shur, A., Waleń, T.: String periods in the order-preserving model. Inf. Comput. 270(104463), 1–22 (2020). https://doi.org/10.1016/j.ic.2019.104463
18. Hendrian, D., Ueki, Y., Narisawa, K., Yoshinaka, R., Shinohara, A.: Permuted pattern matching algorithms on multi-track strings. Algorithms 12(4), 73:1–20 (2019). https://doi.org/10.3390/a12040073
19. Iliopoulos, C., Mouchard, L.: Quasiperiodicity: from detection to normal forms. J. Autom. Lang. Comb. 4, 213–228 (1999)
20. Iliopoulos, C.S., Moore, D.W.G., Park, K.: Covering a string. Algorithmica 16(3), 288–297 (1996). https://doi.org/10.1007/BF01955677
21. Katsura, T., Narisawa, K., Shinohara, A., Bannai, H., Inenaga, S.: Permuted pattern matching on multi-track strings. In: SOFSEM 2013: Theory and Practice of Computer Science, pp. 280–291 (2013). https://doi.org/10.1007/978-3-642-35843-2_25
22. Kim, J., Eades, P., Fleischer, R., Hong, S.H., Iliopoulos, C.S., Park, K., Puglisi, S.J., Tokuyama, T.: Order-preserving matching. Theor. Comput. Sci. 525, 68–79 (2014). https://doi.org/10.1016/j.tcs.2013.10.006
23. Knuth, D.E., Morris Jr., J.H., Pratt, V.R.: Fast pattern matching in strings. SIAM J. Comput. 6(2), 323–350 (1977)
24. Kubica, M., Kulczyński, T., Radoszewski, J., Rytter, W., Waleń, T.: A linear time algorithm for consecutive permutation pattern matching. Inf. Process. Lett. 113(12), 430–433 (2013). https://doi.org/10.1016/j.ipl.2013.03.015
25. Li, Y., Smyth, W.F.: Computing the cover array in linear time. Algorithmica 32(1), 95–106 (2002). https://doi.org/10.1007/s00453-001-0062-2
26. Matsuoka, Y., Aoki, T., Inenaga, S., Bannai, H., Takeda, M.: Generalized pattern matching and periodicity under substring consistent equivalence relations. Theor. Comput. Sci. 656, 225–233 (2016). https://doi.org/10.1016/j.tcs.2016.02.017
27. Moore, D., Smyth, W.: An optimal algorithm to compute all the covers of a string. Inf. Process. Lett. 50(5), 239–246 (1994). https://doi.org/10.1016/0020-0190(94)00045-X
28. Moore, D., Smyth, W.: A correction to "an optimal algorithm to compute all the covers of a string". Inf. Process. Lett. 54(2), 101–103 (1995). https://doi.org/10.1016/0020-0190(94)00235-Q
29. Park, S.G., Amir, A., Landau, G.M., Park, K.: Cartesian tree matching and indexing. In: 30th Annual Symposium on Combinatorial Pattern Matching (CPM 2019), pp. 16:1–16:14 (2019). https://doi.org/10.4230/LIPIcs.CPM.2019.16

Longest Square Subsequence Problem Revisited

Takafumi Inoue[1], Shunsuke Inenaga[1,2(✉)] ⓘ, and Hideo Bannai[3] ⓘ

[1] Department of Informatics, Kyushu University, Fukuoka, Japan
[2] PRESTO, Japan Science and Technology Agency, Kawaguchi, Japan
inenaga@inf.kyushu-u.ac.jp
[3] M&D Data Science Center, Tokyo Medical and Dental University, Tokyo, Japan
hdbn.dsc@tmd.ac.jp

Abstract. The *longest square subsequence* (*LSS*) problem consists of computing a longest subsequence of a given string S that is a square, i.e., a longest subsequence of form XX appearing in S. It is known that an LSS of a string S of length n can be computed using $O(n^2)$ time [Kosowski 2004], or with (model-dependent) polylogarithmic speed-ups using $O(n^2(\log \log n)^2 / \log^2 n)$ time [Tiskin 2013]. We present the first algorithm for LSS whose running time depends on other parameters, i.e., we show that an LSS of S can be computed in $O(r \min\{n, M\} \log \frac{n}{r} + n + M \log n)$ time with $O(M)$ space, where r is the length of an LSS of S and M is the number of matching points on S.

1 Introduction

Subsequences of a string S with some interesting properties have caught much attention in mathematics and algorithmics. The most well-known of such kinds is the *longest increasing subsequence* (*LIS*), which is a longest subsequence of S whose elements appear in lexicographically increasing order. It is well known that an LIS of a given string S of length n can be computed in $O(n \log n)$ time with $O(n)$ space [9]. Other examples are the *longest palindromic subsequence* (*LPS*) and the *longest square subsequence* (*LSS*). Since an LPS of S is a *longest common subsequence* (*LCS*) of S and its reversal, an LPS can be computed by a classical dynamic programming for LCS, or by any other LCS algorithms.

Computing an LSS of a string is not as easy, because a reduction from LSS to LCS essentially requires to consider $n-1$ partition points on S. Kosowski [6] was the first to tackle this problem, and showed an $O(n^2)$-time $O(n)$-space LSS algorithm. Computing LSS can be motivated by e.g. finding an optimal partition point on a given string so that the corresponding prefix and suffix are most similar. Later, Tiskin [10] presented a (model-dependent) $O(n^2(\log \log n)^2 / \log^2 n)$-time LSS algorithm, based on his semi-local string comparison technique applied to the $n-1$ partition points (i.e. $n-1$ pairs of prefixes and suffixes.) Since strongly sub-quadratic $O(n^{2-\epsilon})$-time LSS algorithms do not exist for any $\epsilon > 0$ unless the SETH is false [2], the aforementioned solutions are almost optimal in terms of n.

© Springer Nature Switzerland AG 2020
C. Boucher and S. V. Thankachan (Eds.): SPIRE 2020, LNCS 12303, pp. 147–154, 2020.
https://doi.org/10.1007/978-3-030-59212-7_11

In this paper, we present the first LSS algorithm whose running time depends on other parameters, i.e., we show that an LSS of S can be computed in $O(r \min\{n, M\} \log \frac{n}{r} + n + M \log n)$ time with $O(M)$ space, where r is the length of an LSS of S and M is the number of matching points on S. This algorithm outperforms Tiskin's $O(n^2 (\log \log n)^2 / \log^2 n)$-time algorithm when $r = o(n(\log \log n)^2 / \log^3 n)$ and $M = o(n^2 (\log \log n)^2 / \log^3 n)$.

Our algorithm is based on a reduction from computing an LCS of two strings of total length n to computing an LIS of an integer sequence of length at most M, where M is roughly n^2 / σ for uniformly distributed random strings over alphabets of size σ. We then use a slightly modified version of the dynamic LIS algorithm [3] for our LIS instances that dynamically change over $n - 1$ partition points on S. A similar but more involved reduction from LCS to LIS is recently used in an intermediate step of a reduction from dynamic time warping (DTW) to LIS [8]. We emphasize that our reduction (as well as the one in [8]) from LCS to LIS should not be confused with a well-known folklore reduction from LIS to LCS.

Soon after the submission of this paper, independently to this work, Russo and Francisco [7] showed a very similar algorithm to solve the LSS problem, also based on a reduction to LIS. Their algorithm runs in $O(r \min\{n, M\} \log \min\{r, \frac{n}{r}\} + rn + M)$ time and $O(M)$ space.

2   Preliminaries

Let Σ be an alphabet. An element S of Σ^* is called a string. The length of a string S is denoted by $|S|$. For any $1 \leq i \leq |S|$, $S[i]$ denotes the ith character of S. For any $1 \leq i \leq j \leq |S|$, $S[i..j]$ denotes the substring of X beginning at position i and ending at position j.

A string X is said to be a *subsequence* of a string S if there exists a sequence $1 \leq i_1 < \cdots < i_{|X|} \leq |S|$ of increasing positions of S such that $X = S[i_1] \cdots S[i_{|X|}]$. Such a sequence $i_1, \ldots, i_{|X|}$ of positions in S is said to be an *occurrence* of X in S.

A non-empty string Y of form XX is called a *square*. A square Y is called a *square subsequence* of a string S if square Y is a subsequence of S. Let $\mathsf{LSS}(S)$ denote the length of a *longest square subsequence* (*LSS*) of string S. This paper deals with the problem of computing $\mathsf{LSS}(S)$ for a given string S of length n.

For strings A, B, let $\mathsf{LCS}(A, B)$ denote the length of the *longest common subsequence* (*LCS*) of A and B. For a sequence T of numbers, a subsequence X of T is said to be an *increasing subsequence* of T if $X[i] < X[i+1]$ for $1 \leq i < |X|$. Let $\mathsf{LIS}(T)$ denote the length of the *longest increasing subsequence* (*LIS*) of T.

A pair (i, j) of positions $1 \leq i < j \leq |S|$ is said to be a *matching point* if $S[i] = S[j]$. The set of all matching points of S is denoted by $\mathcal{M}(S)$, namely, $\mathcal{M}(S) = \{(i, j) \mid 1 \leq i < j \leq |S|, S[i] = S[j]\}$. Let $M = |\mathcal{M}(S)|$.

3   Algorithm

We begin with a simple folklore reduction of computing $\mathsf{LSS}(S)$ to computing the LCS of $n-1$ pairs of the prefix and the suffix of S.

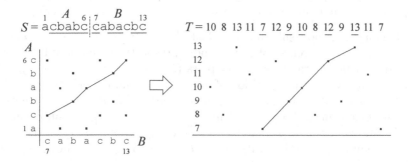

Fig. 1. Correspondence between an LCS of $A = \mathtt{acbabc}$, $B = \mathtt{cabacbc}$ and an LIS of T.

Lemma 1 [6]. $\mathsf{LSS}(S) = 2\max_{1\leq p<n} \mathsf{LCS}(S[1..p], S[p+1..n])$.

Following Lemma 1, one can use the *decremental* LCS algorithm by Kim and Park [5] for computing $\mathsf{LSS}(S)$. Given two strings A and B of length n, Kim and Park proposed how to update, in $O(n)$ time, an $O(n^2)$-space representation for the dynamic programming table for $\mathsf{LCS}(A, B)$ when the leftmost character is deleted from B. Since their algorithm also allows to append a character to A in $O(n)$ time, it turns out that $\mathsf{LSS}(S)$ can be computed in $O(n^2)$ time and space. Kosowski [6] presented an $O(n^2)$-time $\Theta(n)$-space algorithm for computing $\mathsf{LSS}(S)$, which can be seen as a space-efficient version of an application of Kim and Park's algorithm to this specific problem of computing $\mathsf{LSS}(S)$. Tiskin [10] also considered the problem of computing $\mathsf{LSS}(S)$, and showed that using his semi-local LCS method, $\mathsf{LSS}(S)$ can be computed in $O(n^2(\log\log n)^2/\log^2 n)$ time. We remark that the log-shaving factor is model-dependent (i.e., Tiskin's method uses the so-called "Four-Russian" technique).

Let $A = S[1..p]$, $A' = S[1..p+1]$, $B = S[p+1..n]$ and $B' = S[p+2..n]$. For ease of explanations, suppose that the indices on B and B' begin with $p+1$ and $p+2$, respectively. Next, we further reduce computing $\mathsf{LCS}(A', B')$ from (a representation of) $\mathsf{LCS}(A, B)$, to computing an LIS of a dynamic integer sequence of length at most $M = |\mathcal{M}(S)|$.

For any integer pairs (u, v) and (x, y), let $(u, v) \prec (x, y)$ if (i) $u < x$, or (ii) $u = x$ and $v < y$. Consider the following integer sequence T: Let $\mathcal{P}$ be the set of integer pairs $(i, n - j)$ such that $S[i] = A[i] = B[p+j] = B[|A| + j] = S[j]$. Then, we set $T[q] = j$ iff the integer pair $(i, n - j)$ is of rank q in $\mathcal{P}$ w.r.t. $\prec$. See Fig. 1 for an example. Intuitively, T is an integer sequence representation of the (transposed) matching points between A and B, obtained by scanning the matching points between A and B from the bottom row to the top row, where each row is scanned from right to left. Clearly, the length of the integer sequence T is bounded by M.

Lemma 2. *Any common subsequence of A and B corresponds to an increasing subsequence of T of the same length. Also, any increasing subsequence of T corresponds to a common subsequence of A and B of the same length.*

Proof. For any common subsequence C of A and B, let $i_1 < \cdots < i_{|C|}$ and $j_1 < \cdots < j_{|C|}$ be occurrences of C in A and B, respectively. For any $1 \le k < |C|$, let q_k and q_{k+1} be the ranks of integer pairs $(i_k, n - j_k)$ and $(i_{k+1}, n - j_{k+1})$ in the set $\mathcal{P}$ w.r.t. $\prec$. By the definition of T, $q_k < q_{k+1}$ and $T[q_k] < T[q_{k+1}]$ hold. Hence, C corresponds to an increasing subsequence of T of the same length.

For any increasing subsequence I in T, let $t_1 < \cdots < t_{|I|}$ be an occurrence of I in T. For any $1 \le k < |I|$, let $(i_k, n - j_k)$ and $(i_{k+1}, n - j_{k+1})$ be the integer pairs corresponding to $I[k] = T[t_k]$ and $I[k + 1] = T[i_{k+1}]$, respectively. Since $j_k = T[t_k] < T[t_{k+1}] = j_{k+1}$, we have

$$n - j_{k+1} < n - j_k. \tag{1}$$

Since $(i_k, n - j_k) \prec (i_{k+1}, n - j_{k+1})$, either (i) $i_k < i_{k+1}$ or (ii) $i_k = i_{k+1}$ and $n - j_k < n - j_{k+1}$ must hold. By inequality (1), (ii) cannot hold, and thus (i) holds. Hence $A[i_k]A[i_{k+1}] = B[j_k]B[j_{j+1}]$ is a common subsequence of A and B. Hence, I corresponds to a common subsequence of A and B of the same length. □

By Lemma 2, computing $\mathsf{LCS}(A, B)$ can be reduced to computing $\mathsf{LIS}(T)$.

Let T' be the integer sequence for A' and B' defined analogously to T for A and B. Now the task is how to compute $\mathsf{LIS}(T')$ from (a data structure that represents) $\mathsf{LIS}(T)$. See Fig. 2 for an example. Observe that when the leftmost character is deleted from B (upper part of Fig. 2), then the lowest points are deleted from the 2D plane, and thus all the elements with minimum value are deleted from T. Also, when the leftmost character of B is appended to A (upper part of Fig. 2), which gives us $A' = S[1..p+1]$, then a new point for every j with $A'[|A'|] = B'[j]$ is inserted to the right end of the 2D plane in decreasing order of j, and thus j is appended to the right end of T in decreasing order of j, one by one. Thus, computing $\mathsf{LCS}(A', B')$ from $\mathsf{LCS}(A, B)$ reduces to the following sub-problem:

Problem 1. For a dynamic integer sequence T, maintain a data structure that supports the following operations and queries:

- Insertion: Insert a new element to the right-end of T;
- Batched Deletion: Delete all the elements with minimum value from T;
- Query: Return $\mathsf{LIS}(T)$.

We can use Chen et al.'s algorithm [3] for insertions. Let $\ell = \mathsf{LIS}(T)$. Their algorithm supports insertions at the right-end of T in $O(\log |T|)$ time each. Since $|T| \le M \le n^2$, insertions at the right-end can be done in $O(\log n)$ time.

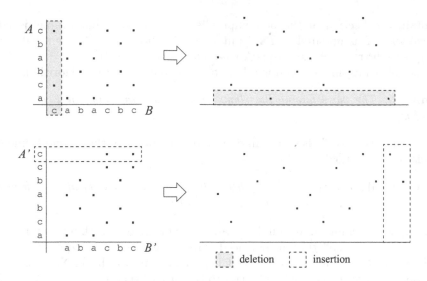

Fig. 2. Illustration on how points in the 2D plane (and elements in T) are to be deleted or inserted when A and B are updated to A' and B', respectively.

Next, let us consider batched deletions. Chen et al. [3] showed that an insertion or deletion of a single element at an arbitrary position of T can be supported in $O(\ell \log \frac{|T|}{\ell}) \subseteq O(\ell \log \frac{n}{\ell})$ time each. However, since our batched deletion may contain $O(|T|) \subseteq O(M)$ characters in the worst case, a naïve application of a single-element deletion only leads to an inefficient $O(\ell|T| \log \frac{n}{\ell}) \subseteq O(\ell M \log \frac{n}{\ell})$ batched deletion. In what follows, we show how to support batched deletions in $O(\ell \log \frac{n}{\ell})$ time each, using Chen et al.'s data structure.

For any position $1 \le t \le |T|$ in sequence T, let $l(t)$ denote the length of an LIS of $T[1..t]$ that has an occurrence $i_1 < \cdots < i_{l(t)} = t$, namely, an occurrence that ends at position t in T. The following observations are immediate:

Lemma 3 [3]. *Let q be the second to last position of any occurrence of a length-$l(t)$ LIS of $T[1..t]$ ending at position t. Then, $l(q) = l(t) - 1$.*

Lemma 4 [3]. *If $q < t$ and $l(q) = l(t)$, then $T[q] \ge T[t]$.*

For any $1 \le k \le \ell$, let $\mathcal{L}_k$ be a list of pairs $\langle t, T[t] \rangle$ such that $l(t) = k$, sorted in increasing order of the first elements t. See Fig. 3 for an example. It follows from Lemma 4 that this list is also sorted in non-increasing order of the second elements $T[t]$. It is clear that $\mathsf{LIS}(T) = \max\{k \mid \mathcal{L}_k \ne \emptyset\}$. It is also clear that for any $k > 1$, if $\mathcal{L}_k \ne \emptyset$, then $\mathcal{L}_{k-1} \ne \emptyset$. Thus, our task is to

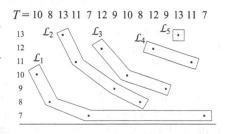

Fig. 3. Lists $\mathcal{L}_k$ for pairs $\langle t, T[t] \rangle$.

maintain a collection of the non-empty lists $\mathcal{L}_1, \ldots, \mathcal{L}_\ell$ that are subject to change when T is updated to T'. As in [3], we maintain each $\mathcal{L}_k$ by a balanced binary search tree such as red-black trees [4] or AVL trees [1].

The following simple claim is a key to our batched deletion algorithm:

Lemma 5. *The pairs having the elements of minimum value in T are at the tail of $\mathcal{L}_1$.*

Proof. Since the list $\mathcal{L}_1$ is sorted in non-increasing order of the second elements, the claim clearly holds. ☐

Lemma 6. *We can perform a batched deletion of all elements of T with minimum value in $O(\ell \log \frac{n}{\ell})$ time, where $\ell = \mathsf{LIS}(T)$.*

Proof. Due to Lemma 5, we can delete all the elements of T with minimum value from the list $\mathcal{L}_1$ by splitting the balanced search tree into two, in $O(\log |\mathcal{L}_1|)$ time.

The rest of our algorithm follows Chen et al.'s approach [3]: Note that the split operation on $\mathcal{L}_1$ can incur changes to the other lists $\mathcal{L}_2, \ldots, \mathcal{L}_\ell$. Let $l'(t)$ be the length of an LIS of $T'[1..t]$ that has an occurrence ending at position t in T', and let $\mathcal{L}'_k$ be the list of pairs $\langle t, T'[t] \rangle$ such that $l'(t) = k$ sorted in increasing order of the first elements t. Let $\mathcal{Q}_1$ be the list of deleted pairs corresponding to the smallest elements in T, and let $\mathcal{Q}_k = \{t \mid l(t) = k, l'(t) = k - 1\}$ for $k \geq 2$. Then, it is clear that $\mathcal{L}'_k = (\mathcal{L}_k \setminus \mathcal{Q}_k) \cup \mathcal{Q}_{k+1}$. Chen et al. [3] showed that $\mathcal{Q}_{k+1}$ can be found in $O(\log |\mathcal{L}_{k+1}|)$ time for each k, provided that $\mathcal{Q}_k$ has been already computed. Since $\mathcal{Q}_k$ is a consecutive sub-list of $\mathcal{L}_k$ (c.f. [3]), the split operation for $\mathcal{L}_k \setminus \mathcal{Q}_k$ can be done in $O(\log |\mathcal{L}_k|)$ time, and the concatenation operation for $(\mathcal{L}_k \setminus \mathcal{Q}_k) \cup \mathcal{Q}_{k+1}$ can be done in $O(\log |\mathcal{L}_k| + \log |\mathcal{L}_{k+1}|)$ time, by standard split and concatenation algorithms on balanced search trees. Thus our batched deletion takes $O(\sum_{1 < k \leq \ell} \log |\mathcal{L}_k|) = O(\log(|\mathcal{L}_1| \cdots |\mathcal{L}_\ell|))$ time, where $\ell = \mathsf{LIS}(T)$. Since $\sum_{1 \leq k \leq \ell} |\mathcal{L}_k| = |T|$ and $\log(|\mathcal{L}_1| \cdots |\mathcal{L}_\ell|)$ is maximized when $|\mathcal{L}_1| = \cdots = |\mathcal{L}_\ell|$, the above time complexity is bounded by $O(\ell \log \frac{|T|}{\ell}) \subseteq O(\ell \log \frac{n}{\ell})$ time. ☐

We are ready to show our main theorem.

Theorem 1. *An LSS of a string S can be computed in $O(r \min\{n, M\} \log \frac{n}{r} + n + M \log n)$ time with $O(M)$ space, where $n = |S|$, $r = \mathsf{LSS}(S)$, and $M = |\mathcal{M}(S)|$.*

Proof. By Lemma 1 and Lemma 2, it suffices to consider the total number of insertions, batched deletions, and queries of Problem 1 for computing an LIS of our dynamic integer sequence T. Since each matching point in $\mathcal{M}(S)$ is inserted to the dynamic sequence exactly once, the total number of insertions is exactly M. The total number of batched deletions is bounded by the number $n - 1$ of partition points p that divide S into $S[1..p]$ and $S[p + 1..n]$. Also, it is clearly bounded by the number M of matching points. Thus, the total number of batched deletions is at most $\min\{n, M\}$. We perform queries $n - 1$ times for all $1 \leq p < n$. Each query for $\mathsf{LIS}(T)$ can be answered in $O(1)$ time, by explicitly maintaining and storing the value of $\mathsf{LIS}(T)$ each time the dynamic integer sequence T is updated. Thus, it follows from Lemma 6 that our algorithm returns $\mathsf{LSS}(S)$ in $O(r \min\{n, M\} \log \frac{n}{r} + M \log n)$ time. By keeping the lists $\mathcal{L}_k$ for a partition point p that gives $2\ell = r = \mathsf{LSS}(S)$, we can also return an LSS (as a string) in $O(r \log \frac{n}{r})$ time, by finding an optimal sequence elements from $\mathcal{L}_\ell, \mathcal{L}_{\ell-1}, \ldots, \mathcal{L}_1$. The additive n term in our $O(r \min\{n, M\} \log \frac{n}{r} + n + M \log n)$ time complexity is for testing whether the input string S consists of n distinct characters (if so, then we can immediately output $r = 0$ in $O(n)$ time).

The space complexity is clearly linear in the total size of the lists $\mathcal{L}_1, \ldots \mathcal{L}_\ell$, which is $|T| \in O(M)$. □

When $r = o(n(\log\log n)^2/\log^3 n)$ and $M = o(n^2(\log\log n)^2/\log^3 n)$, our $O(r \min\{n, M\} \log \frac{n}{r} + n + M \log n)$-time algorithm outperforms Tiskin's solution that uses $O(n^2(\log\log n)^2/\log^2 n)$ time [10]. The former condition $r = o(n(\log\log n)^2/\log^3 n)$ implies that our algorithm can be faster than Tiskin's algorithm (as well as Kosowski's algorithm [6]) when the length r of the LSS of the input string S is relatively short. For uniformly distributed random strings of length n over an alphabet of size σ, we have $M \approx n^2/\sigma$. Thus, for alphabets of size $\sigma = \omega(\log^3 n/(\log\log n)^2)$, the latter condition $M = o(n^2(\log\log n)^2/\log^3 n)$ is likely to be the case for the majority of inputs.

Acknowledgments. This work was supported by JSPS KAKENHI Grant Numbers JP17H01697 (SI), JP20H04141 (HB), and JST PRESTO Grant Number JPMJPR1922 (SI).

References

1. Adelson-Velsky, G., Landis, E.: An algorithm for the organization of information. Proc. USSR Acad. Sci. **146**, 263–266 (1962). (in Russian). English translation by Myron J. Ricci in Soviet Mathematics - Doklady, 3:1259–1263, 1962
2. Bringmann, K., Künnemann, M.: Quadratic conditional lower bounds for string problems and dynamic time warping. In: FOCS 2015, pp. 79–97 (2015). https://doi.org/10.1109/FOCS.2015.15, full version https://arxiv.org/abs/1502.01063
3. Chen, A., Chu, T., Pinsker, N.: Computing the longest increasing subsequence of a sequence subject to dynamic insertion. CoRR abs/1309.7724 (2013). http://arxiv.org/abs/1309.7724
4. Guibas, L.J., Sedgewick, R.: A dichromatic framework for balanced trees. In: FOCS 1978, pp. 8–21 (1978)
5. Kim, S.R., Park, K.: A dynamic edit distance table. J. Discrete Algorithms **2**, 302–312 (2004)
6. Kosowski, A.: An efficient algorithm for the longest tandem scattered subsequence problem. In: Apostolico, A., Melucci, M. (eds.) SPIRE 2004. LNCS, vol. 3246, pp. 93–100. Springer, Heidelberg (2004). https://doi.org/10.1007/978-3-540-30213-1_13
7. Russo, L.M.S., Francisco, A.P.: Small longest tandem scattered subsequences. CoRR abs/2006.14029 (2020). https://arxiv.org/abs/2006.14029

8. Sakai, Y., Inenaga, S.: A reduction of the dynamic time warping distance to the longest increasing subsequence length. CoRR abs/2005.09169 (2020). https://arxiv.org/abs/2005.09169

9. Schensted, C.: Longest increasing and decreasing subsequences. Can. J. Math. **13**, 179–191 (1961). https://doi.org/10.4153/CJM-1961-015-3

10. Tiskin, A.: Semi-local string comparison: algorithmic techniques and applications. CoRR abs/0707.3619 (2013). http://arxiv.org/abs/0707.3619

Adaptive Exact Learning in a Mixed-Up World: Dealing with Periodicity, Errors and Jumbled-Index Queries in String Reconstruction

Ramtin Afshar[1], Amihood Amir[2], Michael T. Goodrich[1] ,
and Pedro Matias[1(✉)]

[1] Department of Computer Science, University of California Irvine, Irvine, USA
{afsharr,goodrich,pmatias}@uci.edu
[2] Department of Computer Science, Bar Ilan University, Ramat Gan, Israel
amir@cs.biu.ac.il

Abstract. We study the query complexity of exactly reconstructing a string from adaptive queries, such as substring, subsequence, and jumbled-index queries. Such problems have applications, e.g., in computational biology. We provide a number of new and improved bounds for exact string reconstruction for settings where either the string or the queries are "mixed-up".

Keywords: Exact learning · String reconstruction · Jumbled-index queries · Periodicity · DNA sequencing · Stringology · Substrings · Hybridization · Information security

1 Introduction

Exact learning involves asking a series of queries so as to learn a configuration or concept uniquely and without errors, e.g., see [12]. For example, imagine a game where a player, Alice, is trying to exactly learn a secret string, S, such as $S = $ "rumpelstiltskin", which is known only to a magic fairy. Alice may ask the fairy questions about S, but only if they are in a form allowed by the fairy, such as "Is X a substring of S?". Any allowable question that Alice asks must be answered truthfully by the fairy. Alice's goal is to learn S by asking the fewest number of allowable questions. Her strategy is ***adaptive*** if her questions can depend on the answers to previous queries. This exact-learning string-reconstruction problem might at first seem like a contrived game, but it actually has a number of applications. For instance, in interactive DNA sequencing, the fairy's string is an unknown DNA sequence, S, and allowable queries are "Is X a substring of S?" Each such question can be answered by a hybridization experiment that exposes copies of S to a mixture containing specific primers to see

---

The full version of this paper is available in [5].

© Springer Nature Switzerland AG 2020
C. Boucher and S. V. Thankachan (Eds.): SPIRE 2020, LNCS 12303, pp. 155–174, 2020.
https://doi.org/10.1007/978-3-030-59212-7_12

which ones bind to S, e.g., see [73]. Thus, we are interested in the exact-learning complexity of adaptively learning an unknown string via queries of various given types, that is, for exactly reconstructing a string from queries. Formally, we are interested in minimizing a *query-complexity* measure, $Q(n)$, which, in our case, is the number of queries of certain types needed in order to exactly learn a string, S. This query-complexity concept comes from machine-learning and complexity theory, e.g., see [3,12,18,25,32,76,83].

1.1   Related Work

Motivated by DNA sequencing, Skiena and Sundaram [73] were the first to study exact string reconstruction from adaptive queries. For *substring queries*, of the form "Is X a substring of S?", they give a bound for $Q(n)$ of $(\sigma - 1)n + 2\log n + O(\sigma)$, where σ is the alphabet size. For *subsequence queries*, of the form "Is X a subsequence of S?", they prove a bound for $Q(n)$ of $\Theta(n\log\sigma + \sigma\log n)$. Recently, Iwama *et al.* [44] study the problem for binary alphabets, which removes the additive logarithmic term in this case. These papers do not consider "mixed-up" strings, however, such as strings that are periodic or periodic with errors. The abundance of repetitions and periodic runs in genomic sequences is well known and has been exploited in the last decades for biologic and medical information (see e.g. [15,16,30,33,35,53,65,66,74,82]). It is somewhat surprising that this phenomenon has not been used to achieve more efficient algorithms. Margaritis and Skiena [60] study a parallel version of exact string reconstruction from queries, which are hybrids of adaptive and non-adaptive strategies, showing, e.g., that a length-n string can be reconstructed in $O(\log^2 n)$ rounds using n substring queries per round. Tsur [77] gives a polynomial approximation algorithm for the 1-round case. As in [73], these papers do not consider bounds for $Q(n)$ based on properties of the string such as its periodicity. Cleve *et al.* [28] study string reconstruction in a quantum-computing model, showing, for example, that a sublinear number of queries are sufficient for a binary alphabet. This result does not seem to carry over to a classical computing model, however, which is the subject of our paper.

Another type of query we consider is the *jumbled (or histogram)-index* query, first considered in [20,21,26,37] and studied more recently in, e.g. [4,7,9,10,52,62]. Jumbled indexing has many applications. It can be used as a tool for de novo peptide identification (as in e.g. [45,50,51]), and has been used as a filter for searching an image database [27,31,75,81,85]. In this query, which has received much study of late, but has not been studied before for adaptive string reconstruction, one is given a Parikh vector, i.e., a vector of frequency counts for each character in an alphabet, and asked if there is a substring of the reference string, S, having these frequency counts and, if so, where it occurs in S. Such reconstruction may aid in narrowing down peptide identification, or focusing on image retrieval.

Another model for string reconstruction, tangential to ours and studied extensively, is the one defined by a non-adaptive oracle, e.g., see [1,2,13,14,19–22,24,26,29,34,36–38,40–43,47–49,54,56,58,59,63,64,67–72,78,79,84]. In this

model we are given a set of answers to queries in advance, and we aim to understand sufficient and necessary conditions on the answers that enable the exact
reconstruction of the string. This model differs from the adaptive one considered in this paper in that it focuses on the study of combinatorial properties
of strings, rather than on minimizing the number of queries. We review existing
literature for non-adaptive string reconstruction in more detail in the full version
of the paper [5].

1.2   Our Results

We provide new and improved results for exactly reconstructing strings from
adaptive substring, subsequence, and jumbled-index queries. For example, we
believe we are the first to characterize query complexities for exactly reconstructing periodic strings from adaptive queries, including the following results
for reconstructing a length-n periodic (i.e., "mixed-up") string, $S = p^k p'$, of
smallest period p, where p' is a prefix of p and the alphabet has size σ:

- It requires at least $|p| \lg \sigma$ substring or subsequence queries.
- It can be done with $\sigma |p| + \lceil \lg |p| \rceil$ substring queries, if n is known.
- It can be done with $O(\sigma |p| + \lg n)$ substring queries, if n is unknown.
- It can be done with $\sigma \lceil \lg n \rceil + 2|p| \lceil \lg \sigma \rceil$ subsequence queries, for known n.
- It can be done with $2\sigma \lceil \lg n \rceil + 2|p| \lceil \lg \sigma \rceil$ subsequence queries, if n is unknown.

Perhaps our most technical result is that we show that we can reconstruct
a length-n string, S, within Hamming distance d of a periodic string $S' = p^k p'$,
of smallest period p, using $O(\min(\sigma n, d\sigma |p| + d|p| \lg \frac{n}{d+1}))$ substring queries, if
n is unknown. We also show that we can exactly reconstruct a general length-n
string, S, using $2\sigma \lceil \lg n \rceil + n \lceil \lg \sigma \rceil$ subsequence queries, if n is unknown. Such
queries are another "mixed-up" setting, since there can be multiple subsequence
matches for a given string. Our bound improves the previous best, decades-old
result, by Skiena and Sundaram [73], who prove a query complexity of $2\sigma \lg n +$
$1.59n \lg \sigma + 5\sigma$ for this case. If n is known, then $\sigma \lceil \lg n \rceil + n \lceil \lg \sigma \rceil$ subsequence
queries suffice. We believe we are the first to study string reconstruction using
jumbled-index queries, which are yet another "mixed-up" setting, since they
simply count the frequency of each character occurring in a substring. We prove
the following results:

- We can reconstruct a length-n string with $O(\sigma n)$ yes/no extended jumbled-
  index queries, which include a count for an end-of-string character, $\$$.
- For jumbled-index queries that return an index of a matching substring, string
  reconstruction is not possible if this index is chosen adversarially, but is possible using $O(\sigma + n \lg n)$ queries if it is chosen uniformly at random.

1.3   Preliminaries

We consider strings over the alphabet $\Sigma = \{a_1, a_2, \ldots, a_\sigma\}$ of σ letters. The size
of a string X is denoted by $|X|$. We use $X[i]$ to denote the ith letter of X and

$X[i..j]$ to refer to the substring of X starting at its ith and ending at its jth letter (e.g., $X = X[1..|X|]$). We may ignore i when expressing a prefix $X[..j]$ of X. Similarly, $X[i..]$ is a suffix of X. Occasionally, we will express concatenation of strings X and Y by $X \cdot Y$ (instead of XY) to emphasize some property of the string. A string X concatenated with itself k (resp. infinitely many) times can be expressed as X^k (resp. X^∞). The reversal of a string X is denoted by X^R.

A string, S, has **period** p if $S = p^k p'$, such that $k > 0$ is an integer and p' is a (possibly empty) prefix of p. Further, a string S is **periodic** if it has a period that repeats at least twice, i.e. $S = p^k p'$ and $k > 1^1$. The following is a well known result concerning the periodicity of a string, due to Fine and Wilf [39], which we will need later on.

Lemma 1 (Periodicity Lemma [39]). *If p, q are periods of a string X of length $|X| \geq |p| + |q| - \gcd(|p|, |q|)$, then X also has a period of size $\gcd(|p|, |q|)$.*

A **doubling search** is the operation used to determine a number n from a (typically unbounded) range of possibilities. It involves doubling a query value, m, until it is greater than n, followed by a binary search to determine n itself. Its time complexity is $2\lfloor \lg n \rfloor + 1^2$.

Due to space constraints, we defer proofs of Lemmas and Theorems marked with ⊛ to the full version of the paper [5], where we also include pseudo-code for our algorithms.

2 Substring Queries

In this section, we study query complexities for a string, S, subject to yes/no **substring** queries, IsSubstr, i.e. queries of "Is X a substring of S?". We focus on the cases where S corresponds to an originally periodic string, that may have lost its periodicity property due to error corruption. The nature of the errors is context-dependent. For example, corruption may be caused by transmission errors or measurement errors.

There are multiple ways to model errors in strings (see [8, 11, 23, 46, 55, 57, 80]). In this paper, we consider Hamming distance. We say that S is a d-**corrupted periodic string** if there exists a periodic string S' of period p, such that $|S| = |S'|$ and $\delta(S', S) \leq d$, where δ is the Hamming distance. We refer to p as an **approximate period** of S. Notice that, depending on d, there might exist multiple possible strings S' that originate S.

Our main result in this section is the following.

---

[1] Our algorithms assume that S is periodic ($k > 1$), while the Periodicity Lemma (1) only requires a string to have a period ($k > 0$).

[2] A more sophisticated version of this procedure exists (see [17]) that actually improves the constant in the time complexity, but for simplicity, we use the traditional algorithm, which is asymptotically equivalent.

Theorem 1. *We can reconstruct a length-n d-corrupted periodic string S using*

$$O\left(\min\left(\sigma n, d\sigma|p| + d|p|\lg\frac{n}{d+1}\right)\right) \quad queries,$$

for known d, unknown $|p|$, regardless of whether we know n, where p is a smallest approximate period of S.

The algorithm of Theorem 1 is a more elaborate version of a reconstruction algorithm for the special case of $d = 0$, i.e. when no errors occurred and $S = S'$, and when n is not known in advance.

Theorem 2. *We can reconstruct a length-n periodic string, $S = p^k p'$, of smallest period p, using $O(\sigma|p| + \lg n)$ substring queries, assuming both n and $|p|$ are unknown in advance.*

The algorithm of Theorem 2, in turn, builds from a simple reconstruction algorithm that handles the case where n is known in advance and $d = 0$.

For clarity, we will present our results in increasing order of complexity, from the least general result of $d = 0$ and known n, to the most general result of arbitrary d and unknown n.

2.1 Uncorrupted Periodic Strings of Known Size

We first give a simple algorithm to reconstruct a periodic string $S = p^k p'$ of smallest period p and known size with query complexity $O(\sigma|p|)$, and then show how to improve this algorithm to have query complexity $\sigma|p|$ plus lower-order terms. Our algorithms use a primitive developed by Skiena and Sundaram [73], which we call "**append** (resp., **prepend**) a letter." In the append (resp., prepend) primitive, we start with a known substring q of S, and we ask queries IsSubstr(qa_i) (resp., IsSubstr(a_iq)), for each $a_i \in \Sigma$. Note that if we know that one of the qa_i (resp., a_iq) strings must be a substring, we can save one query, so that appending or prepending a letter uses at most $\sigma - 1$ queries in this case.

In our simple algorithm[3], we iteratively grow a candidate period, q, using the append primitive until $q^{g(q)-1}$ is a substring, where $g(x) = \lfloor n/|x| \rfloor$. Notice that q may be an "unlucky" cyclic rotation of p, which only repeats $g(p) - 1$ times, and we need to account for this possibility. Thus, once we get a substring corresponding to $q^{g(q)-1}$, we then append/prepend letters until we recover all of S.

Theorem 3. ⊛ *We can reconstruct a length-n periodic string $S = p^k p'$, of smallest period p, using $O(\sigma|p|)$ substring queries, assuming n is known in advance and $|p|$ is unknown.*

---

[3] Pseudo-code can be found in the full version of the paper [5], where the number of queries is also shown for each step involving queries.

With a little more effort, we can improve the constant factor in the query complexity, by showing that, for $k = \lfloor n/|p| \rfloor > 3$, the following implication holds: if $q^{g(q)-1}$ is a substring, then q must be a cyclic rotation of p.

Theorem 4. ⊛ *We can reconstruct a length-n periodic string $S = p^k p'$, of smallest period p, using at most $\sigma|p| + \lceil \lg |p| \rceil$ substring queries, assuming that: n is known in advance, $k > 3$ and $|p|$ is unknown.*

Notice that any reconstruction algorithm requires at least $|p| \lg \sigma$ queries.

Theorem 5. *Reconstructing a length-n string, $S = p^k p'$, of smallest period p, requires at least $|p| \lg \sigma$ IsSubstr queries, even if n and $|p|$ are known.*

Proof. There are $\sigma^{|p|}$ possible periods for S. Since each period corresponds to a different output of a reconstruction algorithm, A, and each query is binary, we can model any such algorithm, A, as a binary decision tree, where each internal node corresponds to an IsSubstr query. Each of the $\sigma^{|p|}$ possible periods must correspond to at least one leaf of A; hence, the minimum height of A is $\lg(\sigma^{|p|})$. □

2.2 Uncorrupted Periodic Strings of Unknown Size

As in Sect. 2.1, we iteratively grow a candidate period q and attempt to recover S by concatenating q with itself in the appropriate way. The difficulty when n is unknown is that we can no longer confidently predict $g(q)$. Thus, we can no longer issue a single query to test if q is the right period. An immediate solution is to use a doubling search. Unfortunately, this introduces a multiplicative $O(\lg n)$ term into the query complexity. To avoid it, we show how we can take advantage of the Periodicity Lemma (1) to amortize the extra work needed to recover S.

Let us describe the algorithm[4]. We start with an empty candidate period q. At each iteration, we add a letter to q, using the append primitive and, using a doubling search, determine the ***run-length*** t of q, i.e. the maximum integer t such that q^t is a substring of S. If $t = 1$, we advance to the next iteration and repeat this process. If, on the other hand, $t > 1$, we use q to determine the largest substring T that has a period of size $|q|$. This can be done efficiently, using doubling searches, by determining the largest suffix l of q and the largest prefix r of q, such that $\mathsf{IsSubstr}(l \cdot q^t \cdot r)$. Once T is determined, we check whether it corresponds to S by checking if there is any letter preceding and succeeding T. If T corresponds to S, we output it. Otherwise, we update q to be any largest substring of T whose size is assuredly less than $|p|$: using Periodicity Lemma (1), we argue in Lemma 2 below that, if q is not a cyclic rotation of p, then p must be as large as *almost* the entire substring T; more specifically, it must be the case that $|p| > |T| - |q| + 1$. Thus, we update q to be a length-$(|T| - |q| + 1)$ prefix of T (any other substring of T would also work). We use this fact to get

---

[4] Pseudo-code can be found in the full version of the paper [5], where the number of queries is also shown for each step involving queries.

a faster convergence to a cyclic rotation of p, while making sure that we do not overshoot $|p|$. Indeed, this observation will enable us to incur a $O(\lg n)$ additive factor, instead of a multiplicative one. After updating q, we advance to the next iteration, where a new letter is appended to q, and repeat this process until $T = S$.

Lemma 2. *Let T be the largest proper substring of $S = p^k p'$, of smallest period p, such that: $|q|$ is the length of the smallest period of T. Then, $|p| > |T| - |q| + 1$.*

Proof. Let us assume, by contradiction, that $|p| \leq |T| - |q| + 1$. Then, $|T| \geq |q| + |p| - 1$ and, thus, $|T| \geq |q| + |p| - \gcd(|q|, |p|)$. In addition, if p is a period of S, then T must have a period of size $|p|$. So, by the Periodicity Lemma (1), T also has a period of size $\gcd(|q|, |p|)$. Moreover, since T is the largest proper substring of S, $|p|$ is not a multiple of $|q|$. Therefore, T must have a period shorter than $|q|$, a contradiction. □

Let $q_1, q_2, \ldots, q_m$ be the sequence of m candidate periods of increasing length, each of which is the result of the append/prepend primitive at the beginning of every iteration, e.g. $|q_1| = 1$. In addition, let us use t_i to denote the run-length of q_i. Correctness of our algorithm follows from the following two lemmas.

Lemma 3. *The algorithm successfully returns $S = p^k p'$, of smallest period p, if there exists an iteration $i \in \{1, 2, \ldots, m\}$, such that q_i is a cyclic rotation of p.*

Proof. If $t_i > 1$, then it is easy to see that the string T computed at iteration i, must correspond to S. If $f_i = 1$, then the algorithm essentially switches to the letter-by-letter algorithm, appending or prepending letters until the end, when $q_m = S$. □

Lemma 4. *There exists an iteration $i \in \{1, 2, \ldots, m\}$, such that q_i is a cyclic rotation of p.*

Proof. Let us assume that there is no such iteration i. Then, since all the q_i's are increasing in length, it must be the case that there exists an iteration $j \in \{1, 2, \ldots, m-1\}$, such that: $|q_j| < |p|$, but $|q_{j+1}| > |p|$. However, it follows from Lemma 2 (when $f_t > 1$) and the fact that we add a single letter to q_j (when $f_t = 1$) that p must be at least as large as q_{j+1}, a contradiction. □

The following lemma shows that we can charge the logarithmic factors, incurred in each iteration j, to the work that would have been required to find the letters introduced in q_{j+1}. This establishes the amortization in query complexity.

Lemma 5. ✹ *The number of queries performed in the jth iteration is at most $\sigma(|q_{j+1}| - |q_j|) + O(\sigma)$, for $j < m$, or $O(\sigma + \lg n)$, for $j = m$.*

Theorem 2 follows from Lemmas 3 to 5. A detailed proof can be found in the full version of the paper [5].

2.3   Corrupted Periodic Strings

Let us assume throughout the remainder of this section that S is a d-corrupted periodic string of approximate period p. Again, the main idea of the algorithm described in this section consists of: (1) determining a cyclic rotation of a true period (in this case, there might be multiple true periods), by iteratively growing a candidate period q, and (2) using q to recover S accordingly. However, in the presence of errors, each of these steps becomes more difficult to realize efficiently. For example, in the first step, we might be growing a candidate period q that includes an error. So, in order to rightfully reject the hypothesis that q is at most as large as some approximate period p, our algorithm should be able to tell the difference between (i) $|p| = |q|$ and q includes an error and (ii) $|p| > |q|$. Otherwise, the algorithm will keep on growing q until it is equal to S, possibly incurring σn queries. In addition, the second step of using q to determine S requires more work, since the presence of errors discards the possibility of simply concatenating q with itself the required number of times. Because of these issues, it is crucial that our algorithm understands when a candidate period is or not free of errors. Thus, the algorithm relies on the following.

Lemma 6. *Let A be any length-$(2d + 1)|p|$ substring of a d-corrupted periodic string S of approximate period p, corresponding to the concatenation of length-$|p|$ substrings $q_1, q_2, \ldots, q_{2d+1}$. Then, a cyclic rotation of p must be the only substring q_j appearing at least $d + 1$ times in $q_1, q_2, \ldots, q_{2d+1}$.*

Proof. Clearly, there is some q_i that is a cyclic rotation of p. Moreover, there is some q_j that appears at least $d + 1$ times in $q_1, q_2, \ldots, q_{2d+1}$, or the number of errors would exceed d, by the pigeonhole principle. If $i \neq j$, then each occurrence of q_j, contributes at least 1 error, resulting in at least $d+1$ errors, a contradiction. Finally, q_j must be the only string with $d + 1$ appearances in $q_1, q_2, \ldots, q_{2d+1}$, by the pigeonhole principle.   □

Let us give the details for our algorithm[5], which is able to recover S, even when its size n is unknown. We maintain an initially empty substring, A, of S, by extending it with $2d + 1$ letters in each iteration, using the append and prepend primitives (as described in Sect. 2.1), potentially incurring an extra σ queries for detecting a left or right endpoint of S. In the case that $n = |S| < |p|(2d+1)$, the last iteration requires only $\min(2d + 1, |S| - |A|)$ new letters. Thus, after adding letters to A in the ith iteration, A is a substring of S of size at most $i(2d + 1)$. Before advancing to the next iteration, we determine the only possible length-i candidate period q that could have originated A with at most d errors (by Lemma 6). At this point we do not know if some approximate period p has size $|p| = i$, so we try to use q to recover the rest of the string, halting whenever the total number of errors exceeds d, in which case we advance to the next iteration and repeat this process for a new candidate period of size $i + 1$. This logic is in

---

[5] Pseudo-code can be found in the full version of the paper [5], where the number of queries is also shown for each step involving queries.

the subroutine Expand(q), described next(See footnote 5). It initializes a string T to q and expands it by doing the following at each iteration:

1. Appending to T the largest periodic substring of period $\overrightarrow{q}$, where $\overrightarrow{q}$ is the appropriate cyclic rotation of q that aligns with the right-endpoint of T. This can be done efficiently by determining the maximum value of x, using a doubling search, for which

$$\mathsf{IsSubstr}(T \cdot (\overrightarrow{q}^{\infty}[..\ x])),$$

incurring $2\lfloor \lg x \rfloor + 1$ queries. The cyclic rotation $\overrightarrow{q}$ can be determined with no additional queries, by maintaining the value x', which is the value of x in the previous iteration, i.e. $\overrightarrow{q}$ is the cyclic rotation of q starting at the index $(x' \mod |q| + 2)$ of q.

2. Prepending to T the largest periodic substring of period $\overleftarrow{q}$, where $\overleftarrow{q}$ is the appropriate cyclic rotation of q that aligns with the left-endpoint of T. This can be done efficiently by determining the maximum value of y, using a doubling search, for which

$$\mathsf{IsSubstr}(((\overleftarrow{q}^{R})^{\infty}[..\ y])^{R} \cdot T),$$

incurring $2\lfloor \lg y \rfloor + 1$ queries. The cyclic rotation $\overleftarrow{q}$ can be determined with no additional queries in a similar fashion to $\overrightarrow{q}$.

3. Determining, if they exist, the letters immediately to the left and to the right of T, using 2σ queries, and adding them to T.

The expansion process in Expand(q) halts when either the total number of errors with respect to q, $\delta(T, q^{\infty}[..|T|])$, exceeds d (in which case we advance to the next iteration), or when $T = S$ (in which case we return T).

Remark 1. Expand(q) successfully returns S if and only if q is a cyclic rotation of some approximate period.

Lemma 7. *The number of queries performed during any call to* Expand *is* $O(d\sigma + d\lg \frac{n}{d+1})$.

Proof. Each call to Expand uses at most $2(d+1)\sigma$ queries to determine the corrupted letters, as well as the left/right endpoints of S – the total number of iterations of the while loop in Expand is $d+1$, since every iteration except the last introduces at least 2 errors in T, and each iteration incurs 2σ queries.

In addition, the number of queries used by Expand(q) during the doubling searches is

$$\sum_{j=1}^{|q|} \left(2\lfloor \lg x_j \rfloor + 2\lfloor \lg y_j \rfloor + 2\right),$$

where x_j and y_j denote, respectively, the lengths of the substrings determined via doubling searches in steps 1 and 2, during the jth call to Expand. Since the total number of iterations is $d + 1$, there is at most $d + 2$ such x_j's and y_j's. Moreover, the above summation is maximized when all the x_j's and y_j's have the same average value of at most $(n - d)/(d + 1)$. This follows from Jensen's inequality and concavity of log. Thus, the overall time complexity is

$$O\left(d\sigma + d\lg\frac{n}{d+1}\right).$$

□

Correctness and query complexity of our algorithm follows from Remark 1 and Lemmas 6 and 7, giving us:

Theorem 6. ⊛ *We can reconstruct a length-n d-corrupted periodic string S using $O(d\sigma|p| + d|p|\lg\frac{n}{d+1})$ queries, for known d, unknown $|p|$, regardless of whether we know n, where p is a smallest approximate period of S.*

If n is known, we could save the queries used to check the left and right endpoints of S, but this does not alter the query complexity asymptotically.

We assume a small enough number of errors, following [6]. In particular, if $d = O(k/(1 + \lg n))$, our algorithm is an improvement to the $O(\sigma n)$ letter-by-letter algorithm of Skiena and Sundaram [73] for general strings, where $k = \lfloor n/|p| \rfloor$. Thus, our algorithm performs better if there is, on average, at most 1 error in every other $O(1 + \lg n)^{\text{th}}$ non-overlapping occurrence of p. If the number of errors is not small enough, then one should run the letter-by-letter algorithm intercalated with ours, to get an upper bound of $O(\sigma n)$ queries, giving us Theorem 1, introduced at the beginning of this section.

3   Subsequence Queries

We study the query complexity for a length-n string, S, subject to yes/no **subsequence** queries, IsSubseq, i.e., queries of the form "Is X a subsequence of S?" We begin with a simple lower bound.

Theorem 7. ⊛ *Reconstructing a length-n periodic string, $S = p^k p'$, of smallest period p, requires at least $|p|\lg\sigma$ IsSubseq queries, even if n and $|p|$ are known.*

Let us next describe an algorithm for reconstructing a periodic length-n periodic string, $S = p^k p'$, of smallest period p. We begin by performing either binary searches (if n is known) or doubling search (if n is unknown), using queries of the form IsSubseq(a^i) to determine the number of a's in S, for each $a \in \Sigma$. From all of these queries, we can determine the value of n if it was previously unknown. This part of our algorithm requires either $\sigma\lceil\lg n\rceil$ or $2\sigma\lceil\lg n\rceil$ queries in total, depending on whether we knew n at the outset.

If the number of a's in S is n, for any $a \in \Sigma$, then we are done, so let us assume the number of a's in S is less than n, for each $a \in \Sigma$. Thus, when we complete all our doubling/binary searches, for each letter, $a \in \Sigma$ that occurs a nonzero number of times in S, we have a maximal subsequence, S_a, of S, consisting of a's. Moreover, since S is periodic with a period that repeats k times, each S_a is periodic with a period that repeats k times. Unfortunately, at this point in the algorithm, we may not be able to determine k. So next we create a binary merge tree, T, with each of its leaves associated with a nonempty subsequence, S_a, much in the style of the well-known merge-sort algorithm, so that T has height $\lceil \lg \sigma \rceil$. We then perform a bottom-up merge-like procedure in T using IsSubseq queries, as follows.

Let v be an internal node in T, with children x and y for which we have inductively determined periodic subsequences, S_x and S_y, respectively, of S. Let $n_x = |S_x|$ and $n_y = |S_y|$. To create the subsequence, S_v, for v, we need to perform a merge procedure to interleave S_x and S_y. To do this, we maintain indices i and j in S_x and S_y, respectively, such that we have already determined an interleaving, $S_v[..i + j]$, of $S_x[..i]$ and $S_y[..j]$. Initially, $i = j = 0$. We then perform the query IsSubseq($S_v[..i+j] \cdot S_x[i+1] \cdot S_y[j+1..n_y]$). Suppose the answer to this query is "yes". In this case, we set $S_v[..i+j+1] = S_v[..i+j] \cdot S_x[i+1]$ and we increment i. If, on the other hand, the answer to the above query is "no", then we set $S_v[..i + j + 1] = S_v[..i + j] \cdot S_y[j + 1]$, because in this case we know that IsSubseq($S_v[..i+j] \cdot S_y[j+1] \cdot S_x[i+1..n_x]$) would return "yes". If this latter condition occurs, then we increment j.

Let q_v denote this new interleaving prefix, $S_v[..i + j]$, and let $\hat{k} = \lfloor n/|q_v| \rfloor$. If $q_v{}^{\hat{k}} q_v'$ is a plausible interleaving of S_x and S_y, where q_v' is a prefix of q_v, then we next ask the query IsSubseq($q_v{}^{\hat{k}} q_v'$). If the answer is "yes", then we set $S_v = q_v{}^{\hat{k}} q_v'$ and this completes the merge. Otherwise, we continue incrementally interleaving S_x and S_y, using the current values of i and j, by iterating the procedure described above. Clearly, this merge procedure asks at most $2|q_v|$ queries in total.

Theorem 8. ⊛ *We can determine a length-n periodic string, $S = p^k p'$, of smallest period p of unknown size, using $2\sigma \lceil \lg n \rceil + 2|p| \lceil \lg \sigma \rceil$ IsSubseq queries, if n is unknown. If n is known, then $\sigma \lceil \lg n \rceil + 2|p| \lceil \lg \sigma \rceil$ IsSubseq queries suffice.*

A simple modification of our algorithm also implies the following.

Theorem 9. ⊛ *We can determine a length-n string, S, using $2\sigma \lceil \lg n \rceil + n \lceil \lg \sigma \rceil$ IsSubseq queries, without knowing the value of n in advance. If n is known, then $\sigma \lceil \lg n \rceil + n \lceil \lg \sigma \rceil$ IsSubseq queries suffice.*

This latter theorem improves a result of Skiena and Sundaram [73], who prove a query bound of $2\sigma \lg n + 1.59 n \lg \sigma + 5\sigma$ when n is unknown.

4   Jumbled-Index Queries

Jumbled-indexing involves preprocessing a given string, S, so as to determine whether there exists a substring of S whose letter frequencies match the given **Parikh vector**, i.e., a vector $\psi = (f_1, \ldots, f_\sigma)$ such that f_i is the number of occurrences in S of $a_i \in \Sigma$, e.g., see [4,7,9,10,52,62]. In this section, we study the query complexity for reconstructing an unknown length-n string, S, using jumbled-index queries. As observed by Acharya *et al.* [1,2], strings and their reversals have the same "composition multiset". This immediately implies the following negative result.

Lemma 8. ⊛ *If S is not a palindrome, then S cannot be reconstructed by yes/no jumbled-index queries, which return whether there is a substring in S with a given Parikh vector.*

Given that simple yes/no jumbled-index queries are not sufficient for string reconstruction, let us consider an extended type of yes/no jumbled-index query.

- *Jumbled-Indexing with End-of-string symbol "$\$$"* (JIE): given an **extended** Parikh vector, $\psi = (f_1, \ldots, f_\sigma, f_\$)$, for the letters in Σ and an end-of-string symbol, $\$$, which is not in Σ, this query returns a yes/no response as to whether there is a substring of $S\$$ with extended Parikh vector ψ.

Unlike the yes/no jumbled-index queries, this variant enables full reconstruction.

Theorem 10. *We can reconstruct a length-n string, S, using $(\sigma - 1)n$ JIE queries, if n is known, or $\sigma(n + 1)$ JIE queries, if n is unknown.*

Proof. Our method is to use a letter-by-letter reconstruction algorithm via an adaption of the prepend-a-letter primitive for substring queries. Suppose n is unknown. Let ψ be an extended Parikh vector for a known suffix, s, of $S\$$; initially, $\psi = (0, 0, \ldots, 0, 1)$ and $s = \$$. Then we perform a jumbled-index query for ψ_i, for each $a_i \in \Sigma$, where $\psi_i = \psi$ except that ψ_i adds 1 to the f_i value in ψ. If one of these, say, ψ_i, returns "yes", then we prepend a_i to our known suffix and we repeat this procedure using ψ_i for ψ. If all of these queries return "no", then we are done. If n is known, on the other hand, then we can skip this last test of all-no responses and we can also save at least one query with each iteration, with the algorithm otherwise being the same. □

We can also consider jumbled-index queries that return an index of a matching substring for a given Parikh vector, if such a substring exists. Though related, notice that this type of query is not subsumed by the query studied in Acharya *et al.* [1,2], which returns the number of occurrences (instead of position) of matching substrings in S. There is some ambiguity, however, if there is more than one matching substring; hence, we should consider how to handle such multiple matches. For example, if a jumbled-index query returns the indices of all matching substrings, then σ queries are clearly sufficient to reconstruct any length-n string, for any n, without knowing the value of n in advance. Thus, let us consider two more-interesting types of jumbled-index queries.

- *Adversarial Jumbled-Indexing* (AJI): given a Parikh vector, $\psi = (f_1, \ldots, f_\sigma)$, this query returns, in an adversarial manner, one of the starting indices of a matching substring, if such a string exists. If there is no matching substring, this query returns False.

- **Random Jumbled-Indexing** (RJI): given a Parikh vector, $\psi = (f_1, \ldots, f_\sigma)$, this query returns, uniformly at random, one of the indices of a substring with Parikh vector ψ if such a substring exists in S. If there is no such substring, this query returns False.

Unfortunately, for the AJI variant, there are some strings that cannot be fully reconstructed, but this is admittedly not obvious. In fact, the unreconstructability characterization of [1,2] fails for AJI queries, because the symmetry property used in their construction of pairwise "equicomposable" strings inherently yields matching substrings with symmetric (e.g. different) positions in S.

Nevertheless, we give a construction of an infinite family of pairwise undistinguishable strings, i.e. two strings such that, for every possible query, there exists an answer (positive or negative) that is common to both strings. Clearly, the adversarial strategy is to output these common answers when given either of these strings. In particular, for all $b \geq 1$, consider the two binary strings of length $4b + 14$ given below, which differ only in the middle section, consisting of 01 in the first string and 10 in the second:

$$S_1 = 101101(10)^b 01(10)^b 010010$$
$$S_2 = 101101(10)^b 10(10)^b 010010$$

Theorem 11. ⊛ *The strings S_1 and S_2 cannot be distinguished using AJI queries, for $b \geq 1$.*

In contrast, the query variant RJI can be used to reconstruct any length-n string, S, without knowing the value of n in advance. In particular, it is possible to reconstruct any length-n string, S, using $O(\sigma + n \log n)$ RJI queries with high probability. Our algorithm for doing this involves a reduction to a multi-window coupon-collector problem.

Let ψ_i be a Parikh vector that is all 0's except for a count of 1 for the letter $a_i \in \Sigma$. Note that an RJI query using ψ_i will return one of the n_i locations in S with an a_i uniformly at random (if $n_i > 0$). If $n_i = 0$, for any $i = 1, 2, \ldots, \sigma$, we learn this fact immediately after one RJI query for ψ_i, so let us assume, w.l.o.g., that $n_i > 0$, for all $i = 1, 2, \ldots, \sigma$, after performing an initial σ number of RJI queries.

Recall that in the **coupon-collector** problem, a collector visits a coupon window each day and requests a coupon from an agent, who chooses one of n coupons uniformly at random and gives it to the collector, e.g., see [61]. The expected number of days required for the collector to get al.l n coupons is nH_n, where H_n is the nth Harmonic number. But this assumes the collector knows when they have received all n coupons (i.e., the collector knows the value of n).

In a coupon-collector formulation of our reconstruction problem, we instead have σ coupon windows, one for each letter $a_i \in \Sigma$, where each window i has n_i coupons that differ from the coupons for the other windows, and we do not know the value of any n_i. Each day the collector must choose one of the coupon windows, i, and request one of its coupons (corresponding to an RJI query for ψ_i), which is chosen uniformly at random from the n_i coupons for window i. We are interested in a strategy and analysis for the collector to collect all $n = n_1 + n_2 + \cdots + n_\sigma$ coupons, with high probability (i.e., with probability at least $1 - 1/n$).

Note that although we do not know the value of any n_i, we can nonetheless test whether the collector has collected all n coupons. In particular, suppose we have received RJI responses for all indices, $1, 2, \ldots, n$, for letters in S, and let n_i be the number of a_i's we have found so far. Let $\psi' = (n_1, n_2, \ldots, n_\sigma)$, and let ψ'_i be equal to ψ' except that we increment n_i by 1. If an RJI query for each ψ'_i returns False, then we know we have fully reconstructed S. Thus, if $n = 1$, then we can determine this and S after 2σ RJI queries, so let us assume that $n \geq 2$. Further, we can assume we have a bound, $N \geq 2$, which is at least n and at most twice n, by a simple doubling strategy, where we double N any time a test for n fails and we set N equal to any RJI query response that is larger than N. Therefore, the remaining problem is to solve the multi-window coupon-collector problem.

Our strategy for the multi-window coupon-collector problem is simply to visit the coupon windows in phases, so that in phase i we repeatedly visit window i until we are confident we have all of its n_i coupons, for which the following lemma will prove useful.

Lemma 9. ⊛ *Let T_i be the number of trips to window i needed to collect all its $n_i \geq 1$ coupons. Then, for any real number β:*

$$\Pr\left(T_i > \beta n_i \ln N\right) \leq \frac{n_i}{N^\beta}.$$

Our strategy, then, is to let $\beta \geq 2$ be constant, and in phase i, implement a doubling strategy where we perform $\beta N_i \log N$ RJI queries for ψ_i, such that N_i is an upper bound estimate for n_i, which we double each time we get more than N_i distinct responses to our queries in this phase. So by the end of the phase i, $n_i \leq N_i \leq 2n_i$. This gives us:

Theorem 12. ⊛ *A string, S, of unknown size, n, can be reconstructed using $O(\sigma + n \log n)$ RJI queries, with high probability.*

5   Conclusion and Open Questions

We have studied the reconstruction of strings under the following settings, by giving efficient reconstruction algorithms and proving lower bounds: (i) periodic strings of known and unknown sizes, with and without mismatch errors, using substring queries; (ii) periodic strings of known and unknown sizes, using subsequence queries and (iii) general strings, using variations of jumbled-indexing queries. For the non-optimal algorithms given here, it would be nice to know whether there exist matching lower bounds, or whether there exist faster algorithms. We mention additional possible future work in the full version of the paper [5].

Acknowledgments. This research was funded in part by the U.S. National Science Foundation under grant 1815073. Amihood Amir was partly supported by BSF grant 2018141 and ISF grant 1475-18.

References

1. Acharya, J., Das, H., Milenkovic, O., Orlitsky, A., Pan, S.: Quadratic-backtracking algorithm for string reconstruction from substring compositions. In: 2014 IEEE International Symposium on Information Theory, Honolulu, HI, USA, 29 June–4 July 2014, pp. 1296–1300. IEEE (2014). https://doi.org/10.1109/ISIT.2014.6875042
2. Acharya, J., Das, H., Milenkovic, O., Orlitsky, A., Pan, S.: String reconstruction from substring compositions. SIAM J. Discrete Math. **29**(3), 1340–1371 (2015). https://doi.org/10.1137/140962486
3. Afshani, P., Agrawal, M., Doerr, B., Doerr, C., Larsen, K.G., Mehlhorn, K.: The query complexity of finding a hidden permutation. In: Brodnik, A., López-Ortiz, A., Raman, V., Viola, A. (eds.) Space-Efficient Data Structures, Streams, and Algorithms. LNCS, vol. 8066, pp. 1–11. Springer, Heidelberg (2013). https://doi.org/10.1007/978-3-642-40273-9_1
4. Afshani, P., van Duijn, I., Killmann, R., Nielsen, J.S.: A lower bound for jumbled indexing. In: 2020 ACM-SIAM Symposium on Discrete Algorithms (SODA), pp. 592–606 (2020). https://doi.org/10.1137/1.9781611975994.36
5. Afshar, R., Amir, A., Goodrich, M.T., Matias, P.: Adaptive exact learning in a mixed-up world: dealing with periodicity errors, and jumbled-index queries in string reconstruction. arXiv preprint arXiv:2007.08787 (2029). https://arxiv.org/abs/2007.08787
6. Amir, A., Eisenberg, E., Levy, A., Porat, E., Shapira, N.: Cycle detection and correction. ACM Trans. Alg. **9**(1) (2012). Article no. 13
7. Amir, A., Apostolico, A., Hirst, T., Landau, G.M., Lewenstein, N., Rozenberg, L.: Algorithms for jumbled indexing, jumbled border and jumbled square on run-length encoded strings. Theor. Comput. Sci. **656**, 146–159 (2016). https://doi.org/10.1016/j.tcs.2016.04.030. http://www.sciencedirect.com/science/article/pii/S030439751630069X
8. Amir, A., et al.: Pattern matching with address errors: rearrangement distances. J. Comput. Syst. Sci. **75**(6), 359–370 (2009). https://doi.org/10.1016/j.jcss.2009.03.001
9. Amir, A., Butman, A., Porat, E.: On the relationship between histogram indexing and block-mass indexing. Philos. Trans. Roy. Soc. Math. Phys. Eng. Sci. **372**(2016) (2014). https://doi.org/10.1098/rsta.2013.0132. https://royalsocietypublishing.org/doi/abs/10.1098/rsta.2013.0132
10. Amir, A., Chan, T.M., Lewenstein, M., Lewenstein, N.: On hardness of jumbled indexing. In: Esparza, J., Fraigniaud, P., Husfeldt, T., Koutsoupias, E. (eds.) ICALP 2014. LNCS, vol. 8572, pp. 114–125. Springer, Heidelberg (2014). https://doi.org/10.1007/978-3-662-43948-7_10
11. Amir, A., Hartman, T., Kapah, O., Levy, A., Porat, E.: On the cost of interchange rearrangement in strings. In: Arge, L., Hoffmann, M., Welzl, E. (eds.) ESA 2007. LNCS, vol. 4698, pp. 99–110. Springer, Heidelberg (2007). https://doi.org/10.1007/978-3-540-75520-3_11
12. Angluin, D.: Queries and concept learning. Mach. Learn. **2**(4), 319–342 (1988). https://doi.org/10.1023/A:1022821128753

13. Arratia, R., Martin, D., Reinert, G., Waterman, M.S.: Poisson process approximation for sequence repeats and sequencing by hybridization. J. Comput. Biol. **3**(3), 425–463 (1996). https://doi.org/10.1089/cmb.1996.3.425

14. Batu, T., Kannan, S., Khanna, S., McGregor, A.: Reconstructing strings from random traces. In: Munro, J.I. (ed.) Proceedings of the Fifteenth Annual ACM-SIAM Symposium on Discrete Algorithms, SODA 2004, New Orleans, Louisiana, USA, 11–14 January 2004, pp. 910–918. SIAM (2004). http://dl.acm.org/citation.cfm?id=982792.982929

15. Benson, G.: Tandem repeats finder: a program to analyze DNA sequence. Nucleic Acids Res. **27**(2), 573–580 (1999)

16. Benson, G., Waterman, M.: A method for fast database search for all k-nucleotide repeats. Nucleic Acids Res. **22**, 4828–4836 (1994)

17. Bentley, J.L., Yao, A.C.: An almost optimal algorithm for unbounded searching. Inf. Process. Lett. **5**(3), 82–87 (1976). https://doi.org/10.1016/0020-0190(76)90071-5

18. Bernasconi, A., Damm, C., Shparlinski, I.: Circuit and decision tree complexity of some number theoretic problems. Inf. Comput. **168**(2), 113–124 (2001). https://doi.org/10.1006/inco.2000.3017. http://www.sciencedirect.com/science/article/pii/S0890540100930177

19. Bresler, G., Bresler, M., Tse, D.: Optimal assembly for high throughput shotgun sequencing. BMC Bioinform. **14**(2013). Article number. S18. https://doi.org/10.1186/1471-2105-14-S5-S18

20. Burcsi, P., Cicalese, F., Fici, G., Lipták, Z.: Algorithms for jumbled pattern matching in strings. Int. J. Found. Comput. Sci. **23**(2), 357–374 (2012). https://doi.org/10.1142/S0129054112400175

21. Butman, A., Eres, R., Landau, G.M.: Scaled and permuted string matching. Inf. Process. Lett. **92**(6), 293–297 (2004). https://doi.org/10.1016/j.ipl.2004.09.002

22. Carpi, A., de Luca, A.: Words and special factors. Theor. Comput. Sci. **259**(1–2), 145–182 (2001). https://doi.org/10.1016/S0304-3975(99)00334-5

23. Cayley, A.: LXXVII. Note on the theory of permutations. Lond. Edinb. Dublin Philos. Mag. J. Sci. **34**(232), 527–529 (1849)

24. Chang, Z., Chrisnata, J., Ezerman, M.F., Kiah, H.M.: Rates of DNA sequence profiles for practical values of read lengths. IEEE Trans. Inf. Theory **63**(11), 7166–7177 (2017). https://doi.org/10.1109/TIT.2017.2747557

25. Choi, S.S., Kim, J.H.: Optimal query complexity bounds for finding graphs. Artif. Intell. **174**(9), 551–569 (2010). https://doi.org/10.1016/j.artint.2010.02.003. http://www.sciencedirect.com/science/article/pii/S0004370210000251

26. Cicalese, F., Fici, G., Lipták, Z.: Searching for jumbled patterns in strings. In: Holub, J., Žďárek, J. (eds.) Proceedings of the Prague Stringology Conference 2009, Prague, Czech Republic, 31 August–2 September 2009, pp. 105–117. Prague Stringology Club, Department of Computer Science and Engineering, Faculty of Electrical Engineering, Czech Technical University in Prague (2009). http://www.stringology.org/event/2009/p10.html

27. Cieplinski, L.: MPEG-7 color descriptors and their applications. In: Skarbek, W. (ed.) CAIP 2001. LNCS, vol. 2124, pp. 11–20. Springer, Heidelberg (2001). https://doi.org/10.1007/3-540-44692-3_3

28. Cleve, R., et al.: Reconstructing strings from substrings with quantum queries. In: Fomin, F.V., Kaski, P. (eds.) SWAT 2012. LNCS, vol. 7357, pp. 388–397. Springer, Heidelberg (2012). https://doi.org/10.1007/978-3-642-31155-0_34

29. Dakic, T.: On the turnpike problem. Simon Fraser University BC, Canada (2000)

30. Deininger, P.: SINEs: short interspersed repeated DNA elements in higher eukary-otes. In: Berg, D., Howe, M. (eds.) Mobile DNA, Chap. 27, pp. 619–636. American Society for Microbiology (1989)

31. Deselaers, T., Keysers, D., Ney, H.: Features for image retrieval: an experimental comparison. Inf. Retrieval **11**(2), 77–107 (2008). https://doi.org/10.1007/s10791-007-9039-3

32. Dobzinski, S., Vondrak, J.: From query complexity to computational complexity. In: Proceedings of the Forty-Fourth Annual ACM Symposium on Theory of Computing, STOC 2012, pp. 1107–1116. ACM, New York (2012). https://doi.org/10.1145/2213977.2214076

33. Domaniç, N.O., Preparata, F.P.: A novel approach to the detection of genomic approximate tandem repeats in the levenshtein metric. J. Comput. Biol. **14**(7), 873–891 (2007)

34. Dudík, M., Schulman, L.J.: Reconstruction from subsequences. J. Comb. Theory Ser. A **103**(2), 337–348 (2003). https://doi.org/10.1016/S0097-3165(03)00103-1

35. Dudley, J., Lin, M.T., Le, D., Eshleman, J.R.: Microsatellite instability as a biomarker for PD-1 blockade. Clin. Cancer Res. **22**(4), 813–820 (2016)

36. Elishco, O., Gabrys, R., Médard, M., Yaakobi, E.: Repeat-free codes. In: IEEE International Symposium on Information Theory, ISIT 2019, Paris, France, 7–12 July 2019, pp. 932–936. IEEE (2019). https://doi.org/10.1109/ISIT.2019.8849483

37. Eres, R., Landau, G.M., Parida, L.: Permutation pattern discovery in biosequences. J. Comput. Biol. **11**(6), 1050–1060 (2004). https://doi.org/10.1089/cmb.2004.11.1050

38. Fici, G., Mignosi, F., Restivo, A., Sciortino, M.: Word assembly through minimal forbidden words. Theor. Comput. Sci. **359**(1–3), 214–230 (2006). https://doi.org/10.1016/j.tcs.2006.03.006

39. Fine, N.J., Wilf, H.S.: Uniqueness theorems for periodic functions. Proc. Am. Math. Soc. **16**(1), 109–114 (1965)

40. Gabrys, R., Milenkovic, O.: The hybrid k-Deck problem: reconstructing sequences from short and long traces. In: 2017 IEEE International Symposium on Information Theory, ISIT 2017, Aachen, Germany, 25–30 June 2017, pp. 1306–1310. IEEE (2017). https://doi.org/10.1109/ISIT.2017.8006740

41. Gabrys, R., Milenkovic, O.: Unique reconstruction of coded sequences from multiset substring spectra. In: 2018 IEEE International Symposium on Information Theory, ISIT 2018, Vail, CO, USA, 17–22 June 2018, pp. 2540–2544. IEEE (2018). https://doi.org/10.1109/ISIT.2018.8437909

42. Ganguly, S., Mossel, E., Rácz, M.Z.: Sequence assembly from corrupted shotgun reads. In: IEEE International Symposium on Information Theory, ISIT 2016, Barcelona, Spain, 10–15 July 2016, pp. 265–269. IEEE (2016). https://doi.org/10.1109/ISIT.2016.7541302

43. Holenstein, T., Mitzenmacher, M., Panigrahy, R., Wieder, U.: Trace reconstruction with constant deletion probability and related results. In: Teng, S. (ed.) Proceedings of the Nineteenth Annual ACM-SIAM Symposium on Discrete Algorithms, SODA 2008, San Francisco, California, USA, 20–22 January 2008, pp. 389–398. SIAM (2008). http://dl.acm.org/citation.cfm?id=1347082.1347125

44. Iwama, K., Teruyama, J., Tsuyama, S.: Reconstructing strings from substrings: optimal randomized and average-case algorithms (2018)

45. Jeong, K., Bandeira, N., Kim, S., Pevzner, P.A.: Gapped spectral dictionaries and their applications for database searches of tandem mass spectra. Mol Cell Proteomics (2011). https://doi.org/10.1074/mcp.M110.002220

46. Jerrum, M.: The complexity of finding minimum-length generator sequences. Theor. Comput. Sci. **36**, 265–289 (1985). https://doi.org/10.1016/0304-3975(85)90047-7
47. Kalashnik, L.: The reconstruction of a word from fragments. In: Numerical Mathematics and Computer Technology, pp. 56–57 (1973)
48. Kannan, S., McGregor, A.: More on reconstructing strings from random traces: insertions and deletions. In: Proceedings of the 2005 IEEE International Symposium on Information Theory, ISIT 2005, Adelaide, South Australia, Australia, 4–9 September 2005, pp. 297–301. IEEE (2005). https://doi.org/10.1109/ISIT.2005.1523342
49. Kiah, H.M., Puleo, G.J., Milenkovic, O.: Codes for DNA sequence profiles. IEEE Trans. Inf. Theory **62**(6), 3125–3146 (2016). https://doi.org/10.1109/TIT.2016.2555321
50. Kim, S., Bandeira, N., Pevzner, P.A.: Spectral profiles: a novel representation of tandem mass spectra and its applications for de novo peptide sequencing and identification. Mol. Cell. Proteomics **8**, 1391–1400 (2009)
51. Kim, S., Gupta, N., Bandeira, N., Pevzner, P.A.: Spectral dictionaries: integrating de novo peptide sequencing with database search of tandem mass spectra. Mol. Cell. Proteomics **8**(1), 53–69 (2009)
52. Kociumaka, T., Radoszewski, J., Rytter, W.: Efficient indexes for jumbled pattern matching with constant-sized alphabet. In: Bodlaender, H.L., Italiano, G.F. (eds.) ESA 2013. LNCS, vol. 8125, pp. 625–636. Springer, Heidelberg (2013). https://doi.org/10.1007/978-3-642-40450-4_53
53. Kolpakov, R., Kucherov, G.: mreps: efficient and flexible detection of tandem repeats in DNA. Nucleic Acids Res. **31**, 3672–3678 (2003). http://www.loria.fr/mreps/
54. Krasikov, I., Roditty, Y.: On a reconstruction problem for sequences. J. Comb. Theory Ser. A **77**(2), 344–348 (1997). https://doi.org/10.1006/jcta.1997.2732
55. Levenshtein, V.I.: Binary codes capable of correcting, deletions, insertions and reversals. Soviet Phys. Dokl. **10**, 707–710 (1966)
56. Levenshtein, V.I.: Efficient reconstruction of sequences. IEEE Trans. Inf. Theory **47**(1), 2–22 (2001). https://doi.org/10.1109/18.904499
57. Lowrance, R., Wagner, R.A.: An extension of the string-to-string correction problem. J. ACM **22**(2), 177–183 (1975). https://doi.org/10.1145/321879.321880
58. Manvel, B., Meyerowitz, A., Schwenk, A.J., Smith, K., Stockmeyer, P.K.: Reconstruction of sequences. Discrete Math. **94**(3), 209–219 (1991). https://doi.org/10.1016/0012-365X(91)90026-X
59. Marcovich, S., Yaakobi, E.: Reconstruction of strings from their substrings spectrum. CoRR abs/1912.11108 (2019). http://arxiv.org/abs/1912.11108
60. Margaritis, D., Skiena, S.S.: Reconstructing strings from substrings in rounds. In: IEEE 36th Symposium on Foundations of Computer Science (FOCS), pp. 613–620, October 1995. https://doi.org/10.1109/SFCS.1995.492591
61. Mitzenmacher, M., Upfal, E.: Probability and Computing: Randomized Algorithms and Probabilistic Analysis, 2nd edn. Cambridge University Press, Cambridge (2017)
62. Moosa, T.M., Rahman, M.S.: Indexing permutations for binary strings. Inf. Process. Lett. **110**(18), 795–798 (2010). https://doi.org/10.1016/j.ipl.2010.06.012. http://www.sciencedirect.com/science/article/pii/S0020019010002012
63. Motahari, A.S., Bresler, G., Tse, D.N.C.: Information theory of DNA shotgun sequencing. IEEE Trans. Inf. Theory **59**(10), 6273–6289 (2013). https://doi.org/10.1109/TIT.2013.2270273

64. Motahari, A.S., Ramchandran, K., Tse, D., Ma, N.: Optimal DNA shotgun sequencing: noisy reads are as good as noiseless reads. In: Proceedings of the 2013 IEEE International Symposium on Information Theory, Istanbul, Turkey, 7–12 July 2013, pp. 1640–1644. IEEE (2013). https://doi.org/10.1109/ISIT.2013.6620505
65. Parisi, V., Fonzo, V.D., Aluffi-Pentini, F.: STRING: finding tandem repeats in DNA sequences. Bioinformatics **19**(14), 1733–1738 (2003)
66. Pellegrini, M., Renda, M.E., Vecchio, A.: TRStalker: an efficient heuristic for finding fuzzy tandem repeats. Bioinformatics [ISMB] **26**(12), 358–366 (2010)
67. Sala, F., Gabrys, R., Schoeny, C., Mazooji, K., Dolecek, L.: Exact sequence reconstruction for insertion-correcting codes. In: IEEE International Symposium on Information Theory, ISIT 2016, Barcelona, Spain, 10–15 July 2016, pp. 615–619. IEEE (2016). https://doi.org/10.1109/ISIT.2016.7541372
68. Scott, A.D.: Reconstructing sequences. Discrete Math. **175**(1–3), 231–238 (1997). https://doi.org/10.1016/S0012-365X(96)00153-7
69. Shomorony, I., Courtade, T.A., Tse, D.N.C.: Do read errors matter for genome assembly? In: IEEE International Symposium on Information Theory, ISIT 2015, Hong Kong, China, 14–19 June 2015, pp. 919–923. IEEE (2015). https://doi.org/10.1109/ISIT.2015.7282589
70. Shomorony, I., Kamath, G.M., Xia, F., Courtade, T.A., Tse, D.N.C.: Partial DNA assembly: a rate-distortion perspective. In: IEEE International Symposium on Information Theory, ISIT 2016, Barcelona, Spain, 10–15 July 2016, pp. 1799–1803. IEEE (2016). https://doi.org/10.1109/ISIT.2016.7541609
71. Simon, I.: Piecewise testable events. In: Brakhage, H. (ed.) GI-Fachtagung 1975. LNCS, vol. 33, pp. 214–222. Springer, Heidelberg (1975). https://doi.org/10.1007/3-540-07407-4_23
72. Skiena, S., Smith, W.D., Lemke, P.: Reconstructing sets from interpoint distances (extended abstract). In: Seidel, R. (ed.) Proceedings of the Sixth Annual Symposium on Computational Geometry, Berkeley, CA, USA, 6–8 June 1990, pp. 332–339. ACM (1990). https://doi.org/10.1145/98524.98598
73. Skiena, S., Sundaram, G.: Reconstructing strings from substrings. J. Comput. Biol. **2**(2), 333–353 (1995). https://doi.org/10.1089/cmb.1995.2.333
74. Sokol, D.: TRedD - a database for tandem repeats over the edit distance. Database J. Biol. Databases Curation **2010**(baq003) (2010). https://doi.org/10.1093/database/baq003
75. Tan, K., Ooi, B.C., Yee, C.Y.: An evaluation of color-spatial retrieval techniques for large image databases. Multimed. Tools Appl. **14**(1), 55–78 (2001). https://doi.org/10.1023/A:1011359607594
76. Tardos, G.: Query complexity, or why is it difficult to separate $NP^A \cap coNP^A$ from P^A by random oracles A? Combinatorica **9**(4), 385–392 (1989). https://doi.org/10.1007/BF02125350
77. Tsur, D.: Tight bounds for string reconstruction using substring queries. In: Chekuri, C., Jansen, K., Rolim, J.D.P., Trevisan, L. (eds.) APPROX/RANDOM -2005. LNCS, vol. 3624, pp. 448–459. Springer, Heidelberg (2005). https://doi.org/10.1007/11538462_38
78. Ukkonen, E.: Approximate string matching with q-grams and maximal matches. Theor. Comput. Sci. **92**(1), 191–211 (1992). https://doi.org/10.1016/0304-3975(92)90143-4
79. Viswanathan, K., Swaminathan, R.: Improved string reconstruction over insertion-deletion channels. In: Teng, S. (ed.) Proceedings of the Nineteenth Annual ACM-SIAM Symposium on Discrete Algorithms, SODA 2008, San Francisco, California,

USA, 20–22 January 2008, pp. 399–408. SIAM (2008). http://dl.acm.org/citation. cfm?id=1347082.1347126

80. Wagner, R.A.: On the complexity of the extended string-to-string correction problem. In: Rounds, W.C., Martin, N., Carlyle, J.W., Harrison, M.A. (eds.) Proceedings of the 7th Annual ACM Symposium on Theory of Computing, Albuquerque, New Mexico, USA, 5–7 May 1975, pp. 218–223. ACM (1975). https://doi.org/10. 1145/800116.803771

81. Wang, J., Hua, X.: Interactive image search by color map. ACM Trans. Intell. Syst. Technol. **3**(1), 12:1–12:23 (2011)

82. Wexler, Y., Yakhini, Z., Kashi, Y., Geiger, D.: Finding approximate tandem repeats in genomic sequences. In: RECOMB, pp. 223–232 (2004)

83. Yao, A.C.C.: Decision tree complexity and Betti numbers. In: Proceedings of the Twenty-Sixth Annual ACM Symposium on Theory of Computing, STOC 1994, pp. 615–624. ACM, New York (1994). https://doi.org/10.1145/195058.195414

84. Zenkin, A., Leont'ev, V.K.: On a non-classical recognition problem. USSR Comput. Math. Math. Phys. **24**(3), 189–193 (1984)

85. Zhou, W., Li, H., Tian, Q.: Recent advance in content-based image retrieval: a literature survey. CoRR abs/1706.06064 (2017). http://arxiv.org/abs/1706.06064

Information Retrieval

Pre-indexing Pruning Strategies

Soner Altin[1]([✉])[iD], Ricardo Baeza-Yates[1,2][iD], and B. Barla Cambazoglu[3][iD]

[1] Web Science and Social Computing Research Group, DTIC, Universitat Pompeu Fabra, Barcelona, Spain
sonersukru.altin01@estudiant.upf.edu

[2] Khoury College of Computer Sciences, Northeastern University at Silicon Valley, San Jose, USA

[3] RMIT University, Melbourne, Australia

Abstract. We explore different techniques for pruning an inverted index in advance, that is, without building the full index. These techniques provide interesting trade-offs between index size, answer quality and query coverage. We experimentally analyze them in a large public web collection with two different query logs. The trade-offs that we find range from an index of size 4% and 35% of precision@10 to an index of size 46% and 90% of precision@10, with respect to the full index case. In both cases we cover almost 97% of the query volume. We also do a relative relevance analysis with a smaller private web collection and query log, finding that some of our techniques allow a reduction of almost 40% the index size by losing less than 2% for NDCG@10.

Keywords: Web search · Inverted index · Index pruning · Search efficiency

1 Introduction

Commercial web search engines evaluate queries by processing a very large inverted index built using pages crawled from the Web. Storing and maintaining such an index as well as providing low-latency query processing requires a large amount of hardware investments. Therefore, performance optimization in the context of web search engine indexes has been a very active research area in the last couple of decades.

Among the possible optimizations, a relatively well studied one is static index pruning. The main idea behind this optimization is to create a so-called pruned inverted index which stores less information than the full web index while attaining the search result quality obtained with a full index as much as possible. A pruned index has lower space requirements and leads to faster query processing since fewer postings are stored and processed. The main challenge is to prevent degradation in search quality and query coverage due to the absence of potentially useful indexed content in the pruned index.

All existing approaches so far assume the presence of a full web index to facilitate the construction of the pruned index. That is, a pruned index is created

© Springer Nature Switzerland AG 2020
C. Boucher and S. V. Thankachan (Eds.): SPIRE 2020, LNCS 12303, pp. 177–193, 2020.
https://doi.org/10.1007/978-3-030-59212-7_13

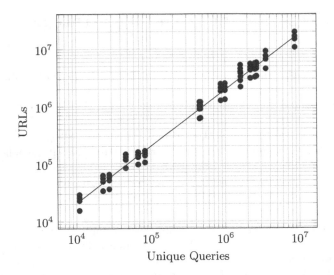

Fig. 1. Relation between unique URLs and unique queries and the simple model found.

by removing postings from the inverted lists in the full web index, entirely or selectively, by using statistical techniques, with the objective of maintaining the search quality. In certain scenarios, however, the resource constraints are really tight (*e.g.*, a low-budget web search engine), and it may not be feasible to build and/or store the full web index. Therefore, pruning decisions need to be made without constructing a full web index first, perhaps even at crawling time to reduce the storage requirements. Hence, how much we lose if we decide *a priori* which documents or parts of them should be indexed?

Another motivation comes from [4], that studied the minimal set of URLs U from a web inverted index that are needed to cover a maximal number of unique queries. They consider that a query is covered if at least one clicked top-k result for that query belongs to U (for k varying from 10 to 1,000). To find U they only used queries with a minimal frequency ranging from 5 to 50 occurrences and pairs (query, clicked result) that appeared at least 5 to 20 times in a period ranging from one day to six months. This setup generated 91 different data sets reaching up to 10 million queries and 50 million URLs. Using these data sets characteristics we can find a simple model between the relation of unique queries and unique clicked URLs given by $U = 2.308 \, Q^{0.987}$ with $R = 0.986$ where U is the number of unique URLs and Q is the number of unique queries (see Fig. 1).

Although the model is simple, the error is low, and says that U grows almost linearly with Q, but for all practical purposes, always there are more URLs than queries (*e.g.*, for 1B unique queries we have 1.76B unique clicked URLs). As unique long tail queries grow, this implies that the index grows even faster. Hence, it makes sense to select documents that cover more queries than others, because as query coverage follows a power law, with a small fraction of documents we can cover an even larger fraction of queries. This is even more important in the long tail as the coverage there is lower. Indeed, in [4] they found that 0.7% of the web pages covered 25% of the queries. Although this result was for a static set of queries, in [5] they found that the correlation of query terms for segments of 3 weeks across a period of 15 weeks was above 0.995. That is, query distribution changes slowly in time. On the other hand, the former result can only be found if the whole index is available. Hence, can a similar result be achieved without having the full index?

Motivated by the aforementioned scenarios, here we design document pruning strategies which do not require the presence of a full web index. That is, the pruning decisions are made in a document-centric manner, *i.e.*, the indexed content is determined as the documents flow in the ingestion pipeline (the sequence of software modules that convert crawled documents to an indexable form). We propose various heuristics to exclude documents' content, partially or entirely, from the index based on the size of the documents, their query-independent summaries, and query terms statistics.

We perform experiments using a public web collection of 50M pages and a private web collection of 18M pages as well as three different search query logs, one public and two private. As main evaluation metrics, we report the reduction attained in the index size, the loss in search quality with respect to a search system relying on a full web index as well as the unique and overall query coverage. To measure search quality we use precision@10 and NDCG@10 with respect to the full index.

The obtained results are encouraging as the index size can be significantly reduced without hurting the search quality too much and without having to build the full index. The trade-offs that we find range from an index of size 4% and 35% of precision@10 to an index of size 46% and 90% of precision@10, with respect to the full index case. In both cases we cover 83% of all unique queries and almost 97% of the query volume. In the case of NDCG@10, some techniques loose less than 2% providing almost 40% reduction in the index size.

Summarizing, our contributions are two fold: (1) we introduce a new pruning method based in summarization, which performs better than most traditional pruning methods (detailed in Sect. 3.2); and (2) we perform a thorough experimental comparison of all known pre-indexing pruning techniques.

The rest of the paper is organized as follows. Section 2 gives an overview of the related work. The pruning heuristics are described in Sect. 3. The experimental setup is explained in Sect. 4 while the results of the experiments are presented in Sect. 5. Section 6 gives our conclusions and future work.

2   Related Work

Unlike this work, which assumes that the content of the documents are pruned (partially or entirely) without constructing a full index, a number of works in literature considered making the pruning decisions after constructing such an index. This line of research is known as static index pruning. The earliest work belongs to Carmel *et al.* [12], who adopted a term pruning approach where an inverted list is entirely removed from the full index if the score contribution of the corresponding term is lower than a certain threshold. Three different strategies are evaluated to determine the threshold: a uniform thresholding strategy that applies to all list entries and two others, list-specific thresholding strategies. De Moura *et al.* [14] argue that the techniques used by Carmel *et al.* [12] are not very effective when queries are processed in conjunctive mode or contain phrases. To alleviate this issue, they exploit the occurrence of terms in important sentences in pages. A page entry is preserved in the pruned index only if the respective term for the list appears in at least one of the important sentences of the page. Büttcher *et al.* [10] use a language model to determine each page's most important terms and keep them in the pruned index. Blanco and Barreiro [8] prune entire inverted lists based on the informativeness of their corresponding terms. Ntoulas *et al.* [17] evaluate three different approaches, where pruned items are entire inverted lists, all inverted list entries belonging to a page, or selected inverted list entries only. Some of those approaches provide correctness guarantees, *i.e.*, the search results obtained by evaluating the query on the pruned index are identical to those that would be obtained from the full index. Blanco and Barreiro [9] apply the probability ranking principle to static index pruning. Thota *et al.* [21] exploit the entropy measure to come up with a document-centric pruning approach. Altingovde *et al.* [1] make use of term and page access statistics in query logs to guide the pruning process. In a relatively recent study, Chen and Lee [13] argue that the earliest pruning approach, proposed by Carmel *et al.* [12], is superior to other pruning approaches in web-scale settings. Skobeltsyn *et al.* [20] investigate the interplay between search result caching and static index pruning.

Apart from static index pruning, there are other architectural optimizations, such as tiering [6,16,19] and selective search [3,15,18]. Another line of research with a vast number of works includes dynamic index pruning, where the inverted lists are pruned at query processing time. We do not cover them here as they are orthogonal to static index pruning. A good survey of efficiency optimizations related to search engines can be found in [11].

Other related problems include document selection in federated search and other types of distributed search (see Chap. 10 of [7]).

3   Pruning Heuristics

Based on the granularity at which document content is pruned, we propose three types of heuristics: document-, sentence, and term-level pruning. Document-level pruning heuristics include or exclude the entire document without investigating its constituting syntactical units. Sentence-level heuristics select a (potentially empty) set of sentences to be pruned from the document content. Finally, term-level heuristics deal with individual terms statistics taken from query logs or consider removing/transforming specific sets of words, providing the finest granularity of pruning. All the strategies presented in this section are summarized in Table 1.

3.1   Document-Level Pruning

Document Size. Very large documents may take too much space in the index. Yet, their contribution to search quality is often not likely to be very different from medium-size documents. Based on this idea, we exclude largest $L\%$ of documents in the collection from the index and we index the remaining documents. The size of a document is determined based on the number of bytes in the textual content that remains after removing the document's boilerplate.

Set Cover. Documents vary in their likelihood of being a good match to a web query. Also, indexing certain documents is relatively more important as they may match many queries, while certain documents are relatively less important since they are seldom queried. The set cover heuristic which was previously proposed in [2] exploits this idea. Essentially, this heuristic tries to select a minimal subset D of documents from a given collection such that the number of queries whose ideal top k results (obtained using a full web index) contain at least one document from D is maximized. In other words, this heuristic tries to minimize the number of queries whose top-k results (obtained using a pruned index) do not contain any ideal results. Once we find a set cover D, we index all documents in D.

Access-Based Document-Centric Pruning. Altingovde *et al.* [1] proposed an access based pruning strategy to prune documents directly from the collection. Documents with low access count are removed from the collection until a fraction of pruned documents is reached.

3.2   Sentence-Level Pruning

Summarization. Certain sentences in a document are relatively more important or represent the document better than the others. It may be more beneficial to index such sentences as they are more likely to be of interest to users and match their queries. To this end, we apply the Textteaser summarizer[1] to the content of documents to obtain the most important or representative S sentences and then index only the terms occurring in those sentences. If the document contains less than S sentences, we do not apply summarization, *i.e.*, we index all of the terms in the document.

3.3   Term-Level Pruning

Query Terms Popularity. It is usually important to index terms that often appear in web queries. In this heuristic, we extract the set of terms occurring in a web query log to obtain a representative set of such useful terms. When processing documents, we index only the terms which appear in this set.

---

[1] https://github.com/MojoJolo/textteaser.

Table 1. Properties of the heuristics.

Index	Uses training		Index	Uses training	
	Queries	Documents		Queries	Documents
Full	No	No	TermPopularity	Yes	No
DocumentSize	No	No	Stemming	No	No
SetCover	Yes	Yes	Stopwords	No	No
aDCP	Yes	Yes	StemStop	No	No
Summarization	No	No			

Stopword Removal and Stemming. We also considered these standard term processing operations as baseline cases for term-level pruning. We have three cases, just stopwords, just stemming, or use both of them.

4   Experimental Setup

4.1   Document Collection

As web document collection, we mostly use the open source web collection provided by Common Crawl, CC, in November 2017.[2] In total, the CC collection takes 4 TB of disk space when compressed. The full web index constructed using this collection contains 50.3M English documents (see end of next section). The average document size is 3.5 KB before parsing the HTML content, and after processing the average number of sentences and terms in a document is around 26 and 628, respectively. This collection is quite diverse in that it contains pages crawled from more than 1.3M different web domains.

For the relevance evaluation we use an ad-hoc web collection A of almost 19M English web documents, also crawled in 2017. This collection is even more diverse as it contains content from 1.6M different domains.

4.2   Document Processing

Before indexing, we have a document processing pipeline that includes HTML parsing, boilerplate extraction, sentence extraction, summarization, tokenization, and language detection. For consistency, we apply the same pipeline to documents and queries.

We first use the open-source Boilerpipe library to remove the boilerplate of the web document (*e.g.*, headers, footers, menus, and ads in the document).[3] Accurate removal of the boilerplate is important since failure to remove the boilerplate may affect the quality of the succeeding text processing steps, such as language detection or summarization. We then use the Jsoup library to extract the

---

[2] Common Crawl web collection, http://commoncrawl.org/2017/11/november-2017-crawl-archive-now-available/.

[3] Boilerpipe, https://github.com/robbypond/boilerpipe.

textual content of the document from the remaining HTML content as well as the document's title. For sentence extraction and tokenization, we use OpenNLP's English models.[4] To identify the language of a document, we use Fasttext's open-source language detection model, which has support for 176 languages.[5] We use only documents whose language is detected as English and when the likelihood estimated by the model is larger than 0.8.

4.3 Indexing

We index documents using an open source version of Elasticsearch.[6] The documents are indexed in an incremental fashion. All indexes were created with 15 shards with a replication factor of 3. `Full` is the full index constructed using the entire collection for comparison purposes. Most of the evaluation metrics computed are relative to this index.

We used Elasticsearch's built-in stopword remover and its built-in stemmer, which removes only possessive suffixes, to obtain our three baselines: `Stemming`, `Stopwords`, and `StemStop`). In all other cases, we do not perform stopword removal nor stemming.

For the `DocumentSize` heuristic, we set L to 1% and 10%. These thresholds result in pruning of documents that are larger than 7 KB and 33 KB, respectively. In the `SetCover` heuristic, we set the C parameter to 1, 5, or 10, resulting in approximately 3.1, 1.7 and 1.3 million documents being indexed, respectively. In the `Summarization` heuristic, we set the S parameter to 10, 20, 40, 80, and 160. For the aDCP heuristic, we set the μ parameter to 0.1, 0.2 and 0.3 resulting in 5, 10 and 15 million documents being indexed, respectively. In the case of the `Term Popularity` index, we prune all terms which do not appear in a given query log.

Table 2 summarizes the size properties of the constructed indexes.

4.4 Ranking

We use two different ranking techniques in the experimental comparison. The simplest one just uses the well-known BM25 technique (see [7], Section 3.5.1) which is native to ElasticSearch, as it is one of the best baselines based just in textual content. A more sophisticated version uses a two phase ranking approach, first using BM25 to obtain a pool of 2,000 candidates and then using learning-to-rank (LTR), LambdaMart [22], to do the final ranking. The LTR variant uses more than 200 features that include query-document similarity (44%), link analysis (20%), query-document relevance (16%), URL name features (10%) and textual content (10%) features.

---

[4] http://opennlp.sourceforge.net/models-1.5/.
[5] https://fasttext.cc/docs/en/language-identification.html.
[6] https://www.elastic.co/products/elasticsearch.

4.5   Query Logs

In most experiments, we use a public query log A, which contains 7.3 million queries submitted in 2006. The query log is split, in temporal order, into training and test sets, which contain 6.4 million and 900K queries, respectively. The training set is used to compute the set cover as well as terms' query frequencies. The test queries are used for evaluation.

Table 2. Index sizes and corresponding pruning ratios (PR) shown as percentages.

Index	Parameter	Size (GB)	PR (%)
Full	–	180.200	0.00
DocumentSize (DS#)	$L = 1\%$	54.147	69.95
DocumentSize	$L = 10\%$	32.226	82.12
SetCover (SC#)	$C = 1$	9.400	94.78
SetCover	$C = 5$	7.000	96.12
SetCover	$C = 10$	6.600	**96.34**
aDCP (aDECP#)	$\mu = 0.1$	22.900	87.29
aDCP	$\mu = 0.2$	31.600	81.90
aDCP	$\mu = 0.3$	41.100	77.19
Summarization (S#)	$S = 10$	23.725	86.83
Summarization	$S = 20$	32.930	81.73
Summarization	$S = 40$	42.526	76.40
Summarization	$S = 80$	50.713	71.86
Summarization	$S = 160$	82.58	**54.17**
TermPopularity (TP)	–	52.014	71.14
Stemming (ST)	–	69.791	61.27
Stopwords (SW)	–	64.338	64.30
StemStop (BOTH)	–	60.907	66.20

To measure the relevance of the results as well as the temporal robustness of them, we also use two newer (2017) but smaller query logs, B1 and B2, obtained from a commercial search engine. Query log B1 contains about 300K queries with relevance judgments for the top 50 results of our LTR model, where 125K queries are used for training the model and the 175K others are left for relevance evaluation. Query log B2 contains almost 2.7M queries and they are used to study how sensitive are our pruning techniques to a query log from a different search engine and from a different time (notice that the year of this query log matches the year of the web collection).

4.6   Evaluation

To evaluate the proposed pruning strategies, we use the Elasticsearch API to submit each of the two test sets of queries sequentially, to compute the metrics detailed below. In all experiments, we set k to 100, unless otherwise stated.

Pruning Ratio (PR). For each index, we compute the ratio of its size to the size of the Full index (in bytes) and subtract this ratio from 1 as follows $PR = 1 - \frac{S(I)}{S(I_{\mathrm{Full}})}$, where $S(I)$ denotes the size of index I in bytes. In the rest of the paper this ratio is shown as a percentage.

Average Precision (AP@k). For each query, we compute the precision as the fraction of relevant results within the retrieved top-k results set. We assume that the top-k results obtained by processing the full web index constitute our relevant results. We then average the precision values over all queries (see Table 3, first column) where $R_q(I, k)$ is the set of top-k retrieved results from the pruned index I for a given query q. I_{Full} is the full web index and m is the number of queries in the test query sample.

Table 3. Evaluation metrics for query relevance and similarity.

$AP@k$	$AR@k$	$ARS@k$
$\sum_{q=1}^{m} \dfrac{\frac{R_q(I,k) \cap R_q(I_{\mathrm{Full}},k)}{R_q(I,k)}}{m}$	$\sum_{q=1}^{m} \dfrac{\frac{R_q(I,k) \cap R_q(I_{\mathrm{Full}},k)}{R_q(I_{\mathrm{Full}},k)}}{m},$	$\sum_{q=1}^{m} \dfrac{\frac{|R_q(I,k) \cap R_q(I_{\mathrm{Full}},k)|}{|R_q(I,k) \cup R_q(I_{\mathrm{Full}},k)|}}{m}$

Average Recall (AR@k). For each query, we compute the recall as the fraction of retrieved relevant results (full index case) and then average them over all queries (see Table 3, second column).

Average Result Similarity (ARS@k). This metric measures the similarity of result sets retrieved for all the queries in the test set between the pruned index and the full index, using the Jaccard similarity metric (see Table 3, third column).

Query Coverage (QC@k). This is the ratio between the number of unique test queries which have at least one relevant result in its top-k results set and the total number of unique queries in the test set, $QC@k = \frac{u}{U}$, where u and U are the number of unique queries that are covered and the total number of unique queries in the test set, respectively.

Query Volume Coverage (QVC@k). This is the ratio between the number of test queries which have at least one relevant result in its top-k results set and the total number of queries in the test, $QVC@k = \frac{v}{V}$, where v and V denote the volume of queries that are covered and the total volume of queries in the test set, respectively.

Normalized Discounted Cumulative Gain (NDCG). NDCG is the most used measure to evaluate relevance for web search (see [7], Section 4.3.4) and we measure it for the first ten results, NDCG@10.

5 Experimental Results

5.1 Common Crawl and BM25

In this section we present most of the evaluation measures using the CC collection with the BM25 baseline ranking for query logs A and C.

Table 4. Evaluation metrics in percentages (CC collection, BM25, query log A).

Index	Parameter	PR	$AP@k$	$AR@k$	$ARS@k$	$QC@k$	$QVC@k$
Full			100	100	100	82.91	96.54
DocumentSize	$L = 1\%$	69.95	78.91	94.59	75.21	78.56	95.66
DocumentSize	$L = 10\%$	82.12	58.84	87.27	51.76	72.54	94.19
SetCover	$C = 1$	94.78	41.95	51.99	34.07	82.80	96.53
SetCover	$C = 5$	96.12	36.27	47.46	29.63	82.79	96.52
SetCover	$C = 10$	**96.34**	35.34	46.62	28.99	82.78	96.52
aDCP	$\mu = 0.1$	87.29	51.92	53.19	40.53	82.15	96.15
aDCP	$\mu = 0.2$	81.90	63.32	53.86	51.52	82.52	96.33
aDCP	$\mu = 0.3$	77.19	64.97	64.09	52.28	82.69	96.43
Summarization	$S = 10$	86.83	39.99	65.94	31.31	70.90	93.57
Summarization	$S = 20$	81.73	52.01	75.19	43.14	73.81	94.46
Summarization	$S = 40$	76.40	64.27	83.91	56.64	76.21	95.11
Summarization	$S = 80$	71.86	73.83	90.22	68.22	78.08	95.54
Summarization	$S = 160$	54.17	**89.75**	**97.21**	**87.90**	**82.91**	**96.54**
TermPopularity		71.14	71.57	79.85	61.39	79.69	95.60
Stemming		61.27	66.13	58.42	49.94	**85.92**	**97.13**
Stopwords		64.30	**93.70**	**93.64**	**88.86**	82.91	96.55
StemStop		66.20	64.47	56.73	47.64	**85.92**	**97.13**

Table 5. Evaluation metrics in percentages (CC collection, BM25, query log B2).

Index	Parameter	PR	$AP@k$	$AR@k$	$ARS@k$	$QC@k$	$QVC@k$
Full			100	100	100	73.43	78.09
DocumentSize	$L = 1\%$	69.95	75.16	81.36	69.03	67.39	75.96
DocumentSize	$L = 10\%$	82.12	56.05	66.86	45.99	59.81	72.89
SetCover	$C = 1$	94.78	44.39	47.82	34.08	72.15	77.41
SetCover	$C = 5$	96.12	40.02	43.94	30.55	71.97	77.27
SetCover	$C = 10$	**96.34**	39.33	43.31	30.10	71.95	77.24
aDCP	$\mu = 0.1$	87.29	54.53	55.51	43.65	72.92	77.79
aDCP	$\mu = 0.2$	81.90	63.97	64.37	52.73	73.15	77.92
aDCP	$\mu = 0.3$	77.19	64.69	64.78	53.59	73.27	77.99
Summarization	$S = 10$	86.83	41.61	51.01	30.91	59.37	72.58
Summarization	$S = 20$	81.73	51.48	59.92	40.55	62.16	73.73
Summarization	$S = 40$	76.40	61.87	69.33	52.01	64.74	74.75
Summarization	$S = 80$	71.86	70.44	76.98	62.53	67.01	75.66
Summarization	$S = 160$	54.17	**84.75**	**84.39**	**78.82**	**73.43**	**78.09**
TermPopularity		71.14	69.18	73.12	58.17	69.21	75.80
Stemming		61.27	63.97	60.49	50.91	77.14	**79.45**
Stopwords		64.30	**90.56**	**90.51**	**84.65**	73.44	78.09
StemStop		66.20	61.55	58.04	47.79	**77.15**	**79.45**

Pruning Ratio. Table 2 shows the size of each index we created and the corresponding pruning ratios. According to this table, the most aggressive pruning strategy is the SetCover strategy, which prunes 96% of the full web index, when $C = 10$. Even when C is set to 1, the pruning ratio remains around 95% with this strategy. The remaining strategies are relatively less aggressive and the pruning ratio remains over 60% (excluding the Summarization strategy with $S = 160$).

Precision, Recall, and Result Similarity. As expected, less aggressive pruning strategies tend to yield the highest precision values. For example, with query log A (Table 4), the highest precision value (90%) is obtained using the Summarization strategy with $S = 160$, while the lowest value (35%) is obtained using the SetCover strategy with $C = 10$. In terms of the recall and result similarity metrics, we observe very similar values. The metrics obtained by using the query log B2 (Table 4) confirm the validity of the results since they exhibit similar behavior.

Query Coverage. Coverage metrics are relatively high for all strategies (Table 4). The worst performing strategy (Summarization with $S = 10$) results in 71% of unique test queries being covered. For the same strategy, the coverage goes up to almost 94% when the query volume is considered. Certain strategies attain the same coverage value attained by the full web index or very close values (*e.g.*, `Summarization` with $S = 160$ or the `SetCover` strategy). We observe similar behavior when query log B2 is used (Table 5). Since the aDCP heuristic uses the most popular documents, it is very successful, as expected, with query volume coverage, but still cannot beat the largest summarization index. The best volume coverage is for aDCP using $\mu = 0.3$ with 96%. On the other hand, aDCP is not as good for unique query coverage.

Table 6. Relevance results (collection A, LTR, query log B2).

Index	Parameter	Size (GB)	PR (%)	$NDCG@10$ (%)	$\Delta NDCG@10$ (%)
Full		172.1	0.00	100	0
Summarization	$S = 40$	133.9	22.19	98.97	−1.02
Summarization	$S = 20$	119.6	30.50	98.63	−1.36
Summarization	$S = 10$	107.5	37.53	98.46	−1.53
aDCP	$\mu = 0.3$	47.2	72.57	79.59	−20.41
DocumentSize	$L = 1\%$	145.6	15.39	69.98	−30.02
TermPopularity		95.4	44.56	30.21	−69.79
SetCover	$C = 1$	1.3	99.24	11.54	−88.46

5.2 Relevance

To understand the relevance loss due to the pruned indexes, here we use the web collection A with our LTR ranking variant and the test queries from query log B1. Based on the results of the previous section, we analyze only the heuristics with higher precision score. For example, *DocumentSize* with $L = 1\%$.

We make an exception in the case of the *Summarization* heuristics, to understand better its effect in relevance. In total we try seven cases and we evaluate relevance using NDCG with respect to the full index version. We do not include stemming nor stopwords removal in this case, because the LTR variant has these functionality embedded during the feature extraction process.

The results for this experiment are given in Table 6 and Fig. 2. For the summarization heuristics, the relevance loss is marginal, while for the other techniques there is a significant NDCG loss, from 20% to 89%. We observe that NDCG loss is much higher for the heuristics that uses document pruning (aDCP and Set Cover). Nevertheless, this effect is also augmented by the fact that the relevance judgments do not cover the whole collection.

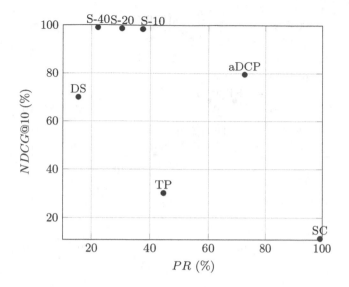

Fig. 2. Pruning ratio versus NDCG@10 (collection A, LTR, query log B1).

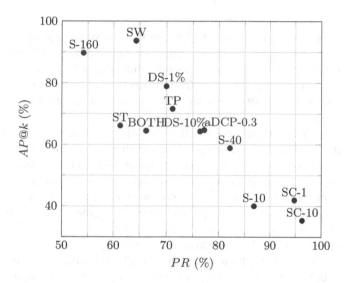

Fig. 3. Pruning ratio vs. precision (CC collection, BM25, query log A).

5.3   Trade-Off Analysis

Figure 3 shows the trade-off between the pruning ratio and the precision attained by different pruning strategies. In general, as the pruning ratio increases the precision is observed to decrease, as expected. The Summarization strategy (with $S = 160$) can be seen to cut the index size by half with around only 10% decrease in precision with respect to the full web index. Therefore, search engines, for

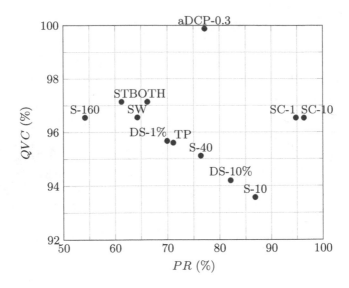

Fig. 4. Pruning ratio vs. query volume coverage (CC collection, BM25, query log A).

which quality is vital, can use this technique to achieve drastic reduction in the hardware needed for storing and processing the web index. The `SetCover` strategy, on the other hand, results in significant quality loss (around 60%), but can yield huge resource savings (around 95%). Therefore, commercial search engines that operate with very limited resources can employ these strategy to cut their operational costs.

Figure 4 shows a similar trade-off between the pruning ratio and query volume coverage. The `Summarization` strategy (with $S = 160$) results in high query volume coverage, again, under the same pruning ratio. The `SetCover` strategy, however, achieves almost the same query volume coverage (around 96.5%), with a drastic reduction in index size. Therefore, a search engine, which aims to satisfy as many queries as possible (rather than the aggregate performance over many queries) may adopt `SetCover` as its pruning strategy.

Regarding relevance, the summarization technique allows to keep parts of all the documents, leading to a marginal NDCG@10 drop, with a reasonable index size reduction. So clearly they are very competitive. The differences that appear in the pruning ratios for some heuristics, such as `Summarization` and `Documentsize`, can be explained by the fact that collection A is of better quality (that is, has much less web spam as has been curated) and also is more homogeneous and hence the fraction of documents removed diminishes.

One way to do a fair comparison of all the cases is to normalize by the index size in percentage. That is, compute the power factor gain of each measure dividing by the index size (the factor would be 1 for the full index in most cases). Larger the power factor, more you gain per space unit used. In Fig. 5 we show these values for the average precision (maximum factor of almost 10), the

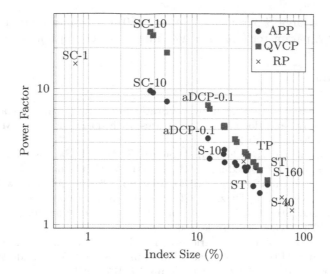

Fig. 5. Power factor for average precision, query volume coverage and NDCG@10.

query volume coverage (maximum factor over 26), and NDCG@10 (maximum factor over 15), using a log-log graph. In both cases, smaller the index, larger the power factor, which is expected given the power law shown in [4] (few documents answers many queries). In this sense, almost all the strategies are similar as there are almost no real outliers. One interesting fact is that the quality measures (APP and RP) are almost in a line while QVCP is almost a perfect line showing an underline power law for this power factor.

6 Conclusions

We have shown different index pruning strategies that aim to reduce the index size significantly, while keeping the search quality and query coverage as similar as possible to the case of the full web index. We conducted large-scale experiments demonstrating the feasibility of these approaches. Our results show that there is no clear dominant technique for different query sets and surely the same is true for different web document collections. We have not included the gains in query processing time, but in all the techniques we get significant faster processing times, from 50% more throughput for S160 to almost a 10 times improvement for the most aggressive pruning techniques.

In practice, each search application has different use cases and different hardware resources. So, we believe that constructing an index that will satisfy all kinds of use cases is not feasible in practice. However, we can recommend different index pruning techniques for different use cases. For example, if we do not have large amounts of hardware and the quality of results is not critical, `SetCover` or `Summarization` with a small S parameter (*e.g.*, $S = 10$) can be a good fit. If we have lots of hardware and search quality is vital, then the `Summarization` strategy with a large S parameter (e.g., $S = 80$) or the DocumentSize strategy with a small L parameter (e.g., $L = 1\%$) might be good options.

References

1. Altingovde, I.S., Ozcan, R., Ulusoy, O.: Static index pruning in web search engines: combining term and document popularities with query views. ACM Trans. Inf. Syst. **30**(1):2:1–2:28 (2012)
2. Anagnostopoulos, A., Becchetti, L., Leonardi, S., Mele, I., Sankowski, P.: Stochastic query covering. In: Proceedings of the Fourth ACM International Conference on Web Search and Data Mining, WSDM 2011, pp. 725–734. ACM, New York (2011)
3. Arguello, J., Callan, J., Diaz, F.: Classification-based resource selection. In Proceedings of the 18th ACM Conference on Information and Knowledge Management, pp. 1277–1286. ACM, New York (2009)
4. Baeza-Yates, R., Boldi, P., Chierichetti, F.: Essential web pages are easy to find. In: Proceedings of the 24th International Conference on World Wide Web, WWW 2015, Florence, Italy, 18–22 May, 2015, pp. 97–107 (2015)
5. Baeza-Yates, R., Gionis, A., Junqueira, F., Murdock, V., Plachouras, V., Silvestri, F. Design trade-offs for search engine caching. TWEB **2**(4):20:1–20:28 (2008)
6. Baeza-Yates, R., Murdock, V., Hauff, C.: Efficiency trade-offs in two-tier web search systems. In: Proceedings of the 32nd International ACM SIGIR Conference on Research and Development in Information Retrieval, pp. 163–170. ACM, New York (2009)
7. Baeza-Yates, R., Ribeiro-Neto, B.: Modern Information Retrieval: The Concepts and Technology Behind Search. Addison-Wesley, Pearson (2011)
8. Blanco, R., Barreiro, Á.: Static pruning of terms in inverted files. In: Amati, Giambattista, Carpineto, Claudio, Romano, Giovanni (eds.) ECIR 2007. LNCS, vol. 4425, pp. 64–75. Springer, Heidelberg (2007). https://doi.org/10.1007/978-3-540-71496-5_9
9. Blanco, R., Barreiro, A.: Probabilistic static pruning of inverted files. ACM Trans. Inf. Syst. **28**(1), 1:1–1:33 (2010)
10. Büttcher, S., Clarke, C.L.A.: A document-centric approach to static index pruning in text retrieval systems. In: Proceedings of the 15th ACM International Conference on Information and Knowledge Management, pp. 182–189. ACM, New York (2006)
11. Cambazoglu, B.B., Baeza-Yates, R.: Scalability Challenges in Web Search Engines. Morgan & Claypool Publishers, San Rafael (2015)
12. Carmel, D., Cohen, D., Fagin, R., Farchi, E., Herscovici, M., Maarek, Y.S., Soffer, A.: Static index pruning for information retrieval systems. In: Proceedings of the 24th Annual International ACM SIGIR Conference on Research and Development in Information Retrieval, pp. 43–50. ACM, New York (2001)
13. Chen, R.-C., Lee, C.-J.: An information-theoretic account of static index pruning. In: Proceedings of the 36th International ACM SIGIR Conference on Research and Development in Information Retrieval, pp. 163–172. ACM, New York (2013)
14. de Moura, E.S., dos Santos, C.F., Fernandes, D.R., Silva, A.S., Calado, P., Nascimento, M.A.: Improving web search efficiency via a locality based static pruning method. In: Proceedings of the 14th International Conference on World Wide Web, pp. 235–244. ACM, New York (2005)

15. Kulkarni, A., Tigelaar, A.S., Hiemstra, D., Callan, J.: Shard ranking and cutoff estimation for topically partitioned collections. In: Proceedings of the 21st ACM International Conference on Information and Knowledge Management, pp. 555–564. ACM, New York (2012)
16. Leung, G., Quadrianto N., Tsioutsiouliklis, K., Smola, A.J.: Optimal web-scale tiering as a flow problem. In: Lafferty, J., Williams, C., Shawe-Taylor, J., Zemel, R., Culotta, A. (eds.) Advances in Neural Information Processing Systems 23, pp. 1333–1341. Curran Associates Inc. (2010)
17. Ntoulas, A., Cho, J.: Pruning policies for two-tiered inverted index with correctness guarantee. In: Proceedings of the 30th Annual International ACM SIGIR Conference on Research and Development in Information Retrieval, pp. 191–198. ACM, New York (2007)
18. Puppin, D., Silvestri, F., Perego, R., Baeza-Yates, R.: Tuning the capacity of search engines: load-driven routing and incremental caching to reduce and balance the load. ACM Trans. Inf. Syst. **28**(2), 1–36 (2010)
19. Risvik, K.M., Aasheim, Y., Lidal, M.: Multi-tier architecture for web search engines. In: Proceedings of the 1st Conference on Latin American Web Congress, p. 132. IEEE Computer Society, Washington (2003)
20. Skobeltsyn, G., Junqueira, F., Plachouras, V., Baeza-Yates, R.: ResIn: a combination of results caching and index pruning for high-performance web search engines. In: Proceedings of the 31st Annual International ACM SIGIR Conference on Research and Development in Information Retrieval, pp. 131–138. ACM, New York (2008)
21. Thota, S.L., Carterette, B.: Within-document term-based index pruning with statistical hypothesis testing. In: Clough, Paul, Foley, Colum, Gurrin, Cathal, Jones, Gareth J.F., Kraaij, Wessel, Lee, Hyowon, Mudoch, Vanessa (eds.) ECIR 2011. LNCS, vol. 6611, pp. 543–554. Springer, Heidelberg (2011). https://doi.org/10.1007/978-3-642-20161-5_54
22. Wu, Q., Burges, C.J.C., Svore, K.M., Gao, J.: Adapting boosting for information retrieval measures. Inf. Retrieval **13**(3), 254–270 (2010)

Measuring Controversy in Social Networks Through NLP

Juan Manuel Ortiz de Zarate[1](✉)(iD), Marco Di Giovanni[2](iD),
Esteban Zindel Feuerstein[1](iD), and Marco Brambilla[2](iD)

[1] Universidad de Buenos Aires, C1053 Buenos Aires, Argentina
{jmoz,efeuerst}@dc.uba.ar
[2] Politecnico di Milano, Milan 20133, Italy
{marco.digiovanni,marco.brambilla}@polimi.it

Abstract. Nowadays controversial topics on social media are often linked to hate speeches, fake news propagation, and biased or misinformation spreading. Detecting controversy in online discussions is a challenging task, but essential to stop these unhealthy behaviours.

In this work, we develop a general pipeline to quantify controversy on social media through content analysis, and we widely test it on Twitter.

Our approach can be outlined in four phases: an initial *graph building* phase, a *community identification* phase through graph partitioning, an *embedding* phase, using language models, and a final *controversy score computation* phase. We obtain an index that quantifies the intuitive notion of controversy.

To test that our method is general and not domain-, language-, geography- or size-dependent, we collect, clean and analyze 30 Twitter datasets about different topics, half controversial and half not, changing domains and magnitudes, in six different languages from all over the world.

The results confirm that our pipeline can quantify correctly the notion of controversy, reaching a ROC AUC score of 0.996 over controversial and non-controversial scores distributions. It outperforms the state-of-the-art approaches, both in terms of accuracy and computational speed.

Keywords: Controversy · Polarization · NLP · Social networks

1 Introduction

Controversy in social networks is a phenomenon with a high social and political impact. Interesting analysis have been performed about presidential elections [44], congress decisions [21], hate spread [9], and harassing [31]. This phenomenon has been broadly studied from the perspective of different disciplines, ranging from the seminal analysis of conflicts within the members of a karate club [54] to political issues in modern times [1,3,10,13,36].

J. M. O. de Zarate and M. Di Giovanni—Equal contribution.

© Springer Nature Switzerland AG 2020
C. Boucher and S. V. Thankachan (Eds.): SPIRE 2020, LNCS 12303, pp. 194–209, 2020.
https://doi.org/10.1007/978-3-030-59212-7_14

The irruption of digital social networks [17] gave raise to new ways of intentional intervention for taking advantages [9,44]. Moreover, highly contrasting points of view in groups tend to provoke conflicts that lead to attacks from one community to the other, such as harassing, "brigading", or "trolling" [31]. The existing literature reports a huge number of issues related to controversy, ranging from the splitting of communities and the biased information spread, to the increase of hate speeches and attacks between groups. For example, Kumar, Srijan, et al. [31] analyze many defense techniques from attacks on *Reddit*[1] while Stewart, et al. [44] insinuate that there was external interference in *Twitter* during the 2016 US presidential elections to benefit one candidate.

As shown in [30,33], detecting controversy also provides the basis to improve the *"news diet"* of readers, offering the possibility to connect users with different points of view by recommending them personalized content to read [37]. Other studies on "bridging echo chambers" [19] and the positive effects of inter group dialogue [4,38] suggest that direct engagement is effective for mitigating conflicts.

An accurate and automatic classifier of controversial topics, therefore, helps to develop quick strategies to prevent miss-information, fights and biases. Moreover, the identification of the main viewpoints and the detection of semantically closer users is also useful to lead people to healthier discussions. *Measuring* controversy is even more powerful, as it can be used to establish controversy levels. For this purpose, we propose a content-based pipeline to measure controversy on social networks, collecting posts' content about a fixed topic (an hashtag or a keyword) as root input.

Controversy quantification through vocabulary analysis also opens several research avenues, such as the analysis whether polarization is being created, maintained or augmented through community's way of talking.

Our main contribution can be summarized as the design of a controversy detection pipeline and the its application to 30 heterogeneous Twitter datasets. We outperform the state-of-the-art approaches, both in terms of accuracy and computational speed.

Our method is tested on datasets from Twitter. This microblogging platform has been widely used to analyze discussions and polarization [36,39,46,50,53]. It is a natural choice for this task, as it represents one of the main fora for public debate [50], it is a common destination for affiliative expressions [23] and it is often used to report and read news about current events [43]. An extra advantage is the availability of real-time data generated by millions of users. Other social media platforms offer similar data-sharing services, but few can match the amount of data and the documentation provided by Twitter. One last asset of Twitter for our work is given by *retweets* (sharing a tweet created by a different user), that typically indicate endorsement [6] and hence they help to model discussions as they can signal "who is with who".

Our paper is organized as follows: in Sect. 2 we list and summarize other works about controversy and polarization on social networks, in Sect. 3 we present the datasets collected for this study, while Sect. 4 contains the

---

[1] https://www.reddit.com/.

step-by-step description of our pipeline. In Sect. 5 we show the results and we conclude with Sect. 6.

2   Related Work

Due to its high social importance, many works focus on polarization measures in online social networks and social media [2,10,11,20,22]. The main characteristic that connects these works is that the measures proposed are based on the structural characteristics of the underlying social-graph. Among them, we highlight the work of Garimella et al. [20] that presents an extensive comparison of controversy measures, different graph-building approaches and data sources, achieving a state-of-the-art performance. We use this approach as a baseline to compare our results.

In [20] the authors propose many metrics to measure polarization on Twitter. Their techniques, based on the structure of the endorsement graph, can successfully detect whether a discussion (represented by a set of tweets), is controversial or not, regardless of the context and, most importantly, without the need of domain expertise. They also include two methods to measure controversy based on the analysis of the posts' contents, both failing. The first of these methods starts with the embedding of tweets in vectors, the clustering of these vectors into two groups and a final computation of KL divergence[2] as a distance measure between clusters, and of I2 measure [27] to quantify the cluster heterogeneity. The second method is based on sentiment analysis. Their hypothesis is that controversial discussions have a higher variance than non-controversial ones. This approach is limited to the fact that it is dependent on language-specific tools that do not work reliably for languages other than English.

Matakos et al. [35] also develop a *polarization index* with a graph-based approach, not including text related features, modelling opinions as real numbers. Their measure successfully captures the tendency of opinions to concentrate in network communities, creating echo-chambers.

Other recent works [34,41,45] prove that communities may express themselves with different terms or ways of speaking, and use different jargon, which can be detected with the use of text-related techniques. Ramponi et al. [40,41] build very efficient classifiers and predictors of account membership within a given community by inspecting the vocabulary used in tweets, for many heterogeneous Twitter communities, such as chess players, fashion designers and members and supporters of political parties [15]. In [45] Tran et al. found that language style, characterized using a hybrid word and part-of-speech tag n-gram language model, is a better indicator of community identity than topic, even for communities organized around specific topics. Finally, Lahoti et al. [34] model the problem of learning the liberal-conservative ideology space of social media users and media sources as a constrained non-negative matrix-factorization problem. They validate their model and solution on a real-world Twitter dataset

---

[2] Kullback–Leibler divergence is a measure of how a probability distribution is different from a reference probability distribution.

consisting of controversial topics, and show that they are able to separate users by ideology with over 90% purity.

Other works for controversy detection through content have been made over Wikipedia [16,26] showing that text contents are good indicatives to estimate polarization. These works are heavily dependent on Wikipedia and can not be extrapolated to social networks.

In her thesis [25], Jang explains controversy via generating a summary of two conflicting stances that build the controversy. Her work shows that a specific sub-set of tweets is enough to represent the two opposite positions in a polarized debate.

A first approach to content-based controversy detection was made in [55]. The main difference between this work and [55] is that the techniques presented here are less dependent on the graph structure. Our new content-based pipeline introduces the possibility of defining and detecting concepts like the "semantic frontier" of a cluster. This opens new ways to activate interventions in the communities, such as the investigation of users lying near that frontier to facilitate a healtier interaction between the communities, or the analysis of users far away from the frontier to understand which aspects establish the real differences. Improvements on [55] (used as a second baseline in this work), include a wider comparison of NLP models and distance measures, a higher heterogeneity of datasets used, and results in better performances both in terms of AUC ROC scores and computational times.

3   Datasets

To test our approach, we collect 30 Twitter datasets in six languages. Each dataset corresponds to a manually selected topic among the trending ones. The collection is performed through the official Twitter API.

3.1   Topic Definition

In the literature, a topic is often defined by a single hashtag. We believe that this might be too restrictive since some discussions may not have a defined hashtag, but they are about a *keyword* that represents the main concept, i.e. a word or expression that is not specifically an hashtag but it is widely used in the discussion. For example during the Brazilian presidential elections in 2018, we collected tweets mentioning to the word *Bolsonaro*, the principal candidate's surname. Thus, in our approach, a topic is defined as a specific hashtag or keyword, depending on the discussion. For each topic we collect all the tweets that contain its hashtag or keyword, posted during a selected observation window. We also check that each topic is associated with a large enough activity volume.

3.2   Description of the Datasets

We collected 30 discussions (50% more than the baseline work [20]) that took place between 2015 and 2020, half of them controversial and half not. We selected discussions in six languages: English, Portuguese, Spanish, French, Korean and Arabic, occurring in five regions over the world: South and North America, Western Europe, Central and Southern Asia. The details of each discussion are described in Table 2. We have chosen discussions clearly recognizable as controversial or not to have an evident groundtruth. Blurry discussions will be analyzed in future works. The encoded datasets are available on github[3].

Since our models require a large amount of text and since a tweet contains no more than 240 characters, we established a threshold of at least 100000 tweets per topic. Topics containing a lower number of tweets were discarded. To select discussions and to determine if they are controversial or not we looked for topics widely covered by mainstream media that have generated ample discussion, both online and offline. For non-controversial discussions we focused both on "soft news" and entertainment, and on events that, while being impactful and/or dramatic, did not generate large controversies. On the other side, for controversial debates we focused on political events such as elections, corruption cases or justice decisions. We validate our intuition by manually checking random samples of tweets.

To furtherly establish the presence or absence of controversy of our datasets, we visualized the corresponding networks through ForceAtlas2 [24], a widely used force-directed layout. This algorithm has been recently found to be very useful at visualizing community interactions [49], as it represents closer users interacting among each other, and farther users interacting less. Figure 1 shows examples of how non-controversial and controversial discussions respectively look like with ForceAtlas2 layout. As we can see in these figures, in a controversial discussion the layout shows two well separated groups, while in a non-controversial one it generates one big cluster.

More information on the datasets is given in Table 2 in Appendix A.

4   Methodology

Our approach to measure controversy can be outlined into four phases, namely *graph building* phase, *community identification* phase, *embedding* phase and *controversy score computation* phase. The final output of the pipeline is a positive value that measures the controversy of a topic, with higher values corresponding to lower degrees of controversy.

Our hypothesis is that using the embeddings generated by an NLP model, we can distinguish different ways of speaking; the more controversial the discussion is, the better differentiation we obtain.

4.1   Graph Building Phase

Firstly, our purpose is to build a conversation graph that represents activities related to a single topic of discussion. For each topic, we build a retweet-graph

---

[3] Code and datasets used in this work are available here: https://github.com/ jmanuoz/Measuring-controversy-in-Social-Networks-through-NLP.

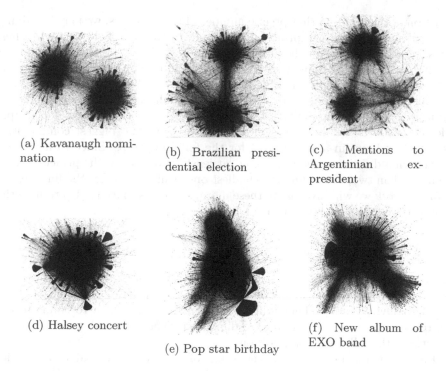

(a) Kavanaugh nomination

(b) Brazilian presidential election

(c) Mentions to Argentinian ex-president

(d) Halsey concert

(e) Pop star birthday

(f) New album of EXO band

Fig. 1. ForceAtlas2 layout for different discussions. (a), (b) and (c) are controversial while (d), (e) and (f) are non-controversial.

G where each user is represented by a vertex, and a directed edge from node u to node v indicates that user u retweeted a tweet posted by user v.

Retweets typically indicate endorsement [6]: users who retweet signal endorsement of the opinion expressed in the original tweet by propagating it further. Retweets are not constrained to occur only between users who are connected in Twitter's social network, but users are allowed to retweet posts generated by every other user. As typically in the literature [7,9,18,32,36,44] we establish that one retweet among a pair of users is enough to define an edge between them. We do not use "quotes" to build the graph since, due to their nature, they can both signal endorsement and opposition, allowing users to comment the quoted tweet.

We remark that the "retweet information" is included in the tweets extracted, allowing us to build the graph without increasing the number of twitter API requests needed. This makes this stage faster than, for example, building a follower graph, another popular alternative.

4.2 Community Identification Phase

To identify the jargon of the community we need to be very accurate at defining its members. If we, in our will of finding two principal communities, force the partition of the graph in that precise number of communities, we may be adding noise in the jargon of the principal communities that are fighting each other. Thus, we decide to cluster the graph using Louvain [8], one of the most popular graph-clustering algorithms. It is a greedy technique that can run over big networks without memory or running time problems, and does not detect a fixed number of clusters. Its output depends on the Modularity Q optimization, resulting in less "noisy" communities. In a polarized context there are two principal sides covering the whole discussion, thus we take the two biggest communities identified by Louvain and use them for the following steps. Since to have controversy in a discussion there must be "at least" two sides, if the principal sides are more than two, discarding the smallest ones will not impact the final result. In future work we will investigate these more complex situations. Up to here the approach we follow is the same as in [55].

4.3 Embedding Phase

In this phase, our purpose is to embed each user into a corresponding vector. These vectors encode syntactic and semantic proprieties of the posts of the corresponding accounts. They will be used in the next phase to compute the controversy score, since we need fixed dimension semantically significant vectors to perform the following computations.

Firstly, tweets belonging to the users of the two principal communities selected in the previous stage are grouped by user and sanitized. We remove duplicates and, from each tweet, we remove user names, links, punctuation, tabs, leading and lagging blanks, general spaces and the retweet keyword "RT", the string that points that a tweet is in fact a retweet. Many sentence embedding techniques have been developed in the literature, ranging from simple bag-of-words models to complex deep language models. To perform this step we selected two models among the most advanced ones, namely Fasttext and BERT, that embed texts into fixed dimension vectors encoding semantically significance and meaning.

Fasttext [28]. This is a tool based on the skipgram model, where each word is represented as a bag of character n-grams. A vector representation is associated to each character n-gram; words being represented as the sum of these representations. This is a fast method that allows to quickly train models on large corpora and to compute word representations also for words that do not appear in the training data. We train this model with tagged data, accordingly to the output of Louvain (previous stage), representing the community of the user. To define the values of the hyper-parameters we use the findings of [52], where the authors investigate the best hyper-parameters to train word embedding models using Fasttext and Twitter data. We use the trained model to compute the text embedding.

BERT. Bidirectional Encoder Representations from Transformers (BERT) [14] is a deep state-of-the-art language representation model based on Transformers [48] pretrained in an unsupervised way on the entire Wikipedia dump for more than 100 languages. The model is designed for transfer learning, so it has to be finetuned for a few epochs for a specific tasks, inserting an additional fully-connected layer on the top, without any substantial task-specific architecture modifications. We use the BASE version of BERT (12 layer, 768 hidden dimension, 12 heads per layer, for a total of $110M$ parameters).

Given a dataset of tweets labeled accordingly to the output of Louvain (previous stage), we finetune BERT on a 2-classes classification task for 6 epochs (learning rate set to 10^{-5}). Since our goal is to obtain embeddings of tweets, after the training procedure we remove the fully-connected layer and we use the outputs of BERT as embeddings. In detail, BERT firstly split a sentence into tokens, adding the *[CLS]* token at the beginning. Then, it embeds each token into a 786-dimensional vector. Since we need a single vector of fixed length to compute our score, we select as aggregator the embedding of the *[CLS]* token. This is the same strategy selected during the fine-tuning step. We perform this stage using bert-as-service GitHub repository [51].

To train Fasttext and BERT in a supervised way, we need to create a training set with its labels. We label each user with its community, namely with tags C_1 and C_2, corresponding respectively to the biggest (Community 1) and second biggest (Community 2) groups. It is important to note that, to prevent bias in the model, we take the same number of users from each community, downsampling the first principal community to the number of users of the second one.

4.4   Controversy Score Computation Phase

To compute the controversy score, we select some users as the best representatives of each side's main point of view. We run the HITS algorithm [29] to estimate the authoritative and hub score of each user. We take the 30% of the users with the highest authoritative score and the 30% with the highest hub score and we call them *central users*.

Finally, we compute the controversy score r, using the embeddings of the central users $x_i \in \mathbb{R}^k$ and the labels $y_i \in \{1,2\}$, imposing their belonging to cluster C_1 or C_2, computed during the community identification phase.

We compute the centroids of each cluster j with Eq. 1, where $|C_j|$ is the magnitude of cluster C_j, and a global centroid c_{glob} with Eq. 2.

$$c_j = \frac{1}{|C_j|} \sum_{i:y_i=j} x_i \tag{1}$$

$$c_{glob} = \frac{1}{|C_1| + |C_2|} \sum_i x_i \tag{2}$$

We define D_j as the sum of distances between the embeddings x_i and their centroids c_j using Eq. 3 for $j = 1, 2$, where *dist* is a generic distance function.

Similarly, D_{glob} is the sum of distances between all the embeddings and the global centroid.

$$D_j = \sum_{i:y_i=j} dist(x_i, c_j) \tag{3}$$

Because of the *curse of dimensionality* [5], measuring distances over big number of dimensions is not a trivial task and the usefulness of a distance measure depends on the sub-spaces that the problem belongs to [42]. For this reason, we select and test four distance measure: L_1 (Manhattan), L_2 (Euclidean), Cosine and Mahalanobis [12] distance (particularly useful when the embedding space is not interpretable and not homogeneous, since it takes into account also correlations of the dataset and reduces to Euclidean distance if the covariance matrix is the identity matrix).

The controversy score r is defined in Eq. 4.

$$r = \frac{D_1 + D_2}{D_{glob}} \tag{4}$$

Intuitively, it represents how much the clusters are separated. We expect that, if the dataset is a single cloud of points, this value should be near 1 since the two centroids c_1 and c_2 will be near each other and near the global centroid c_{glob}. On the contrary, if the embeddings successfully divide the dataset in two clearly separated clusters, their centroids will be far apart and near to the points that belong to their own clusters. Note that r is, by definition, positive, since D_1, D_2 and D_{glob} are positive too.

The datasets and the full code is available on github[4] and the results discussed in the following section are fully reproducible.

5   Results

In this section we collect the results obtained with the different techniques described above and we compare them to the state-of-the-art structured-based method "RW" [20] and our previous work "DMC" [55], a structure and text-based approach. In Fig. 2 we show the distributions of scores of Fasttext and BERT, using the four different distances described before, compared to the baselines "RW" and "DMC". We plot them as beanplots with scores of controversial datasets on the left side and non-controversial ones on the right side. Note that, since by definition "DMC" approach gives higher scores for controversial datasets and lower scores for non-controversial ones, the two distributions are reversed.

The less the two distributions overlap, the better the pipeline works. Thus, to quantify the performance of different approaches, we compute the ROC AUC. By definition, this value is between 0 and 1, where 0.5 means that the curves are perfectly overlapped (i.e. random scoring), while values of 0 and 1 correspond to

---

[4] https://github.com/jmanuoz/Measuring-controversy-in-Social-Networks-through-NLP.

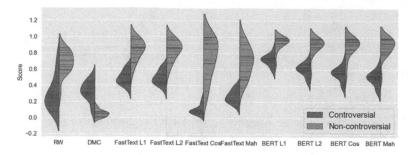

Fig. 2. Scores distributions comparison

perfectly separated distributions. The comparison among the different distance measures is reported in Table 1. As we can see, the best score (the highest value) is obtained by Fasttext model with cosine distance, outperforming the state-of-the-art methods [20,55].

Table 1. ROC AUC scores comparison

Method	L1	L2	Cosine	Mahalanobis	Baseline
FastText	0.987	0.987	**0.996**	0.991	–
BERT	0.942	0.947	0.942	**0.964**	–
DMC	–	–	–	–	**0.982**
RW	–	–	–	-	**0.924**

Even if BERT reached many state-of-the-art results in different NLP tasks [14], FastText suits better in our pipeline. Analyzing the wrongly scored cases we observe that BERT fails mainly with the non-controversial datasets, for example *Feliz Natal* dataset (0.51 controversy score). Our hypothesis is that, since BERT is a bigger and more complex model than FastText, sometimes it overfits the data. BERT is able to separate the two communities' ways of speaking even when they are very similar, not opposite sides of a controversy, exploiting differences that we are not able to perceive. To qualitatively check this behaviour we plot the embeddings produced by each technique by reducing their dimension to 2 with t-SNE algorithm [47] for visualization purposes.

In Fig. 3 we show the reduced embeddings obtained by each method for two non-controversial datasets *Jackson's birthday* and *Feliz Natal*. The first dataset is correctly predicted as non-controversial by both methods and we can see that their embeddings are highly mixed, as expected. However *Feliz Natal* embeddings are mixed when Fasttext is used, while BERT is still able to split them in two separate clusters. This shows that, for the *Feliz Natal* case, BERT is still differentiating two ways of speaking.

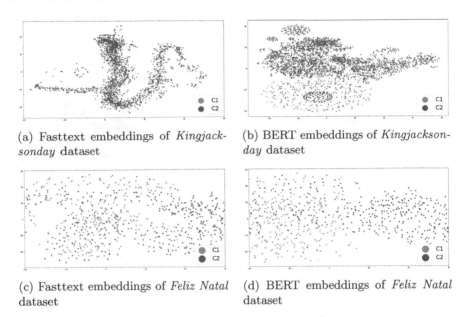

(a) Fasttext embeddings of *Kingjack-sonday* dataset

(b) BERT embeddings of *Kingjackson-day* dataset

(c) Fasttext embeddings of *Feliz Natal* dataset

(d) BERT embeddings of *Feliz Natal* dataset

Fig. 3. t-SNE reduced embeddings produced by Fasttex and Bert

Computational Time. Figure 4 shows the boxplots over the 30 datasets of the total computational times (in seconds) of our two best algorithms, from the beginning (graph building stage) to the end (controversy score computation stage), compared to the baselines. Our approaches are faster than the baseline graph-based method (RW), while DMC approach is only faster than our BERT variant. Fastext approach outperforms both the baselines, allowing a quicker analysis when used in a real-time perspective, since intervention could be necessary for prevention of malicious behaviours, already described in Sect. 1.

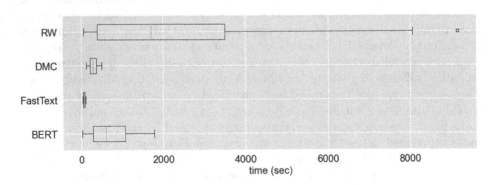

Fig. 4. Computational time comparison

6   Conclusions

In this work we designed an NLP-based pipeline to measure controversy. We test some variants, such as two embedding techniques (using Fasttext and BERT language models) and four distance measures. We applied these approaches on 30 heterogeneous Twitter datasets, and we compared the results. Our best approach, using FastText and cosine distance, outperforms not only the state-of-the-art graph-based method [20], where the authors state that content based techniques do not perform as well as structure based ones, but also our previous work [55], in terms of ROC AUC score and speed, due to the lower dependence on the graph structure and the insertion of a semantic contribute.

Our pipeline involves FastText, a fast model to encode sentences, or BERT, a more accurate language model, slower due to the complex finetuning process required. Fasttext obtains the best performance overall, reaching a ROC AUC score of 0.996. As we reported in the previous section, this is probably because BERT is so strong that it could differentiate ways of speaking even when they are not in controversy. Due to the nature of our pipeline, Fasttext performs better having also a much faster computing time. These results open to a whole new social network analysis to help people participate in healthier discussions, since these approaches allow us to detect faster and better the different points of view.

Since this approach on controversy detection shares similarities with previous works [20,55], we share some limitations too: *Evaluation*, difficulties to establish the ground-truth, *Multisided controversies*, controversy with more than two sides, *Choice of data*, manual collection of topics, and *Overfitting*, small set of experiments, although now we have 10 more discussions, it is still not big enough from a statistical point of view.

Our language-based approach has other limitations. Firstly, training an NLP model that can have a good performance requires significant amount of text, therefore our method works only for "big" enough discussions. However, most interesting controversies are those that have consequences at a society level, in general big enough for our method. Secondly, our findings are based on datasets coming from Twitter. While this is certainly a limitation, Twitter is one of the main venues for online public discussion, and one of the few for which data is easily available. Hence, Twitter is a natural choice. However, Twitter's characteristic limit of 280 characters per message (140 till short time ago) is an intrinsic limitation. We believe that our method, applied to other social networks like Facebook or Reddit, could perform even better, since having more text per user could redound on a more accurate computation of the controversy score.

Future work will involve also user-related analysis, such as the detection of users that are in the "semantic border", on controversial cases, and how they behave over time. This could be useful to find whether there are actors that may help to prevent polarization. We will also analyze which users lay on opposite semantic sides to quickly detect the main differences between two communities.

Finally, we will also detect and analyze the behaviours of users performing mixed interventions on a polarized debate, e.g. posting opinions of both sides of the controversy.

Appendix A   Details on the discussions

Table 2. Datasets statistics, the top group represent controversial topics, while the bottom one represent non-controversial ones

Hashtag/Keywords	#Lang	#Tweets	Description and collection period
#LeadersDebate	EN	250 000	Candidates debate, Nov 11–21,2019
pelosi	EN	252 000	Trump Impeachment, Dec 06,2019
@mauriciomacri	ES	108 375	Macri's mentions, Jan 1–11,2018
@mauriciomacri	ES	120 000	Macri's mentions, Mar 11–18,2018
@mauriciomacri	ES	147 709	Macri's mentions, Mar 20–27,2018
@mauriciomacri	ES	309 603	Macri's mentions, Apr 05–11,2018
@mauriciomacri	ES	254 835	Macri's mentions, May 05–11,2018
Kavanaugh	EN	260 000	Kavanaugh's nomination, Oct 03,2018
Kavanaugh	EN	259 999	Kavanaugh's nomination, Oct 05,2018
Kavanaugh	EN	260 000	Kavanaugh's nomination, Oct 08,2018
Bolsonaro	PT	170 764	Brazilian elections, Oct 27,2018
Bolsonaro	PT	260 000	Brazilian elections, Oct 28,2018
Bolsonaro	PT	260 000	Brazilian elections, 30-10-2018
Lula	PT	250 000	Mentions to Lula the day of Moro chats news, Jun 11-10,2019
Dilma	PT	209 758	Roussef impeachment, 06-11-2015
EXODEUX	EN	179 908	EXO's new album, Nov 07,2019
Thanksgiving	EN	250 000	Thanksgiving day, Nov 28,2019
#Al-HilalEntertainment	AR	221 925	Al-Hilal champion, Dec 01,2019
#MiracleOfChristmasEve	KO	251 974	Segun Woo singer birthday, 23-12-2019
Feliz Natal	PT	305 879	Happy Christmas wishes, Dec 24,2019
#kingjacksonday	EN	186 263	popstar's birthday, Mar 24–27,2019
#Wrestlemania	EN	260 000	Wrestlemania event, Apr 08,2019
Notredam	FR	200 000	Notredam fire, Apr 16,2019
Nintendo	EN	203 992	Nintendo's release, May 19–28,2019
Halsey	EN	250 000	Halsey's concert, Jun 07–08,2019
Bigil	EN	250 000	Vijay's birthday, Jun 21–22,2019
#VanduMuruganAJITH	EN	250 000	Ajith's fans, Jun 23,2019
Messi	ES	200 000	Messi's birthday, Jun 24,2019
#Area51	EN	178 220	Jokes about Area51, Jul 13,2019
#OTDirecto20E	ES	148 061	Event of a Music TV program in Spain, Jan 20,2020

References

1. Adamic, L.A., Glance, N.: The political blogosphere and the 2004 US election: divided they blog. In: Proceedings of the 3rd International Workshop on Link Discovery, pp. 36–43. ACM (2005)
2. Akoglu, L.: Quantifying political polarity based on bipartite opinion networks. In: Eighth International AAAI Conference on Weblogs and Social Media (2014)
3. Al-Ayyoub, M., Rabab'ah, A., Jararweh, Y., Al-Kabi, M.N., Gupta, B.B.: Studying the controversy in online crowds' interactions. Appl. Soft Comput. **66**, 557–563 (2018)
4. Allport, G.W., Clark, K., Pettigrew, T.: The Nature of Prejudice. Addison-Wesley, Reading (1954)
5. Bellman, R.: Dynamic programming. Science **153**(3731), 34–37 (1966)
6. Bessi, A., Caldarelli, G., Del Vicario, M., Scala, A., Quattrociocchi, W.: Social determinants of content selection in the age of (mis)information. In: Aiello, L.M., McFarland, D. (eds.) SocInfo 2014. LNCS, vol. 8851, pp. 259–268. Springer, Cham (2014). https://doi.org/10.1007/978-3-319-13734-6_18
7. Bild, D.R., Liu, Y., Dick, R.P., Mao, Z.M., Wallach, D.S.: Aggregate characterization of user behavior in Twitter and analysis of the retweet graph. ACM Trans. Internet Technol. (TOIT) **15**(1), 1–24 (2015)
8. Blondel, V.D., Guillaume, J.L., Lambiotte, R., Lefebvre, E.: Fast unfolding of communities in large networks. J. Stat. Mech: Theory Exp. **2008**(10), P10008 (2008)
9. Calvo, E.: Anatomía política de Twitter en argentina. Tuiteando# Nisman. Capital Intelectual, Buenos Aires (2015)
10. Conover, M.D., Ratkiewicz, J., Francisco, M., Gonçalves, B., Menczer, F., Flammini, A.: Political polarization on Twitter. In: Fifth International AAAI Conference on Weblogs and Social Media (2011)
11. Dandekar, P., Goel, A., Lee, D.T.: Biased assimilation, homophily, and the dynamics of polarization. Proc. Natl. Acad. Sci. **110**(15), 5791–5796 (2013)
12. De Maesschalck, R., Jouan-Rimbaud, D., Massart, D.L.: The mahalanobis distance. Chcmometr. Intell. Lab. Syst. **50**(1), 1–18 (2000)
13. Del Vicario, M., Zollo, F., Caldarelli, G., Scala, A., Quattrociocchi, W.: Mapping social dynamics on Facebook: the Brexit debate. Soc. Netw. **50**, 6–16 (2017)
14. Devlin, J., Chang, M., Lee, K., Toutanova, K.: BERT: pre-training of deep bidirectional transformers for language understanding. CoRR abs/1810.04805 (2018). http://arxiv.org/abs/1810.04805
15. Di Giovanni, M., Brambilla, M., Ceri, S., Daniel, F., Ramponi, G.: Content-based classification of political inclinations of Twitter users. In: 2018 IEEE International Conference on Big Data (Big Data), pp. 4321–4327 (2018)
16. Dori-Hacohen, S., Allan, J.: Automated controversy detection on the web. In: Hanbury, A., Kazai, G., Rauber, A., Fuhr, N. (eds.) ECIR 2015. LNCS, vol. 9022, pp. 423–434. Springer, Cham (2015). https://doi.org/10.1007/978-3-319-16354-3_46
17. Easley, D., Kleinberg, J., et al.: Networks, Crowds, and Markets, vol. 8. Cambridge University Press, Cambridge (2010)
18. Feng, W., Wang, J.: Retweet or not?: personalized tweet re-ranking. In: Proceedings of the Sixth ACM International Conference on Web Search and Data Mining, pp. 577–586. ACM (2013)
19. Garimella, K., De Francisci Morales, G., Gionis, A., Mathioudakis, M.: Reducing controversy by connecting opposing views. In: Proceedings of the Tenth ACM International Conference on Web Search and Data Mining, pp. 81–90. ACM (2017)

20. Garimella, K., Morales, G.D.F., Gionis, A., Mathioudakis, M.: Quantifying controversy on social media. ACM Trans. Soc. Comput. **1**(1), 3 (2018)
21. Grčar, M., Cherepnalkoski, D., Mozetič, I., Kralj Novak, P.: Stance and influence of Twitter users regarding the Brexit referendum. Comput. Soc. Netw. **4**(1), 1–25 (2017). https://doi.org/10.1186/s40649-017-0042-6
22. Guerra, P.C., Meira Jr., W., Cardie, C., Kleinberg, R.: A measure of polarization on social media networks based on community boundaries. In: Seventh International AAAI Conference on Weblogs and Social Media (2013)
23. Hong, S.: Online news on Twitter: newspapers' social media adoption and their online readership. Inf. Econ. Policy **24**(1), 69–74 (2012)
24. Jacomy, M., Venturini, T., Heymann, S., Bastian, M.: ForceAtlas2, a continuous graph layout algorithm for handy network visualization designed for the Gephi software. PLoS One **9**(6), e98679 (2014)
25. Jang, M.: Probabilistic models for identifying and explaining controversy (2019)
26. Jang, M., Foley, J., Dori-Hacohen, S., Allan, J.: Probabilistic approaches to controversy detection. In: Proceedings of the 25th ACM International on Conference on Information and Knowledge Management, pp. 2069–2072 (2016)
27. Jeh, G., Widom, J.: SimRank: a measure of structural-context similarity. In: Proceedings of the Eighth ACM SIGKDD International Conference on Knowledge Discovery and Data Mining, pp. 538–543. ACM (2002)
28. Joulin, A., Grave, E., Bojanowski, P., Mikolov, T.: Bag of tricks for efficient text classification. arXiv preprint arXiv:1607.01759 (2016)
29. Kleinberg, J.M.: Authoritative sources in a hyperlinked environment. J. ACM (JACM) **46**(5), 604–632 (1999)
30. Kulshrestha, J., Zafar, M.B., Noboa, L.E., Gummadi, K.P., Ghosh, S.: Characterizing information diets of social media users. In: Ninth International AAAI Conference on Web and Social Media (2015)
31. Kumar, S., Hamilton, W.L., Leskovec, J., Jurafsky, D.: Community interaction and conflict on the web. In: Proceedings of the 2018 World Wide Web Conference on World Wide Web, pp. 933–943. International World Wide Web Conferences Steering Committee (2018)
32. Kupavskii, A., et al.: Prediction of retweet cascade size over time. In: Proceedings of the 21st ACM International Conference on Information and Knowledge Management, pp. 2335–2338. ACM (2012)
33. LaCour, M.: A balanced news diet, not selective exposure: evidence from a direct measure of media exposure. In: APSA 2012 Annual Meeting Paper (2015)
34. Lahoti, P., Garimella, K., Gionis, A.: Joint non-negative matrix factorization for learning ideological leaning on Twitter. In: Proceedings of the Eleventh ACM International Conference on Web Search and Data Mining, pp. 351–359. ACM (2018)
35. Matakos, A., Terzi, E., Tsaparas, P.: Measuring and moderating opinion polarization in social networks. Data Min. Knowl. Disc. **31**(5), 1480–1505 (2017). https://doi.org/10.1007/s10618-017-0527-9
36. Morales, A., Borondo, J., Losada, J.C., Benito, R.M.: Measuring political polarization: Twitter shows the two sides of Venezuela. Chaos: Interdisc. J. Nonlinear Sci. **25**(3), 033114 (2015)
37. Munson, S.A., Lee, S.Y., Resnick, P.: Encouraging reading of diverse political viewpoints with a browser widget. In: Seventh International AAAI Conference on Weblogs and Social Media (2013)
38. Pettigrew, T.F., Tropp, L.R.: Does intergroup contact reduce prejudice? Recent meta-analytic findings. In: Reducing Prejudice and Discrimination, pp. 103–124. Psychology Press (2013)

39. Rajadesingan, A., Liu, H.: Identifying users with opposing opinions in Twitter debates. In: Kennedy, W.G., Agarwal, N., Yang, S.J. (eds.) SBP 2014. LNCS, vol. 8393, pp. 153–160. Springer, Cham (2014). https://doi.org/10.1007/978-3-319-05579-4_19

40. Ramponi, G., Brambilla, M., Ceri, S., Daniel, F., Di Giovanni, M.: Vocabulary-based community detection and characterization. In: Proceedings of the 34th ACM/SIGAPP Symposium on Applied Computing. SAC 2019, pp. 1043–1050. Association for Computing Machinery, New York (2019). https://doi.org/10.1145/3297280.3297384

41. Ramponi, G., Brambilla, M., Ceri, S., Daniel, F., Giovanni, M.D.: Content-based characterization of online social communities. Inf. Process. Manag., 102133 (2019). https://doi.org/10.1016/j.ipm.2019.102133, http://www.sciencedirect.com/science/article/pii/S0306457319303516

42. Sapienza, F., Groisman, P.: Distancia de fermat y geodesicas en percolacion euclidea:teoriaa y aplicaciones en machine learning. M.sc. thesis (2018). http://cms.dm.uba.ar/academico/carreras/licenciatura/tesis/2018/Sapienza.pdf

43. Shearer, E., Gottfried, J.: News use across social media platforms 2017. Pew Research Center 7 (2017)

44. Stewart, L.G., Arif, A., Starbird, K.: Examining trolls and polarization with a retweet network. In: Proceedings of the ACM WSDM, Workshop on Misinformation and Misbehavior Mining on the Web (2018)

45. Tran, T., Ostendorf, M.: Characterizing the language of online communities and its relation to community reception. arXiv preprint arXiv:1609.04779 (2016)

46. Trilling, D.: Two different debates? Investigating the relationship between a political debate on TV and simultaneous comments on Twitter. Soc. Sci. Comput. Rev. 33(3), 259–276 (2015)

47. Van Der Maaten, L.: Accelerating t-SNE using tree-based algorithms. J. Mach. Learn. Res. 15(1), 3221–3245 (2014)

48. Vaswani, A., et al.: Attention is all you need. CoRR abs/1706.03762 (2017). http://arxiv.org/abs/1706.03762

49. Venturini, T., Jacomy, M., Jensen, P.: What do we see when we look at networks. An introduction to visual network analysis and force-directed layouts. An introduction to visual network analysis and force-directed layouts, 26 April 2019 (2019)

50. Weller, K., Bruns, A., Burgess, J., Mahrt, M., Puschmann, C.: Twitter and Society, vol. 89. Peter Lang, Bern (2014)

51. Xiao, H.: Bert-as-service (2018). https://github.com/hanxiao/bert-as-service

52. Yang, X., Macdonald, C., Ounis, I.: Using word embeddings in Twitter election classification. Inf. Retrieval J. 21(2–3), 183–207 (2017). https://doi.org/10.1007/s10791-017-9319-5

53. Yardi, S., Boyd, D.: Dynamic debates: an analysis of group polarization over time on Twitter. Bull. Sci. Technol. Soc. 30(5), 316–327 (2010)

54. Zachary, W.W.: An information flow model for conflict and fission in small groups. J. Anthropol. Res. 33(4), 452–473 (1977)

55. de Zarate, J.M.O., Feuerstein, E.: Vocabulary-based method for quantifying controversy in social media. arXiv preprint arXiv:2001.09899 (2020)

Compression

On Repetitiveness Measures
of Thue-Morse Words

Kanaru Kutsukake[1], Takuya Matsumoto[1], Yuto Nakashima[1]<ID>,
Shunsuke Inenaga[1,2]<ID>, Hideo Bannai[3(✉)]<ID>, and Masayuki Takeda[1]<ID>

[1] Department of Informatics, Kyushu University, Fukuoka, Japan
{kutsukake.kanaru,matsumoto.takuya,
yuto.nakashima,inenaga,takeda}@inf.kyushu-u.ac.jp
[2] PRESTO, Japan Science and Technology Agency, Kawaguchi, Japan
[3] M&D Data Science Center, Tokyo Medical and Dental University, Tokyo, Japan
hdbn.dsc@tmd.ac.jp

Abstract. We show that the size $\gamma(t_n)$ of the smallest string attractor of the n-th Thue-Morse word t_n is 4 for any $n \geq 4$, disproving the conjecture by Mantaci et al. [ICTCS 2019] that it is n. We also show that $\delta(t_n) = \frac{10}{3+2^{4-n}}$ for $n \geq 3$, where $\delta(w)$ is the maximum over all $k = 1, \ldots, |w|$, the number of distinct substrings of length k in w divided by k, which is a measure of repetitiveness recently studied by Kociumaka et al. [LATIN 2020]. Furthermore, we show that the number $z(t_n)$ of factors in the self-referencing Lempel-Ziv factorization of t_n is exactly $2n$.

Keywords: String attractors · Thue-Morse words

1 Introduction

Measures which indicate the repetitiveness in a string is a hot and important topic in the field of string compression. For example, given string w, the size $g(w)$ of the smallest grammar that derives solely w [5], the number $z(w)$ of factors in the Lempel-Ziv factorization [13], the number $r(w)$ of runs in the Burrows-Wheeler transform [4] (RLBWT), and the size $b(w)$ of the smallest bidirectional scheme (or macro schemes) [19]. Recently, Kempa and Prezza proposed the notion of *string attractor* [11], and showed that the size $\gamma(w)$ of the smallest string attractor of w is a lower bound on the size of the compressed representation for these dictionary compression schemes. While $z(w)$ and $r(w)$ are known to be computable in linear time, it is NP-hard to compute $g(w), b(w), \gamma(w)$ [7,11,19].

To further understand these measures, Mantaci et al. [14] studied the size of the smallest string attractor in several well-known families of strings. In particular, they showed a size-2 string attractor for standard Sturmian words which is the smallest possible. They further showed a string attractor of size n for the n-th Thue-Morse word t_n, and conjectured it to be the smallest.

In this paper, we continue this line of work, and investigate the exact values of various repetitive measures of the n-th Thue-Morse word t_n. More specifically,

C. Boucher and S. V. Thankachan (Eds.): SPIRE 2020, LNCS 12303, pp. 213–220, 2020.
https://doi.org/10.1007/978-3-030-59212-7_15

we show that the size $\gamma(t_n)$ of the smallest string attractor of t_n is 4 for $n \geq 4$, disproving Mantaci et al.'s conjecture. Furthermore, we give the exact value $\delta(t_n) = \frac{10}{3+2^{4-n}}$ for $n \geq 3$, of the repetitiveness measure recently studied by Kociumaka et al. [12], and the size $z(t_n) = 2n$ of the self-referencing LZ77 factorization.

We note that for any standard Sturmian word s, $z(s) = \Theta(\log|s|)$ [1], while the size $r(s)$ of the RLBWT is always constant [15]. On the other hand, $z(t_n)$ and $r(t_n)$ are both $\Theta(n)$, i.e., logarithmic in the length $|t_n|$ (the former due to [1] as well as this work, and the latter due to [3]). This shows that Thue-Morse words are an example where the size of smallest string attractor is *not* a tight lower bound on the size of the smallest of the known efficiently computable dictionary compressed representations, namely, $\min\{z(w), r(w)\}$. We also conjecture that $b(t_n) = \Theta(n)$, which would seem to imply that the size of the smallest string attractor is not a tight lower bound for *all* currently known dictionary compression schemes.

Let $\ell(w)$ denote the size of the Lyndon factorization [6] of w. It is known that for any w, $\ell(w) = O(g(w))$ [8] and $\ell(w) = O(z(w))$ [10,21], although it can be much smaller. Interestingly, it is also known that $\ell(t_n) = \Theta(n)$ (Theorem 3.1, Remark 3.8 of [9]). Thus, if $b(t_n) = \Theta(n)$, then $\ell(t_n)$ would be an asymptotically tight lower bound for the smallest size of known dictionary compression schemes for t_n, while $\gamma(t_n)$ is not.

Table 1 summarizes what we know so far.

Table 1. Repetitiveness measures for the n-th Thue-Morse word t_n.

Measure	Description	Value	Reference
$z(t_n)$	Size of Lempel-Ziv factorization with self-reference	$2n$	[1], this work
$r(t_n)$	Number of same-character runs in BWT	$2n$	[3]
$\ell(t_n)$	Size of Lyndon factorization	$\left\lfloor \dfrac{3n-2}{2} \right\rfloor$	[9]
$b(t_n)$	Size of smallest bidirectional scheme	Open	N/A
$\gamma(t_n)$	Size of smallest string attractor	4 $(n \geq 4)$	This work
$\delta(t_n)$	Maximum of subword complexity divided by subword length	$\dfrac{10}{3 + 2^{4-n}}$ $(n \geq 3)$	This work

2   Preliminaries

Let Σ denote a set of symbols called the alphabet. An element of Σ^* is called a string. For any $k \geq 0$, let Σ^k denote the set of strings of length exactly k. For any string w, the length of w is denoted by $|w|$. For any $1 \leq i \leq |w|$, let $w[i]$ denote the ith symbol of w, and for any $1 \leq i \leq j \leq |w|$, let $w[i..j] = w[i]w[i+1]\cdots w[j]$.

If $w = xyz$ for strings $x, y, z \in \Sigma^*$, then x, y, z are respectively called a prefix, substring, suffix of w. We denote by $Substr(w)$, the set of substrings of w.

In this paper, we will only consider the binary alphabet $\Sigma = \{a, b\}$. For any string $w \in \Sigma^*$, let $\overline{w}$ denote the string obtained from w by changing all occurrences of a (resp. b) to b (resp. a).

Definition 1. (Thue-Morse Words [16,17,20]). *The n-th Thue-Morse word t_n is a string over a binary alphabet $\{a, b\}$ defined recursively as follows: $t_0 = a$, and for any $n > 0$, $t_n = t_{n-1}\overline{t_{n-1}}$.*

It is a simple observation that $|t_n| = 2^n$ for any $n \geq 0$.

Below, we define the repetitiveness measures used in this paper:

String attractors [11]. For any string w, a set Γ of positions in w is a string attractor of w, if, for any substring x of w, there is an occurrence of x in w that contains a position in Γ. For any string w, we will denote the size of a smallest string attractor of w as $\gamma(w)$.

δ [12,18]

For any string w,

$$\delta(w) = \max_{k=1,..,|w|} \left(|\Sigma^k \cap Substr(w)|/k \right).$$

LZ factorization [13]. For any string w, the LZ factorization of w is the sequence $f_1, \ldots, f_z$ of non-empty strings such that $w = f_1 \cdots f_z$, and for any $1 \leq i \leq z$, f_i is the longest prefix of $f_i \cdots f_z$ which has at least two occurrences in $f_1 \cdots f_i$, or, $|f_i| = 1$ otherwise. We denote the size of the LZ factorization of string w as $z(w)$.

It is known that $\delta(w) \leq \gamma(w) \leq z(w), r(w)$ for any w [7,11].

3   Repetitive Measures of Thue-Morse Words

3.1 $\gamma(t_n)$

Mantaci et al. [14] showed the following explicit string attractor of size n for the n-th Thue-Morse word.

Theorem 1. (Theorem 8 of [14]). *A string attractor of the n-th Thue Morse word, with $n \geq 3$ is*

$$\{2^{n-1} + 1\} \cup \{3 \cdot 2^{i-2} \mid i = 2, \ldots, n\}.$$

To prove our new upperbound of 4 for the smallest string attractor of t_n for $n \geq 4$, we first show the following lemma.

Lemma 1. *Let*

$$N_n = \{t_{n-1}\overline{t_{n-1}}\} \cup \left(\bigcup_{k=0}^{n-2} \{t_k\overline{t_k}, \overline{t_k}t_k\} \right).$$

Then, for any substring $w \in Substr(t_n)$ and $n \geq 2$, there exists $s \in N_n$ such that the occurrence of w in s contains the center of s (i.e., position $|s|/2$).

Proof. Consider the recursively defined perfect binary tree with t_n as the root, with t_{n-1} and $\overline{t_{n-1}}$ respectively as its left and right children (See Fig. 1). The leaves consist of either t_0 or $\overline{t_0}$, each corresponding to a position of t_n. If $|w| = 1$, then, we can choose $t_1 = t_0\overline{t_0} = \text{ab}$ for a and $t_2 = t_1\overline{t_1} = \text{abba}$ for b. For any substring $w = t_n[i..j]$ of length at least 2, consider the lowest common ancestor of leaves corresponding to $t_n[i]$ and $t_n[j]$. Each node of the tree is $t_n = t_{n-1}\overline{t_{n-1}}$ if it is the root, or otherwise, either $t_{k+1} = t_k\overline{t_k}$ or $\overline{t_{k+1}} = \overline{t_k}t_k$ for some $0 \leq k \leq n-2$. Since w is a substring that starts in the left child and ends in the right child of the lowest common ancestor, the occurrence of w must contain the center, and the lemma holds. $\square$

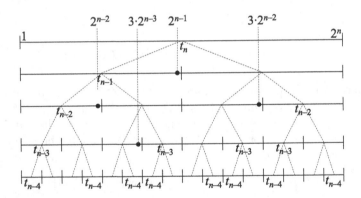

Fig. 1. A representation of t_n as a perfect binary tree (shown to depth 4) introduced in the proof of Lemma 1. For each level where segments are labeled with t_k, non-labeled segments represent $\overline{t_k}$. The black circles depict the four positions in K_n defined in Theorem 2, at the node at which the center of the parent coincides with the position.

Theorem 2. *For any $n \geq 4$, the set*

$$K_n = \{2^{n-2}, 3 \cdot 2^{n-3}, 2^{n-1}, 3 \cdot 2^{n-2}\}$$

is a string attractor of t_n.

Proof. Let w be an arbitrary substring of t_n. From Lemma 1, it suffices to show that any element in N_n has an occurrence in t_n whose center coincides with a position in K_n. $t_{n-1}\overline{t}_{n-1}$, $t_{n-2}\overline{t}_{n-2}$, $\overline{t}_{n-2}t_{n-2}$, and $\overline{t}_{n-3}t_{n-3}$ each have an occurrence whose center coincides respectively with position 2^{n-1}, 2^{n-2}, $3 \cdot 2^{n-2}$, and $3 \cdot 2^{n-3}$ which are all elements of K_n (see Fig. 1). Furthermore, there is an occurrence of $t_{n-3}\overline{t}_{n-3}$ whose center coincides with that of $t_{n-1}\overline{t}_{n-1}$, and thus with an element of K_n. More generally, for any $2 \le k \le n-2$, each occurrence of $t_k\overline{t}_k$ implies an occurrence of $t_{k-2}\overline{t}_{k-2}$ whose centers coincide. This is because

$$t_k\overline{t}_k = t_{k-1}\overline{t}_{k-1}t_{k-1}t_{k-1}$$
$$= t_{k-1}\overline{t}_{k-2}t_{k-2}\overline{t}_{k-2}t_{k-2}t_{k-1}.$$

The same argument holds for $\overline{t}_{k-2}t_{k-2}$ by considering $\overline{t}_k t_k$. The theorem follows from a simple induction. □

Theorem 3. $\gamma(t_n) = 4$ *for any* $n \ge 4$.

Proof. Theorem 2 implies $\gamma(t_n) \le 4$. From Theorem 4 shown in the next subsection, we have $\delta(t_n) > 3$ for $n \ge 6$. Since $\gamma(t_n)$ is an integer which cannot be smaller than $\delta(t_n)$, it follows that $\gamma(t_n) \ge 4$ for $n \ge 6$. For $n = 4, 5$, it can be shown by exhaustive search that there is no string attractor of size 3. □

3.2 $\delta(t_n)$

Brlek [2] investigated the number of distinct substrings of length m in t_n, and gave an exact formula. Below is a summary of his result which will be a key to computing $\delta(t_n)$.

Lemma 2 (Proposition 4.2, Corollary 4.2.1, Proposition 4.4 of [2]). *The number* $P_n(m)$ *of distinct substrings of length* $m \ge 3$ *in* t_n ($n \ge 3$) *is:*

$$P_n(m) = \begin{cases} 2^n - m + 1 & 2^{n-2} + 1 \le m \le 2^n \\ 6 \cdot 2^{q-1} + 4p & 3 \le m \le 2^{n-2}, 0 < p \le 2^{q-1} \\ 8 \cdot 2^{q-1} + 2p & 3 \le m \le 2^{n-2}, 2^{q-1} < p \le 2^q \end{cases}$$

where p, q *are values uniquely determined by* $m = 2^q + p + 1$ *and* $0 < p \le 2^q$.

Theorem 4.

$$\delta(t_n) = \begin{cases} 1 & n = 0 \\ 2 & n = 1, 2 \\ \dfrac{10}{3 + 2^{4-n}} & n \ge 3 \end{cases}$$

Proof. We only consider $n \ge 3$ below. The number of distinct substrings of length 1 and 2 in t_n, are respectively 2 and 4. For $2^{n-2} + 1 \le m \le 2^n$,

$$\max_{2^{n-2}+1 \le m \le 2^n} \frac{P_n(m)}{m} = \max_{2^{n-2}+1 \le m \le 2^n} \left\{ \frac{2^n + 1}{m} - 1 \right\} = \frac{2^n + 1}{2^{n-2} + 1} - 1 = \frac{3}{1 + 2^{2-n}}.$$

For $3 \leq m \leq 2^{n-2}$ and fixed q, it is easy to verify that $P_n(m)/m$ is increasing when $0 < p \leq 2^{q-1}$, and non-increasing when $2^{q-1} < p \leq 2^q$, because

$$\left(\frac{6 \cdot 2^{q-1} + 4p}{2^q + p + 1} \right)' = \frac{4(2^q + p + 1) - (6 \cdot 2^{q-1} + 4p)}{(2^q + p + 1)^2} = \frac{2^q + 4}{(2^q + p + 1)^2} > 0$$

and

$$\left(\frac{8 \cdot 2^{q-1} + 2p}{2^q + p + 1} \right)' = \frac{2(2^q + p + 1) - (8 \cdot 2^{q-1} + 2p)}{(2^q + p + 1)^2} = \frac{(2 - 4 \cdot 2^{q-1})}{(2^q + p + 1)^2} \leq 0.$$

Also note that $6 \cdot 2^{q-1} + 4p = 8 \cdot 2^{q-1} + 2p$ when $p = 2^{q-1}$. Therefore, for a fixed q, the maximum value of $\frac{P_n(m)}{m}$ is obtained when $p = 2^{q-1}$, i.e., $\frac{6 \cdot 2^{q-1} + 4 \cdot 2^{q-1}}{2^q + 2^{q-1} + 1} = \frac{10 \cdot 2^{q-1}}{3 \cdot 2^{q-1} + 1} = \frac{10}{3 + 2^{1-q}}$. Since this is increasing in q, we have that $\max_{3 \leq m \leq 2^{n-2}} \frac{P_n(m)}{m}$ is obtained by choosing the largest possible $q = n - 3$ (where $p = 2^{q-1} = 2^{n-4}$, and thus $m = 2^{n-3} + 2^{n-4} + 1 = 3 \cdot 2^{n-4} + 1 \leq 2^{n-2}$), which gives us the final result $\delta(t_n) = \max\{\frac{2}{1}, \frac{4}{2}, \frac{10}{3 + 2^{4-n}}, \frac{3}{1 + 2^{2-n}}\} = \frac{10}{3 + 2^{4-n}}$. $\square$

3.3   LZ77

We consider the size $z(t_n)$ of the LZ factorization. Although Berstel and Savelli [1] have given a complete characterization of the LZ factorization for the infinite Thue-Morse word, we show an alternate proof in terms of the n-th Thue-Morse word. Below is an important lemma, again by Brlek, we will use.

Lemma 3 (Corollary 4.1.1 of [2]). *The word t_n has one and only one occurrence of every factor w such that $|w| \geq 2^{n-2} + 1$.*

Theorem 5. *For any $n \geq 1$, $z(t_n) = 2n$.*

Proof. Clearly, $z(t_1) = 2$. Since $t_k = t_{k-1}\overline{t_{k-1}} = t_{k-2}\overline{t_{k-2}}\overline{t_{k-2}}t_{k-2}$, it is easy to see that $z(t_k) \leq z(t_{k-1}) + 2$, because $\overline{t_{k-2}}$ and t_{k-2} respectively have earlier occurrences in t_k. Thus, $z(t_n) \leq 2n$. On the other hand, Lemma 3 implies that the substring $t_k[2^{k-1}..3 \cdot 2^{k-2}]$ of length $2^{k-2} + 1$ cannot be a single LZ factor, implying that position $2^{k-1}(= |t_{k-1}|)$ and position $3 \cdot 2^{k-2}(> |t_{k-1}|)$ belong to different factors. Similarly, the substring $t[3 \cdot 2^{k-2}..2^k]$ of length $2^{k-2} + 1$ cannot be a single LZ factor, implying that position $3 \cdot 2^{k-2}$ and position 2^k belong to different factors. Thus, $z(t_k) \geq z(t_{k-1}) + 2$, implying $z(t_n) \geq 2n$. $\square$

Acknowledgments. This work was supported by JSPS KAKENHI Grant Numbers JP18K18002 (YN), JP17H01697 (SI), JP16H02783, JP20H04141 (HB), JP18H04098 (MT), and JST PRESTO Grant Number JPMJPR1922 (SI).

References

1. Berstel, J., Savelli, A.: Crochemore factorization of Sturmian and other infinite words. In: Královič, R., Urzyczyn, P. (eds.) MFCS 2006. LNCS, vol. 4162, pp. 157–166. Springer, Heidelberg (2006). https://doi.org/10.1007/11821069_14

2. Brlek, S.: Enumeration of factors in the Thue-Morse word. Discrete Appl. Math. **24**(1), 83–96 (1989). https://doi.org/10.1016/0166-218X(92)90274-E

3. Brlek, S., Frosini, A., Mancini, I., Pergola, E., Rinaldi, S.: Burrows-Wheeler transform of words defined by morphisms. In: Colbourn, C.J., Grossi, R., Pisanti, N. (eds.) IWOCA 2019. LNCS, vol. 11638, pp. 393–404. Springer, Cham (2019). https://doi.org/10.1007/978-3-030-25005-8_32

4. Burrows, M., Wheeler, D.J.: A block-sorting lossless data compression algorithm. SRC Research Report **124** (1994)

5. Charikar, M., et al.: The smallest grammar problem. IEEE Trans. Inf. Theory **51**(7), 2554–2576 (2005). https://doi.org/10.1109/TIT.2005.850116

6. Chen, K.T., Fox, R.H., Lyndon, R.C.: Free differential calculus, IV. The quotient groups of the lower central series. Ann. Math. **68**(1), 81–95 (1958). http://www.jstor.org/stable/1970044

7. Christiansen, A.R., Ettienne, M.B., Kociumaka, T., Navarro, G., Prezza, N.: Optimal-time dictionary-compressed indexes (2019). http://arxiv.org/abs/1811.12779v6

8. I, T., Nakashima, Y., Inenaga, S., Bannai, H., Takeda, M.: Faster Lyndon factorization algorithms for SLP and LZ78 compressed text. In: Kurland, O., Lewenstein, M., Porat, E. (eds.) SPIRE 2013. LNCS, vol. 8214, pp. 174–185. Springer, Cham (2013). https://doi.org/10.1007/978-3-319-02432-5_21

9. Ido, A., Melançon, G.: Lyndon factorization of the Thue-Morse word and its relatives. Discrete Math. Theor. Comput. Sci. **1**(1), 43–52 (1997). http://dmtcs.episciences.org/233

10. Kärkkäinen, J., Kempa, D., Nakashima, Y., Puglisi, S.J., Shur, A.M.: On the size of Lempel-Ziv and Lyndon factorizations. In: Vollmer, H., Vallée, B. (eds.) 34th Symposium on Theoretical Aspects of Computer Science, STACS 2017. LIPIcs, Hannover, Germany, 8–11 March 2017, vol. 66, pp. 45:1–45:13. Schloss Dagstuhl - Leibniz-Zentrum für Informatik (2017). https://doi.org/10.4230/LIPIcs.STACS.2017.45

11. Kempa, D., Prezza, N.: At the roots of dictionary compression: string attractors. In: Diakonikolas, I., Kempe, D., Henzinger, M. (eds.) Proceedings of the 50th Annual ACM SIGACT Symposium on Theory of Computing (STOC 2018), pp. 827–840. ACM (2018). https://doi.org/10.1145/3188745.3188814

12. Kociumaka, T., Navarro, G., Prezza, N.: Towards a definitive measure of repetitiveness. In: Proceedings of the 14th Latin American Symposium on Theoretical Informatics (LATIN) (2020, to appear). https://arxiv.org/abs/1910.02151

13. Lempel, A., Ziv, J.: On the complexity of finite sequences. IEEE Trans. Inf. Theory **22**(1), 75–81 (1976). https://doi.org/10.1109/TIT.1976.1055501

14. Mantaci, S., Restivo, A., Romana, G., Rosone, G., Sciortino, M.: String attractors and combinatorics on words. In: Proceedings of the 20th Italian Conference on Theoretical Computer Science (ICTCS 2019), pp. 57–71 (2019). http://ceur-ws.org/Vol-2504/paper8.pdf

15. Mantaci, S., Restivo, A., Sciortino, M.: Burrows-Wheeler transform and Sturmian words. Inf. Process. Lett. **86**(5), 241–246 (2003). https://doi.org/10.1016/S0020-0190(02)00512-4

16. Morse, M.: Recurrent geodesics on a surface of negative curvature. Trans. Am. Math. Soc. **22**, 84–100 (1921)

17. Prouhet, E.: Mémoire sur quelques relations entre les puissances des nombres. CR Acad. Sci. Paris Sér. **133**, 225 (1851)

18. Raskhodnikova, S., Ron, D., Rubinfeld, R., Smith, A.D.: Sublinear algorithms for approximating string compressibility. Algorithmica **65**(3), 685–709 (2013). https://doi.org/10.1007/s00453-012-9618-6
19. Storer, J.A., Szymanski, T.G.: Data compression via textual substitution. J. ACM **29**(4), 928–951 (1982). https://doi.org/10.1145/322344.322346
20. Thue, A.: Über unendliche zeichenreihen. Norske vid. Selsk. Skr. Mat. Nat. Kl. **7**, 1–22 (1906)
21. Urabe, Y., Nakashima, Y., Inenaga, S., Bannai, H., Takeda, M.: On the size of overlapping Lempel-Ziv and Lyndon factorizations. In: 30th Annual Symposium on Combinatorial Pattern Matching (CPM 2019), pp. 29:1–29:11 (2019). https://doi.org/10.4230/LIPIcs.CPM.2019.29

Practical Random Access to SLP-Compressed Texts

Travis Gagie[1]([✉])[iD], Tomohiro I[2][iD], Giovanni Manzini[3][iD], Gonzalo Navarro[4][iD],
Hiroshi Sakamoto[2][iD], Louisa Seelbach Benkner[5][iD],
and Yoshimasa Takabatake[2][iD]

[1] Dalhousie University, Halifax, Canada
travis.gagie@dal.ca
[2] Kyushu Institute of Technology, Fukuoka, Japan
[3] University of Eastern Piedmont, Alessandria, Italy
[4] CeBiB & DCC, University of Chile, Santiago, Chile
[5] University of Siegen, Siegen, Germany

Abstract. Grammar-based compression is a popular and powerful approach to compressing repetitive texts but until recently its relatively
poor time-space trade-offs during real-life construction made it impractical for truly massive datasets such as genomic databases. In a recent
paper (SPIRE 2019) we showed how simple pre-processing can dramatically improve those trade-offs, and in this paper we turn our attention to
one of the features that make grammar-based compression so attractive:
the possibility of supporting fast random access. This is an essential
primitive in many algorithms that process grammar-compressed texts
without decompressing them and so many theoretical bounds have been
published about it, but experimentation has lagged behind. We give a
new encoding of grammars that is about as small as the practical state
of the art (Maruyama et al., SPIRE 2013) but with significantly faster
queries.

1 Background

It is widely acknowledged that we now have more data than we can properly
handle, and one possible solution is to compress it in such a way that we can
later process it quickly without decompressing it. Since many of our largest
and most important datasets—such as genomic databases—are highly repetitive texts, grammar-based schemes offer excellent compression ratios while still
admitting algorithms for many natural problems that run in times polynomial
in the sizes of the compressed representations.

Probably the most popular such schemes are those producing straight-line
programs (SLPs), which are context-free grammars in Chomsky normal form
that each generate exactly one string; we refer the reader to Lohrey's [23] and
Navarro's [25] surveys for more details of SLPs, SLP algorithmics, SLP-based
data structures, and related techniques. Since many algorithms that process
SLPs depend on random access to the compressed texts as a primitive operation,

© Springer Nature Switzerland AG 2020
C. Boucher and S. V. Thankachan (Eds.): SPIRE 2020, LNCS 12303, pp. 221–231, 2020.
https://doi.org/10.1007/978-3-030-59212-7_16

there have been several important theoretical papers written about supporting it, which we review in Appendix A.

Unfortunately, there have not been as many breakthroughs about supporting random access to SLP-compressed texts in practice. Block trees [3] are practical, and resemble SLPs in many ways with similar theoretical bounds, but they are not SLPs nor even context-free grammars and so researchers studying SLP algorithmics may wish to avoid them. Variant call format [8] and relative Lempel-Ziv [21] are also practical but even less like SLPs.

In the real world, users still rely on Larsson and Moffat's [22] RePair algorithm, even though the SLPs it produces are not optimal in the worst case and it is not known if they are even always close to optimal.[1] Similarly, users who need random access to SLP-compressed strings often just augment the SLPs produced by RePair and naïvely encode them even though, as far as we are aware, there are no good bounds on their heights and thus no good bounds on the access times (unless we modify the SLPs at the risk of making them impractical). The best encoding we know of is due to Maruyama et al. [24], which is significantly smaller than the naïve encoding but also significantly slower.

Practitioners' main concern about RePair seems to be the large constants in its time-space trade-offs for construction. For example, Navarro's implementation of RePair[2] compresses a 3.0 GB file containing copies of human chromosome 19 from 50 distinct genomes into 23 MB and a 5.9 GB file containing copies from 100 genomes into 24 MB, but on a commodity computer it takes 84 min and 11 GB of workspace for the former and 11 hours and 18 GB of workspace for the latter [13]. Although several alternatives have been proposed [4,11,16,26,30], until recently the most practical option for files of more than a few gigabytes was SOLCA [33], which compresses the 3.0 GB file into 40 MB using 11 min and 310 MB of workspace, and the 5.9 GB file into 45 MB using 22 min and 310 MB of workspace, respectively. In addition to achieving noticeably worse compression than RePair, even SOLCA took over 3.6 h to compress a 59 GB file containing copies of chromosome 19 from 1000 genomes, although it used only 783 MB of workspace and produced an SLP of only 129 MB.

In a recent paper [13] we showed how simple pre-processing with context-triggered piecewise hashing (CTPH) can dramatically improve the trade-offs for both RePair and SOLCA. For CTPH, we run a relatively short sliding window over the text and insert a phrase break whenever the Karp-Rabin hash of the window's contents is 0 modulo some parameter p.[3] Although it works poorly in the worst case even on repetitive texts—for example, the string a^n is either parsed into a single phrase or into nearly n of them—in practice on most

---

[1] RePair is probably most commonly used in natural-language processing, where it is viewed as an implementation of Gage's [12] byte-pair encoding and used for word segmentation in neural machine translation [31]; we refer readers to Gallé's [14] recent survey for more discussion.

[2] https://users.dcc.uchile.cl/~gnavarro/software/repair.tgz.

[3] We realized after [13] went to press that the worst-case approximation ratios in Theorems 1 and 2 should be multiplied by the length of the sliding window, but this does not affect our approach's correctness or practicality.

repetitive texts CTPH produces a dictionary of distinct phrases and a parse that are, together, much smaller than the text. We note in passing the similarity of the high-level ideas behind prefix-free parsing and string synchronizing sets [19], which have good worst-case bounds and seem practical for small files [9] but may not scale as easily to tens or hundreds of gigabytes.

We first experimented with CTPH for building Burrows-Wheeler Transforms (BWT) for massive texts [6,20], because we can quickly build the run-length compressed BWT from the dictionary and the parse in workspace bounded in terms of their combined size. It then occurred to us that, if we build SLPs for the dictionary and the parse, with the SLP for the dictionary restricted such that each phrase is the complete expansion of some non-terminal, then we can easily combine those SLPs to obtain an SLP for the text: we replace each terminal in the SLP for the parse—which is a phrase identifier—by the non-terminal in the SLP for the dictionary whose expansion is that phrase. For example, on the same commodity computer, applying RePair to the dictionary and parse of the 59 GB file containing 1000 copies of chromosome 19, compressed it by a factor of 1000 in 21 min using 7.0 GB of workspace, and applying SOLCA compressed it by a factor of over 400 in 44 min using only 4.6 MB of workspace.

Now that grammar-based compression itself is reasonably scalable, it is time to turn our attention to making SLP algorithmics practical, and an obvious starting place is improving the practicality of random access.

2   Design of the New Grammar Encoding

Random access to an SLP-compressed text works by descending the parse tree and computing the expansion sizes of the non-terminals we visit. In particular, at each non-terminal, we compute the expansion sizes of its children, in order to know to which we should descend. The main idea of our new encoding is that symbols' expansion sizes can tell us a lot about their identities, so we should tightly integrate how we encode these two kinds of information.

If the non-terminals (excluding the start symbol, unless it expands to two symbols in one step) in an SLP have d distinct expansion sizes, then we build a minimal perfect hash function (MPHF) h that maps those sizes bijectively to the numbers in $[0, d - 1]$. In this paper we use Esposito, Graf and Vigna's recent RecSplit [10] MPHF implementation, which occupies only about $1.56d$ bits. We note that we cannot recover the d sizes from the MPHF—given any other size, it will still return a hash value in the range $[0, d - 1]$—so in our algorithm we will be careful to query the MPHF only with numbers we know are non-terminals' expansion sizes in our SLP.

We group the non-terminals by their expansion sizes; sort the groups by the hash values of the expansion sizes of the non-terminals in them; and replace each non-terminal by a triple consisting of the expansion size of its left child, and the offsets of its children in their groups (or, if they are terminals, their offsets in the alphabet). If the start symbol expands to more than one symbol in one step, then we store a bitvector indicating the lengths of the expansions of the symbols

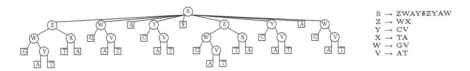

Fig. 1. An SLP (right) for GATTAGATACAT$GATTACATAGAT and its parse tree (left).

it expands to in one step, and we store the offset of each of those symbols in its group (or its offset in the alphabet if it is a terminal).

The random access to the input text T works as follows. Suppose we know $T[i]$ is the jth character in the expansion of the kth non-terminal, say X, in the group of non-terminals with expansion size ℓ. Using some small auxiliary data structures, we can

1. look up X's left child's expansion size ℓ';
2. compute X's right child's expansion size $\ell'' = \ell - \ell'$;
3. look up X's left child's offset k' in the group of non-terminals with expansion size ℓ' (or its offset in the alphabet if $\ell' = 1$ so it is a terminal);
4. look up X's right child's offset k'' in the group of non-terminals with expansion size ℓ'' (or its offset in the alphabet if $\ell'' = 1$ so it is a terminal);
5. if $j \leq \ell'$ then set $j' = j$ and recursively find the j'th character in the expansion of the k'th non-terminal in the group of non-terminals with expansion size ℓ' (or just return the character if it is a terminal);
6. otherwise, $j > \ell'$ and we set $j'' = j - \ell'$ and recursively find the j''th character in the expansion of the k''th non-terminal in the group of non-terminals with expansion size ℓ'' (or just return the character if it is a terminal).

Since $T[i]$ is the $(i+1)$st character in the expansion of the only non-terminal with expansion size n, we can descend down the parse tree in time proportional to its height. If we push the offsets and expansion sizes on a stack as we do so, then we can traverse the parse tree starting from the $(i+1)$st leaf and thus extract subsequent characters of T in constant amortized time per character.

Encoding Example. Consider the SLP for GATTAGATACAT$GATTACATA-GAT that is shown with its parse tree in Fig. 1. The 3 distinct sizes of the non-terminals' expansions (excluding S) are 5 (for Z), 3 (for W and Y) and 2 (for V and X). If we use an MPHF h with $h(5) = 1$, $h(3) = 2$ and $h(2) = 0$, then we can sort the non-terminals into the order V, X; Z; W, Y, with semicolons showing the divisions between the groups.

Assuming the alphabet is $\{\$, A, C, G, T\}$, we replace the non-terminals by the triples $(1, 1, 4), (1, 4, 1); (3, 0, 1); (1, 3, 0), (1, 2, 0)$, with the semicolons again showing the divisions between the groups. For example non-terminal V is represented by $(1, 1, 4)$ since its left child, the terminal A, has expansion size 1, and its offset among the terminals is 1, while the second child, the terminal T, has offset 4. Finally, we encode the rule involving the initial symbol S as the bitvector 0000100110011000010011001, which is the concatenation of the unary representations of the expansion sizes of the symbols on the rule's right-hand side, and the sequence $0, 0, 1, 1, 0, 0, 1, 1, 0$ giving the offset of each symbol in its group.

To extract the 17th character of the text, we start by performing a rank query and two select queries on the bitvector for S, which together tell us that the 17th character is the 4th character in the expansion of the 6th symbol on the right-hand side of the rule for S, and that symbol expands into 5 characters. Checking the sequence for S, we see that the 6th symbol on the right-hand side of the rule for S has rank 0 among all the non-terminals that expand to 5 characters (note there is only one such non-terminal, Z).

We compute $h(5) = 1$ and check the triple with rank 0 in the group with rank 1—i.e., $(3, 0, 1)$—which tells us that Z's left child expands into 3 characters, so its right child X expands into 2 characters and the 4th character in the expansion of Z is the 1st character in the expansion of X, and that X has rank 1 among the non-terminals that expand into 2 characters. Note that we never actually learn or use the identifiers Z or X in the actual data structure: we use them here just to ease the presentation. We compute $h(2) = 0$ and check the triple with rank 1 in the group with rank 0—i.e., $(1, 4, 1)$—which tells us that X's left child expands into 1 character, so it is a terminal, and it has rank 4 in the alphabet, meaning it is a T.

Admittedly, for this small example we do not save space compared to the naïve encoding, but our experiments show that it pays to carefully integrate our encodings of the symbols in the parse and its shape.

3   Experiments

We compared our encoding with the naïve encoding and the state-of-the-art encoding by Maruyama et al. [24]; we refer to these as OURS, NAIVE and MTSS, respectively. For the naïve encoding of an SLP for a string of length n with r rules, we store the following information in plain arrays:

1. the right-hand sides of rules in $2r \lg(r + \sigma)$ bits,
2. the expansion length for every non-terminal in $r \lg n$ bits.

To support random access to the triples in our encoding and to store the bitvector for the start rule, we used SD bitvectors from the SDSL 2.0 library[4]. Our experiments ran on a Xeon E5-1650V3 (6core/12thread 3.5 GHz) machine with 32 GB memory.

In this section we describe only our main experimental results; additional results can be found in Appendix B. For our main experiments, we used the same 59 GB file containing 1000 copies of chromosome 19 that we used in our previous work [13], downloaded from the 1000 Genomes Project [34]; the effective alphabet size was 5. When we compress the dictionary and parse with Navarro's

---

[4] https://github.com/simongog/sdsl-lite.

Table 1. Extraction times in microseconds with the three encodings and various substring length.

substring length	NAIVE (217 MB)	MTSS (86 MB)	OURS (81 MB)
1	1.8	25.9	6.9
10	2.2	29.6	9.3
100	5.2	63.5	31.7
1000	31.6	394.6	249.6

implementation of RePair combined with CTPH, as described in Sect. 1, the resulting 59 MB SLP contains almost 13 million rules with almost 120 000 distinct expansion lengths and almost 4.5 million symbols on the right-hand side of the start rule; the height of the parse tree is 43.

Table 1 shows our main experimental results: for each of the given substring lengths and each of the encodings, we extracted that many consecutive characters from 10000 pseudo-randomly chosen positions in the compressed file and averaged the extraction times. The naïve encoding is obviously the largest but also the fastest: it takes 217 MB, access to a single character taking 1.8 µs, and access to ten consecutive characters taking 2.2 µs. Maruyama et al.'s encoding takes 86 MB—much closer to the size of the unaugmented SLP—but access to one character takes 26 µs and access to ten takes 30 µs. We encode the augmented grammar in 81 MB—even less than Maruyama et al.—with access to one character taking 6.9 µs and access to ten taking 9.3 µs. Although our encoding is still significantly slower than the naïve encoding, it is only a little more than a third of the size. The size difference is particularly pronounced if we compare how much larger the naïve encoding and ours are than the unaugmented SLP: $217/59 \approx 3.7$ versus $81/59 \approx 1.4$. Building our encoding is also reasonably fast, taking only 18 seconds with the source code we have made publicly available at https://github.com/itomomoti/ShapedSlp.

For some applications, we are interested in processing many queries at once, which offers us the opportunity to exploit parallelism. Figure 2 shows the average speedup using up to 8 threads. Since the scale makes it difficult to discern the height of the rightmost points, we note that NAIVE, MTSS and OURS with 8 threads use 0.38, 6.56 and 1.41 µs for length 1; 0.41, 7.01 and 1.86 for length 10; and 0.78, 13.47 and 7.07 for length 100.

Acknowledgements. TG was partly funded by NSERC RGPIN-2020-07185, Canada, and Basal Funds FB0001, Chile. TI, HS and YT were partly funded by JSPS KAKENHI grants 19K20213, 17H01791 and 18K18111, respectively. GM was partly funded by MIUR-PRIN grant 2017WR7SHH. GN was partly funded by Basal Funds FB0001 and Fondecyt grant 1-200038, Chile. LSB was partly funded by DFG project LO 748/10-2 (QUANT-KOMP) and received travel funds from the EU's Horizon 2020 MSC RISE program (grant 690941).

A   Theoretical Bounds

Charikar et al. [7] and Rytter [28, 29] independently showed how, given a text T of length n over an alphabet of size σ whose smallest SLP has g^* rules, in $O(n \log \sigma)$ time we can build an SLP for T with $O(g^* \log(n/g^*))$ rules and height $O(\log n)$. We can augment the non-terminals of this SLP with the sizes of their expansions to obtain an $O(g^* \log(n/g^*))$-space data structure supporting access to any ℓ consecutive characters of T in $O(\log n + \ell)$ time. Bille et al. [5] showed how we can take any SLP for T with g rules, regardless of height, and build a data structure of size $O(g)$ (measured in words of bit length $\log n$) that also supports access to any ℓ consecutive characters in $O(\log n + \ell)$ time, while Verbin and Yu [35] proved we generally cannot support $O(\log^{1-\epsilon} n)$-time random access to T with a poly(g)-space data structure. Belazzougui et al. [2] showed how we can support $O(\log n / \log \log n)$-time random access to T with an $O(g \log^\epsilon n)$-space grammar. Prezza [27] sidestepped Verbin and Yu's lower bound to obtain constant-time random access to T with an $O(gn^\epsilon)$-space grammar (after Belazzougui et al. [3] achieved that tradeoff with block trees). Recently, Ganardi, Jeż and Lohrey [15] showed how we can turn any SLP for T with g rules into an SLP for T with $O(g)$ rules and height $O(\log n)$, thus simplifying many previous proofs.

Regarding SLPs produced with RePair, Charikar et al. [7] showed they can be an $\Omega(\log^{1/2} n)$ factor larger than the smallest possible SLPs, and Hucke, Jeż and Lohrey [1,18] improved that lower bound to $\Omega(\log n / \log \log n)$. Charikar et al. showed they are always within an $O((n/\log n)^{2/3})$-factor of the smallest SLPs and this is still the best upper bound known, although Hucke [17] showed they are within a $\log_2 3$-factor for unary strings.

B   Additional experimental results

We are mainly interested in compressing human DNA but we performed experiments with other datasets to check our approach's robustness: 11264 Salmonella genomes (salx11264) from the GenomeTrakr project [32], and two repetitive files from the Pizza & Chili corpus[5] (einstein.en.txt and kernel).

As can be seen from Tables 2 and 3 below and comparing Fig. 2 to Fig. 3, our results are not as good for the other datasets as for chr19x1000 but our general conclusions are supported: MTSS and OURS are about the same size and several times smaller than NAIVE; NAIVE is by far the fastest to build, with MTSS slower by almost an order of magnitude and OURS slower even than that by a factor of 4 to 7; NAIVE is also the fastest to answer queries, followed by OURS and then MTSS. Since the scale again makes it difficult to discern the height of the rightmost points, we note that NAIVE, MTSS and OURS with 8 threads use 0.53, 9.34 and 3.76 μs for salx11264; 0.15, 6.16 and 1.84 for einstein.en.txt; and 0.53, 22.18 and 12.84 for kernel.

---

[5] http://pizzachili.dcc.uchile.cl/.

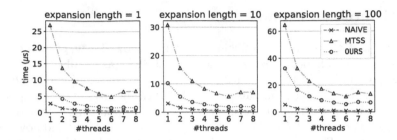

Fig. 2. Average time to answer an expansion query using multiple threads.

Table 2. Statistics of our datasets: name, alphabet size, length (in bytes), number of symbols on the right-hand side of the start rule, number of rules, number of distinct expansion lengths, and height of the grammar.

dataset	σ	n	s	r	d	h
chr19x1000	5	59125115010	4495360	12898128	118889	43
salx11264	4	57033515255	32579379	199121788	332808	18658
einstein.en.txt	139	467626544	62473	100611	17343	1353
kernel	160	257961616	69427	1057914	48453	5820

Table 3. Sizes of the encodings and construction times.

Dataset	Size (bytes)			Construction time (ms)		
	NAIVE	MTSS	OURS	NAIVE	MTSS	OURS
chr19x1000	217418909 (0.37%)	86362255 (0.15%)	80629662 (0.14%)	524	4576	17649
salx11264	2896264885 (5.1%)	799395665 (1.4%)	956575138 (1.7%)	5457	53147	370175
einstein.en.txt	1896040 (0.41%)	674979 (0.14%)	631698 (0.14%)	3	22	92
kernel	12964629 (5.0%)	4473636 (1.7%)	5044020 (2.0%)	30	158	866

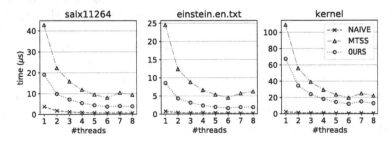

Fig. 3. Average time to answer an expansion query with expansion length 10 using multiple threads.

References

1. Bannai, H., et al.: The smallest grammar problem revisited. CoRR, abs/1908.06428 (2019)
2. Belazzougui, D., Cording, P.H., Puglisi, S.J., Tabei, Y.: Access, rank, and select in grammar-compressed strings. In: Bansal, N., Finocchi, I. (eds.) ESA 2015. LNCS, vol. 9294, pp. 142–154. Springer, Heidelberg (2015). https://doi.org/10.1007/978-3-662-48350-3_13
3. Belazzougui, D., et al.: Queries on LZ-bounded encodings. In: 2015 Data Compression Conference, pp. 83–92. IEEE (2015)
4. Bille, P., Li Gørtz, I., Prezza, N.: Space-efficient re-pair compression. In: 2017 Data Compression Conference (DCC), pp. 171–180. IEEE (2017)
5. Bille, P., Landau, G.M., Raman, R., Sadakane, K., Satti, S.R., Weimann, O.: Random access to grammar-compressed strings and trees. SIAM J. Comput. **44**(3), 513–539 (2015)
6. Boucher, C., Gagie, T., Kuhnle, A., Langmead, B., Manzini, G., Mun, T.: Prefix-free parsing for building big BWTs. Algorithms Mol. Biol. **14**(1), 13 (2019). https://doi.org/10.1186/s13015-019-0148-5
7. Charikar, M., et al.: The smallest grammar problem. IEEE Trans. Inf. Theory **51**(7), 2554–2576 (2005)
8. Danecek, P., et al.: The variant call format and VCFtools. Bioinformatics **27**(15), 2156–2158 (2011)
9. Dinklage, P., Fischer, J., Herlez, A., Kociumaka, T., Kurpicz, F.: Practical performance of space efficient data structures for longest common extensions. In: Proceedings of the Twenty-Eighth European Symposium on Algorithms (ESA) (2020, to appear)
10. Esposito, E., Graf, T.M., Vigna, S.: RecSplit: minimal perfect hashing via recursive splitting. In: 2020 Proceedings of the Twenty-Second Workshop on Algorithm Engineering and Experiments (ALENEX), pp. 175–185. SIAM (2020)
11. Furuya, I., Takagi, T., Nakashima, Y., Inenaga, S., Bannai, H., Kida, T.: MR-RePair: grammar compression based on maximal repeats. In: Data Compression Conference. DCC 2019, Snowbird, UT, USA, 26–29 March 2019, pp. 508–517 (2019)
12. Gage, P.: A new algorithm for data compression. C Users J. **12**(2), 23–38 (1994)
13. Gagie, T., I, T., Manzini, G., Navarro, G., Sakamoto, H., Takabatake, Y.: Rpair: rescaling RePair with Rsync. In: Brisaboa, N.R., Puglisi, S.J. (eds.) SPIRE 2019. LNCS, vol. 11811, pp. 35–44. Springer, Cham (2019). https://doi.org/10.1007/978-3-030-32686-9_3
14. Gallé, M.: Investigating the effectiveness of BPE: the power of shorter sequences. In: Inui, K., Jiang, J., Ng, V., Wan, X. (eds.) Proceedings of the 2019 Conference on Empirical Methods in Natural Language Processing and the 9th International Joint Conference on Natural Language Processing, EMNLP-IJCNLP 2019, Hong Kong, China, 3–7 November 2019, pp. 1375–1381. Association for Computational Linguistics (2019)
15. Ganardi, M., Jeż, A., Lohrey, M.: Balancing straight-line programs. In: 60th IEEE Annual Symposium on Foundations of Computer Science. FOCS 2019, Baltimore, Maryland, USA, 9–12 November 2019, pp. 1169–1183 (2019)
16. Gańczorz, M., Jeż, A.: Improvements on re-pair grammar compressor. In: 2017 Data Compression Conference (DCC), pp. 181–190. IEEE (2017)
17. Hucke, D.: Approximation ratios of RePair, LongestMatch and Greedy on unary strings. In: Brisaboa, N.R., Puglisi, S.J. (eds.) SPIRE 2019. LNCS, vol. 11811, pp. 3–15. Springer, Cham (2019). https://doi.org/10.1007/978-3-030-32686-9_1

18. Hucke, D., Jeż, A., Lohrey, M.: Approximation ratio of RePair. CoRR, abs/1703.06061 (2017)
19. Kempa, D., Kociumaka, T.: String synchronizing sets: sublinear-time BWT construction and optimal LCE data structure. In: Proceedings of the 51st Annual ACM SIGACT Symposium on Theory of Computing, pp. 756–767 (2019)
20. Kuhnle, A., Mun, T., Boucher, C., Gagie, T., Langmead, B., Manzini, G.: Efficient construction of a complete index for pan-genomics read alignment. In: Cowen, L.J. (ed.) RECOMB 2019. LNCS, vol. 11467, pp. 158–173. Springer, Cham (2019). https://doi.org/10.1007/978-3-030-17083-7_10
21. Kuruppu, S., Puglisi, S.J., Zobel, J.: Relative Lempel-Ziv compression of genomes for large-scale storage and retrieval. In: Chavez, E., Lonardi, S. (eds.) SPIRE 2010. LNCS, vol. 6393, pp. 201–206. Springer, Heidelberg (2010). https://doi.org/10.1007/978-3-642-16321-0_20
22. Jesper Larsson, N., Moffat, A.: Offline dictionary-based compression. In: Data Compression Conference. DCC 1999, Snowbird, Utah, USA, 29–31 March 1999, pp. 296–305 (1999)
23. Lohrey, M.: Algorithmics on SLP-compressed strings: a survey. Groups Complex. Cryptol. 4(2), 241–299 (2012)
24. Maruyama, S., Tabei, Y., Sakamoto, H., Sadakane, K.: Fully-online grammar compression. In: Kurland, O., Lewenstein, M., Porat, E. (eds.) SPIRE 2013. LNCS, vol. 8214, pp. 218–229. Springer, Cham (2013). https://doi.org/10.1007/978-3-319-02432-5_25
25. Navarro, G.: Indexing highly repetitive string collections. CoRR, abs/2004.02781 (2020)
26. Ohno, T., Goto, K., Takabatake, Y., I, T., Sakamoto, H.: LZ-ABT: a practical algorithm for α-balanced grammar compression. In: Iliopoulos, C., Leong, H.W., Sung, W.-K. (eds.) IWOCA 2018. LNCS, vol. 10979, pp. 323–335. Springer, Cham (2018). https://doi.org/10.1007/978-3-319-94667-2_27
27. Prezza, N.: Optimal rank and select queries on dictionary-compressed text. In: Pisanti, N., Pissis, S.P. (eds.) 30th Annual Symposium on Combinatorial Pattern Matching. CPM 2019, volume 128 of LIPIcs, Pisa, Italy, 18–20 June 2019, pp. 4:1–4:12. Schloss Dagstuhl - Leibniz-Zentrum für Informatik (2019)
28. Rytter, W.: Application of Lempel-Ziv factorization to the approximation of grammar-based compression. Theoret. Comput. Sci. 302(1–3), 211–222 (2003)
29. Rytter, W.: Grammar compression, LZ-encodings, and string algorithms with implicit input. In: Díaz, J., Karhumäki, J., Lepistö, A., Sannella, D. (eds.) ICALP 2004. LNCS, vol. 3142, pp. 15–27. Springer, Heidelberg (2004). https://doi.org/10.1007/978-3-540-27836-8_5
30. Sakai, K., Ohno, T., Goto, K., Takabatake, Y., I, T., Sakamoto, H.: RePair in compressed space and time. In: 2019 Data Compression Conference (DCC), pp. 518–527. IEEE (2019)
31. Sennrich, R., Haddow, B., Birch, A.: Neural machine translation of rare words with subword units. In: Proceedings of the 54th Annual Meeting of the Association for Computational Linguistics. ACL 2016. Volume 1: Long Papers, Berlin, Germany, 7–12 August 2016. The Association for Computer Linguistics (2016)
32. Stevens, E.L., et al.: The public health impact of a publically available, environmental database of microbial genomes. Front. Microbiol. 8, 808 (2017)
33. Takabatake, Y., I, T., Sakamoto, H.: A space-optimal grammar compression. In: 25th Annual European Symposium on Algorithms. ESA 2017, Vienna, Austria, 4–6 September 2017, pp. 67:1–67:15 (2017)

34. The 1000 Genomes Project Consortium: A global reference for human genetic variation. Nature **526**, 68–74 (2015)
35. Verbin, E., Yu, W.: Data structure lower bounds on random access to grammar-compressed strings. In: Fischer, J., Sanders, P. (eds.) CPM 2013. LNCS, vol. 7922, pp. 247–258. Springer, Heidelberg (2013). https://doi.org/10.1007/978-3-642-38905-4_24

A Comparison of Empirical Tree Entropies

Danny Hucke, Markus Lohrey$^{(\boxtimes)}$, and Louisa Seelbach Benkner

University of Siegen, Siegen, Germany
{hucke,lohrey,seelbach}@eti.uni-siegen.de

Abstract. Whereas for strings, higher-order empirical entropy is the standard entropy measure, several different notions of empirical entropy for trees have been proposed in the past, notably label entropy, degree entropy, conditional versions of the latter two, and empirical entropy of trees (here, called label-shape entropy). In this paper, we carry out a systematic comparison of these entropy measures. We underpin our theoretical investigations by experimental results with real XML data.

1   Introduction

In the area of string compression the notion of higher order empirical entropy yields a well established measure for the compressibility of a string. Roughly speaking, the kth-order empirical entropy of a string is the expected uncertainty about the symbol at a certain position, given the k-preceding symbols. In fact, except for some modifications (as the kth-order modified empirical entropy from [19]) the authors are not aware of any other empirical entropy measure for strings ("empirical" refers to the fact that the entropy is defined for the string itself and not a certain probability distribution on strings). For many string compressors, worst-case bounds on the length of a compressed string in terms of the kth-order empirical entropy are known [11, 19, 20]. For further aspects of higher-order empirical entropy see [8].

If one goes from strings to trees the situation becomes different. Let us first mention that the area of tree compression (and compression of structured data in general) is currently a very active area, which is motivated by the appearance of large tree data in applications like XML processing. Common tree compression techniques are based on succinct tree encodings [5, 6, 12, 17, 21], grammar-based tree compressors [9, 13, 14, 18], directed acyclic graphs [3, 7] and top dags [1, 2]. In recent years, several notions of empirical tree entropy have been proposed with the aim of quantifying the compressibility of a given tree. Let us briefly discuss these entropies in the following paragraphs (all entropies below are unnormalized; the corresponding normalized entropies are obtained by dividing by the tree size).

This work was supported by the DFG research project LO 748/10-2 (Quantitative Aspekte Grammatik-basierter Kompression).

C. Boucher and S. V. Thankachan (Eds.): SPIRE 2020, LNCS 12303, pp. 232–246, 2020.
https://doi.org/10.1007/978-3-030-59212-7_17

Ferragina et al. [5,6] introduced the kth-order label entropy $H_k^\ell(t)$ of a node-labeled unranked[1] tree t. Its normalized version is the expected uncertainty about the label of a node v, given the so-called k-*label-history* of v, which contains the k first labels on the path from v's parent node to the root. The kth-order label entropy is not useful for unlabeled trees since it ignores the tree shape.

In [17], Jansson et al. introduce the *degree entropy* $H^{\deg}(t)$, which is the (unnormalized) 0th-order empirical entropy of the node degrees occurring in the unranked tree t. Degree entropy is mainly made for unlabeled trees since it ignores node labels, but in combination with label entropy it yields a reasonable measure for the compressibility of a tree: every node-labeled unranked tree of size n in which σ many different node labels occur can be stored in $H_k^\ell(t) + H^{\deg}(t) + o(n+n\log\sigma)$ bits if σ is not too big; see Theorem 2. Note that the (unnormalized) degree entropy of a binary tree with n leaves converges to $2n - o(n)$ since a binary tree with n leaves has exactly $n - 1$ nodes of degree 2.

Recently, Ganczorz [12] defined relativized versions of kth-order label entropy and degree entropy: the kth-order degree-label entropy $H_k^{\deg,\ell}(t)$ and the kth-order label-degree entropy $H_k^{\ell,\deg}(t)$. The normalized version of $H_k^{\deg,\ell}(t)$ is the expected uncertainty about the label of a node v of t, given (i) the k-label-history of v and (ii) the degree of v, whereas the normalized version of $H_k^{\ell,\deg}(t)$ is the expected uncertainty about the degree of a node v, given (i) the k-label-history of v and (ii) the label of v. Ganczorz [12] proved that every node-labeled unranked tree of size n can be stored in $H_k^\ell(t) + H_k^{\ell,\deg}(t) + o(n + n\log\sigma)$ bits as well as in $H^{\deg}(t) + H_k^{\deg,\ell}(t) + o(n + n\log\sigma)$ bits (again assuming σ is not too big); see Theorem 2. Note that for unlabeled trees t, we have $H_k^\ell(t) + H_k^{\ell,\deg}(t) = H^{\deg}(t) + H_k^{\deg,\ell}(t) = H^{\deg}(t)$, which for unlabeled binary trees is equal to the information theoretic upper bound $2n - o(n)$ (with n the number of leaves).

Motivated by the inability of the existing entropies for measuring the compressibility of unlabeled binary trees, we introduced in [14] a new entropy for binary trees (possibly with labels) that we called k^{th}-order empirical entropy $H_k(t)$. In order to distinguish it better from the existing tree entropies we prefer the term k^{th}-*order label-shape entropy* in this paper. The main idea is to extend k-label-histories in a binary tree by adding to the labels of the k predecessors of a node v also the k last directions (0 for left, 1 for right) on the path from the root to v. We call this extended label history simply the k-history of v. The normalized version of $H_k(t)$ is the expected uncertainty about the pair consisting of the label of a node and the information whether it is a leaf or an internal node, given the k-history of the node. The main result of [14] states that a node-labeled binary tree t can be stored in $H_k(t) + o(n + n\log\sigma)$ bits using a grammar-based code building on tree straight-line programs. We also defined in [14] the kth-order label-shape entropy of an unranked node-labeled tree t by taking the kth-order label-shape entropy of the first-child next-sibling encoding of t.

---

[1] Unranked means that there is no bound on the number of children of a node. Moreover, we only consider ordered trees, where the children of a node are linearly ordered.

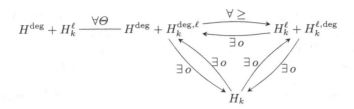

Fig. 1. Comparison of the entropy notions for unranked node-labeled trees. The meaning of the red and green arrows is explained in the main text.

The goal of this paper is to compare the entropy variants $H_k^\ell(t) + H^{\deg}(t)$, $H_k^\ell(t) + H_k^{\ell,\deg}(t)$, $H^{\deg}(t) + H_k^{\deg,\ell}(t)$, and $H_k(t)$. Our results for unranked node-labeled trees are summarized in Fig. 1. Let us explain the meaning of the arrows in Fig. 1: For two entropy notions H and H', an arrow $H \xrightarrow{\exists o} H'$ means that there is a sequence of unranked node-labeled trees t_n ($n \geq 1$) such that (i) the function $n \mapsto |t_n|$ is strictly increasing and (ii) $H(t_n) \leq o(H'(t_n))$ (in most cases we prove an exponential separation). The meaning of the arrow with label $\forall \geq$ is that $H^{\deg}(t) + H_k^{\deg,\ell}(t) \geq H_k^\ell(t) + H_k^{\ell,\deg}(t)$ for every unranked node-labeled tree t, whereas the edge with label $\forall\Theta$ means that $H^{\deg}(t) + H_k^{\deg,\ell}(t)$ and $H^{\deg}(t) + H_k^\ell(t)$ are equivalent up to fixed multiplicative constants (which are 1 and 2).

We also investigate the relationship between the entropies for node-labeled binary trees and unranked unlabeled trees (the case of unlabeled binary trees is not really interesting as explained above). An unranked unlabeled tree t of size n can be represented with $H^{\deg}(t) + o(n)$ bits [17]. Here, we prove that $H_k(t) \leq 2H^{\deg}(t) + 2\log_2(n) + 4$ for every unranked unlabeled tree t.

Finally, we underpin our theoretical results by experimental results with real XML data from XMLCompBench (http://xmlcompbench.sourceforge.net). For each XML document we consider the corresponding tree structure t (obtained by removing all text values and attributes) and compute $H_k^\ell(t) + H^{\deg}(t)$, $H_k^\ell(t) + H_k^{\ell,\deg}(t)$, $H^{\deg}(t) + H_k^{\deg,\ell}(t)$, and $H_k(t)$. The results are summarized in Table 1 on page 14. Our experiments indicate that the upper bound on the number of bits needed by the compressed data structure in [14] is the strongest for real XML data since the kth-order label-shape entropy (for $k > 0$) is significantly smaller than all other entropy values for all XMLs that we have examined.

Let us remark that Ganczorz's succinct tree representations [12] that achieve (up to low-order terms) the entropies $H_k^\ell(t) + H_k^{\ell,\deg}(t)$ and $H^{\deg}(t) + H_k^{\deg,\ell}(t)$, respectively, allow constant query times for a large number of tree queries. For the entropy $H_k(t)$ such a result is not known. The tree representation from [14] is based on tree straight-line programs, which can be queried in logarithmic time (if we assume logarithmic height of the grammar, which can be enforced by [10]).

Missing proofs can be found in the long version [16].

2   Preliminaries

With $\mathbb{N}$ we denote the natural numbers including 0. Let $w = a_1 \cdots a_l \in \Gamma^*$ be a word over an alphabet Γ. With $|w| = l$ we denote the length of w. Let ε denote the empty word. We use the standard $\mathcal{O}$-notation. If $b > 1$ is a constant, then we write $\mathcal{O}(\log n)$ for $\mathcal{O}(\log_b n)$. Moreover, terms $\log_b n$ with $b \geq 1$ are implicitly replaced by $\log_{b'} n$ for $b' = \max\{2, b\}$. We make the convention that $0 \cdot \log(0) = 0$ and $0 \cdot \log(x/0) = 0$ for $x \geq 0$. The well-known log-sum inequality (see e.g. [4, Theorem 2.7.1]) states:

Lemma 1 (Log-Sum inequality). *Let $a_1, a_2, \ldots, a_l, b_1, b_2, \ldots, b_l \geq 0$ be real numbers. Moreover, let $a = \sum_{i=1}^{l} a_i$ and $b = \sum_{i=1}^{l} b_i$. Then*

$$a \log_2 \left(\frac{b}{a} \right) \geq \sum_{i=1}^{l} a_i \log_2 \left(\frac{b_i}{a_i} \right).$$

2.1   Unranked Trees

Let Σ denote a finite alphabet of size $|\Sigma| = \sigma \geq 1$. Later, we need a fixed, distinguished symbol from Σ that we denote with $\square \in \Sigma$. We consider Σ-*labeled unranked ordered trees*, where "Σ-labeled" means that every node is labeled by a symbol from the alphabet Σ, "ordered" means that the children of a node are totally ordered, and "unranked" means that the number of children of a node (also called its *degree*) can be any natural number. In particular, the degree of a node does not depend on the node's label or vice versa. Let us denote by $\mathcal{T}(\Sigma)$ the set of all such trees. Formally, the set $\mathcal{T}(\Sigma)$ is inductively defined as the smallest set of expressions such that if $a \in \Sigma$ and $t_1, \ldots, t_n \in \mathcal{T}(\Sigma)$ then also $a(t_1 \cdots t_n) \in \mathcal{T}(\Sigma)$. This expression represents a tree with an a-labeled root whose direct subtrees are $t_1, \ldots, t_n$. Note that for the case $n = 0$ we obtain the tree $a()$, for which we also write a. The *size* $|t|$ of $t \in \mathcal{T}(\Sigma)$ is the number of occurrences of labels from Σ in t, i.e., $a(t_1 \cdots t_n) = 1 + \sum_{i=1}^{n} |t_i|$. We identify an unranked tree with a graph in the usual way, where each node is labeled with a symbol from Σ. Let $V(t)$ denote the set of nodes of a tree $t \in \mathcal{T}(\Sigma)$. We have $|V(t)| = |t|$. The label of a node $v \in V(t)$ is denoted with $\ell(v) \in \Sigma$. Moreover, we write $\deg(v) \in \mathbb{N}$ for the degree of v. An important special case of unranked trees are *unlabeled unranked trees*: They can be considered as labeled unranked trees over a singleton alphabet $\Sigma = \{a\}$.

For a node $v \in V(t)$ of a tree t, we define its *label-history* $h^{\ell}(v) \in \Sigma^*$ inductively: for the root node v_0, we set $h^{\ell}(v_0) = \varepsilon$ and for a child node w of a node v of t, we set $h^{\ell}(w) = h^{\ell}(v) \ell(v)$. In other words: $h^{\ell}(v)$ is obtained by concatenating the node labels along the unique path from the root to v. The label of v is not part of the label-history of v. The k-*label-history* $h_k^{\ell}(v)$ of a tree node $v \in V(t)$ is defined as the length-k suffix of $\square^k h^{\ell}(v)$, where $\square$ is a fixed dummy symbol in Σ. This means that if the depth of v in t is greater than k, then $h_k^{\ell}(v)$ lists the last k node labels along the path from the root to node v.

If the depth of v in t is at most v, then we pad its label-history $h^\ell(v)$ with the symbol $\square$ such that $h_k^\ell(v) \in \Sigma^k$. For $z \in \Sigma^k$, $a \in \Sigma$ and $i \in \mathbb{N}$ we set

$$n_z^t = |\{v \in V(t) \mid h_k^\ell(v) = z\}|, \tag{1}$$

$$n_i^t = |\{v \in V(t) \mid \deg(v) = i\}|, \tag{2}$$

$$n_{z,i}^t = |\{v \in V(t) \mid h_k^\ell(v) = z \text{ and } \deg(v) = i\}|, \tag{3}$$

$$n_{z,a}^t = |\{v \in V(t) \mid h_k^\ell(v) = z \text{ and } \ell(v) = a\}|, \tag{4}$$

$$n_{z,i,a}^t = |\{v \in V(t) \mid h_k^\ell(v) = z, \ell(v) = a \text{ and } \deg(v) = i\}|. \tag{5}$$

In order to avoid ambiguities in these notations we should assume that $\Sigma \cap \mathbb{N} = \emptyset$. Moreover, when writing $n_{z,i}^t$ (resp., $n_{z,a}^t$) then, implicitly, i (resp., a) always belongs to $\mathbb{N}$ (resp., Σ).

2.2   Binary Trees

An important subset of $\mathcal{T}(\Sigma)$ is the set $\mathcal{B}(\Sigma)$ of *labeled binary trees* over the alphabet Σ. A binary tree is a tree in $\mathcal{T}(\Sigma)$, where every node has either exactly two children or is a leaf. Formally, $\mathcal{B}(\Sigma)$ is inductively defined as the smallest set of terms over Σ such that (i) $\Sigma \subseteq \mathcal{B}(\Sigma)$ and (ii) if $t_1, t_2 \in \mathcal{B}(\Sigma)$ and $a \in \Sigma$, then $a(t_1 t_2) \in \mathcal{B}(\Sigma)$. An *unlabeled binary tree* can be considered as a binary tree over the singleton alphabet $\Sigma = \{a\}$. The *first-child next-sibling encoding* (or shortly *fcns-encoding*) transforms a tree $t \in \mathcal{T}(\Sigma)$ into a binary tree $t \in \mathcal{B}(\Sigma)$. We define it more generally for an ordered sequence of unranked trees $s = t_1 t_2 \cdots t_n$ (a so-called forest) inductively as follows (recall that $\square \in \Sigma$ is a fixed distinguished symbol in Σ): $\mathrm{fcns}(s) = \square$ for $n = 0$ and if $n \geq 1$ and $t_1 = a(t_1' \cdots t_m')$ then $\mathrm{fcns}(s) = a(\mathrm{fcns}(t_1' \cdots t_m') \mathrm{fcns}(t_2 \cdots t_n))$. Thus, the left (resp. right) child of a node in $\mathrm{fcns}(s)$ is the first child (resp., right sibling) of the node in s or a $\square$-labeled leaf, if it does not exist.

For the special case of binary trees, we extend the label history of a node to its full history, which we just call its history. Intuitively, the history of a node v records all information that can be obtained by walking from the root of the tree straight down to the node v. In addition to the node labels this also includes the directions (left/right) of the descending edges. For an integer $k \geq 0$ let

$$\mathcal{L}_k = (\Sigma\{0,1\})^k = \{a_1 i_1 a_2 i_2 \cdots a_k i_k \mid a_j \in \Sigma, i_j \in \{0,1\} \text{ for } 1 \leq j \leq k\}.$$

For a node v of a binary tree t, we define its *history* $h(v) \in (\Sigma\{0,1\})^*$ inductively as follows: For the root node v_0, we set $h(v_0) = \varepsilon$. For a left child node w of a node v of t, we set $h(w) = h(v)\ell(v)0$ and for a right child node w of v, we set $h(w) = h(v)\ell(v)1$ (recall that $\ell(v)$ is the label of v). That is, in order to obtain $h(v)$, while descending in the tree from the root node to the node v, we alternately write down the current node label from Σ and the direction into which we descend (0 if we descend to a left child, 1 if we descend to a right child). Note that the symbol that labels v is not part of the history $h(v)$. The *k-history* of a node v is then defined as the length-$2k$ suffix of the word $(\square 0)^k h(v)$, where

$\square$ is again a fixed dummy symbol in Σ. This means that if the depth of v in t is greater than k, then $h_k(v)$ describes the last k directions and node labels along the path from the root to node v. If the depth of v in t is at most k, then we pad the history of v with $\square$'s and zeroes such that $h_k(v) \in \mathcal{L}_k$. For a node v of a binary tree we define $\lambda(v) = (\ell(v), \deg(v)) \in \Sigma \times \{0, 2\}$. For $z \in \mathcal{L}_k$ and $\tilde{a} \in \Sigma \times \{0, 2\}$, we finally define

$$m_z^t = |\{v \in V(t) \mid h_k(v) = z\}|, \tag{6}$$

$$m_{z,\tilde{a}}^t = |\{v \in V(t) \mid h_k(v) = z \text{ and } \lambda(v) = \tilde{a}\}|. \tag{7}$$

3   Empirical Entropy for Trees

In this section we formally define the various entropy measures that were mentioned in the introduction. Note that in all cases we define so-called unnormalized entropies, which has the advantage that we do not have to multiply with the size of the tree in bounds for the encoding size of a tree. Note that in [5,6,12,17] the authors define normalized entropies. In each case, one obtains the normalized entropy by dividing the corresponding unnormalized entropy by the tree size.

Label Entropy. The first notion of empirical entropy for trees was introduced in [5]. In order to distinguish notions, we call the entropy from [5] *label entropy*. It is defined for unranked labeled trees $t \in \mathcal{T}(\Sigma)$: the kth-order *label entropy* $H_k^\ell(t)$ of t is defined as follows, where n_z^t and $n_{z,a}^t$ are from (1) and (4), respectively:

$$H_k^\ell(t) = \sum_{z \in \Sigma^k} \sum_{a \in \Sigma} n_{z,a}^t \log_2 \left(\frac{n_z^t}{n_{z,a}^t} \right). \tag{8}$$

We remark that in [5], it is not explicitly specified how to deal with nodes, whose label-history is shorter than k. There are three natural variants: (i) padding label-histories with a symbol $\square \subset \Sigma$ (this is our choice), (ii) padding label-histories with a fresh symbol $\diamond \notin \Sigma$, or equivalently, allowing label-histories of length smaller than k, and (iii) ignoring nodes whose label-history is shorter than k. However, similar considerations as in the appendix of [15] show that these approaches yield the same k^{th}-order label entropy up to an additional additive term of at most $m^<(1 + 1/\ln(2) + \log_2(\sigma|t|/m^<))$, where $m^<$ is the number of nodes at depth less than k in t.

Degree Entropy. Another notion of empirical entropy for trees is the entropy measure from [17], which we call *degree entropy*. Degree entropy is primarily made for unlabeled unranked trees, as it completely ignores node labels. Nevertheless the definition works for trees $t \in \mathcal{T}(\Sigma)$ over any alphabet Σ. For a tree $t \in \mathcal{T}(\Sigma)$, the degree entropy $H^{\deg}(t)$ is the 0th-order entropy of the node degrees (n_i^t is from (2)):

$$H^{\deg}(t) = \sum_{i=0}^{|t|} n_i^t \log_2 \left(\frac{|t|}{n_i^t} \right). $$

For the special case of unlabeled trees the following result was shown in [17]:

Theorem 1 ([17, Theorem 1]). *Let t be an unlabeled unranked tree. Then t can be represented with $H^{\deg}(t) + \mathcal{O}(|t| \log \log(|t|)/\log|t|)$ bits.*

Label-Degree Entropy and Degree-Label Entropy. Recently, two combinations of the label entropy from [5] and the degree entropy from [17] were proposed in [12]. We call these two entropy measures *label-degree entropy* and *degree-label entropy*. Both notions are defined for unranked node-labeled trees. Let $t \in \mathcal{T}(\Sigma)$ be such a tree. The kth-order *label-degree entropy* $H_k^{\ell,\deg}(t)$ of t from [12] is defined as follows, where $n_{z,a}^t$ and $n_{z,i,a}^t$ are from (4) and (5), respectively:

$$H_k^{\ell,\deg}(t) = \sum_{z \in \Sigma^k} \sum_{a \in \Sigma} \sum_{i=0}^{|t|} n_{z,i,a}^t \log_2 \left(\frac{n_{z,a}^t}{n_{z,i,a}^t} \right).$$

The kth-order *degree-label entropy* $H_k^{\deg,\ell}(t)$ of t from [12] is defined as follows, where $n_{z,i}^t$ and $n_{z,i,a}^t$ are from (3) and (5), respectively:

$$H_k^{\deg,\ell}(t) = \sum_{z \in \Sigma^k} \sum_{i=0}^{|t|} \sum_{a \in \Sigma} n_{z,i,a}^t \log_2 \left(\frac{n_{z,i}^t}{n_{z,i,a}^t} \right).$$

In order to deal with nodes whose label-history is shorter than k one can again choose one of the three alternatives (i)–(iii) that were mentioned after (8). In [12], variant (ii) is chosen, while the above definitions correspond to choice (i). However, as for the label entropy one can show that these variants only differ by a small additive term of at most $m^<(1/\ln(2) + \log_2(\sigma|t|/m^<))$ in the case of the degree-label entropy, respectively, $m^<(1/\ln(2) + \log_2|t|)$ in the case of the label-degree entropy, where $m^<$ is the number of nodes at depth less than k.

By [12], the following inequalities hold:

Lemma 2. *For every $t \in \mathcal{T}(\Sigma)$, $H_k^{\ell,\deg}(t) \le H^{\deg}(t)$ and $H_k^{\deg,\ell}(t) \le H_k^{\ell}(t)$.*

Moreover, one of the main results of [12] states the following bounds:

Theorem 2 ([12, Theorem 12]). *Let $t \in \mathcal{T}(\Sigma)$, with $\sigma \le |t|^{1-\alpha}$ for some $\alpha > 0$. Then t can be represented in*

$$H + \mathcal{O}\left(\frac{|t|k \log \sigma + |t| \log \log_\sigma |t|}{\log_\sigma |t|} \right),$$

bits, where H is one of $H^{\deg}(t) + H_k^{\ell}(t)$, $H_k^{\ell}(t) + H_k^{\ell,\deg}(t)$, or $H^{\deg}(t) + H_k^{\deg,\ell}(t)$.

Label-Shape Entropy. Another notion of empirical entropy for trees which incorporates both node labels and tree structure was recently introduced in [14]: Let us start with a binary tree $t \in \mathcal{B}(\Sigma)$. The kth-order label-shape entropy $H_k(t)$ of t (in [14] it is simply called the kth-order empirical entropy of t) is

$$H_k(t) = \sum_{z \in \mathcal{L}_k} \sum_{\tilde{a} \in \Sigma \times \{0,2\}} m_{z,\tilde{a}}^t \log_2 \left(\frac{m_z^t}{m_{z,\tilde{a}}^t} \right), \tag{9}$$

where m_z^t and $m_{z,\bar{a}}^t$ are from (6) and (7), respectively. Now let $t \in \mathcal{T}(\Sigma)$ be an unranked tree and recall that $\mathrm{fcns}(t) \in \mathcal{B}(\Sigma)$. The kth-order label-shape entropy $H_k(t)$ of t is defined as

$$H_k(t) = H_k(\mathrm{fcns}(t)). \tag{10}$$

The following result is shown in [14] using a grammar-based encoding of trees:

Theorem 3. *Every tree $t \in \mathcal{T}(\Sigma)$ can be represented within the following bound (in bits):*

$$H_k(t) + \mathcal{O}\left(\frac{k|t|\log\sigma}{\log_\sigma |t|}\right) + \mathcal{O}\left(\frac{|t|\log\log_\sigma |t|}{\log_\sigma |t|}\right) + \sigma.$$

Note that for binary trees, there are two possibilities how to compute the label-shape entropy $H_k(t)$. The first is to compute the label-shape entropy as defined in (9), the second is to consider the binary tree as an unranked tree and compute the label-shape entropy of its first-child next-sibling encoding as defined in (10). The following lemma from [15] states that if we consider the first-child next-sibling encoding of the binary tree instead of the binary tree itself, the kth-order label-shape entropy does not increase if we double the value of k:

Lemma 3. *Let $t \in \mathcal{B}(\Sigma)$ be a binary tree with first-child next-sibling encoding $\mathrm{fcns}(t) \in \mathcal{B}(\Sigma)$. Then $H_{2k}(\mathrm{fcns}(t)) \leq H_{k-1}(t)$ for $1 \leq k \leq n$.*

In contrast to Lemma 3, there are families of binary trees t_n where $H_k(t_n) \in \Theta(n - k)$ and $H_k(\mathrm{fcns}(t_n)) \in \Theta(\log(n - k))$ [15].

4   Comparison of the Empirical Entropy Notions

As we have seen in Theorems 2 and 3, entropy bounds for the number of bits needed to represent an unranked labeled tree t are achievable by $H_k(t)$, $H_k^\ell(t) + H_k^{\ell,\deg}(t)$, $H^{\deg}(t) + H_k^{\deg,\ell}(t)$, and $H^{\deg}(t) + H_k^\ell(t)$, where in all cases we have to add a low-order term. The term $H^{\deg}(t) + H_k^\ell(t)$ is lower-bounded by $H_k^\ell(t) + H_k^{\ell,\deg}(t)$ and $H^{\deg}(t) + H_k^{\deg,\ell}(t)$ by Lemma 2. For the special case of unlabeled unranked trees, $H^{\deg}(t)$ (plus low-order terms) is an upper bound on the encoding length (see Theorem 1) as well. Let us also remark that $H_{k'}(t) \leq H_k(t)$ for $k < k'$ and analogously for H_k^ℓ, $H_k^{\ell,\deg}$, and $H_k^{\deg,\ell}$.

4.1   Binary Trees

Let us start with unlabeled binary trees, i.e., trees $t \in \mathcal{B}(\{a\})$ over the unary alphabet $\Sigma = \{a\}$. As $\Sigma = \{a\}$, the fixed dummy symbol used to pad k-histories and k-label-histories is $\square = a$. The following lemma follows from the fact that every binary tree of size $2n - 1$ consists of n nodes of degree 0 and $n - 1$ nodes of degree 2:

Lemma 4. *Let t be an unlabeled binary tree with n leaves and thus $|t| = 2n-1$. Then $H^{\deg}(t) = H_k^{\ell,\deg}(t) = (2 - o(1))n$.*

For the following lower bound one can take for t_n a left-degenerate chain of height n (formally: $t_1 = a$ and $t_n = a(t_{n-1}\, a)$ for $n \geq 2$).

Lemma 5. *There exists a family of unlabeled binary trees $(t_n)_{n \geq 1}$ such that $|t_n| = 2n - 1$ and $H_k(t_n) \leq \log_2(en)$ for all $n \geq 1$ and $1 \leq k \leq n$.*

Lemmas 4 and 5 already indicate that all entropies considered in this paper except for the label-shape entropy are not interesting for unlabeled binary trees. For every unlabeled binary tree t with n leaves (and $2n - 1$ nodes) we have: $H_k^{\ell}(t) = H_k^{\deg,\ell} = 0$, as every node of t has the same label, and $H_k^{\ell}(t) + H_k^{\ell,\deg}(t) = H^{\deg}(t) + H_k^{\deg,\ell}(t) = H_k^{\ell}(t) + H^{\deg}(t) = H^{\deg}(t)$, and these values are lower bounded by $2n(1 - o(1))$ (Lemma 4). In contrast, the label-shape entropy (9) is able to capture regularities in unlabeled binary trees (and attains different values for different binary trees of the same size).

Let us now look at binary trees $t \in \mathcal{B}(\Sigma)$, where Σ is arbitrary. As in the special case of unlabeled binary trees, we find that $H^{\deg}(t) = 2n(1 - o(1))$ for every binary tree t of size $2n - 1$ (the node labels do not influence $H^{\deg}(t)$), which implies $H^{\deg}(t) + H_k^{\deg,\ell}(t) \geq 2n(1 - o(1))$. The following lemma shows that $H_k(t)$ is always bounded by $H_k^{\ell}(t) + H_k^{\ell,\deg}(t)$ and $H^{\deg}(t) + H_k^{\deg,\ell}(t)$ (and hence also $H_k^{\ell}(t) + H^{\deg}(t)$) for $t \in \mathcal{B}(\Sigma)$.

Lemma 6. *Let $t \in \mathcal{B}(\Sigma)$ be a binary tree. Then (i) $H_k(t) \leq H_k^{\ell}(t) + H_k^{\ell,\deg}(t)$ and (ii) $H_k(t) \leq H^{\deg}(t) + H_k^{\deg,\ell}(t)$.*

Proof. We start with proving statement (i): We have

$$H_k(t) = \sum_{z \in \mathcal{L}_k} \sum_{a \in \Sigma} \sum_{i \in \{0,2\}} m_{z,(a,i)}^t \log_2\left(\frac{m_z^t}{m_{z,(a,i)}^t}\right)$$

$$= \sum_{z \in \mathcal{L}_k} \sum_{a \in \Sigma} \left(m_{z,(a,0)}^t + m_{z,(a,2)}^t\right) \log_2\left(\frac{m_z^t}{m_{z,(a,0)}^t + m_{z,(a,2)}^t}\right)$$

$$+ \sum_{z \in \mathcal{L}_k} \sum_{a \in \Sigma} \sum_{i \in \{0,2\}} m_{z,(a,i)}^t \log_2\left(\frac{m_{z,(a,0)}^t + m_{z,(a,2)}^t}{m_{z,(a,i)}^t}\right)$$

$$\leq \sum_{z \in \Sigma^k} \sum_{a \in \Sigma} n_{z,a}^t \log_2\left(\frac{n_z^t}{n_{z,a}^t}\right) + \sum_{z \in \Sigma^k} \sum_{a \in \Sigma} \sum_{i \in \{0,2\}} n_{z,i,a}^t \log_2\left(\frac{n_{z,a}^t}{n_{z,i,a}^t}\right)$$

$$= H_k^{\ell}(t) + H_k^{\ell,\deg}(t),$$

where the inequality in the second last line follows from the log-sum inequality (Lemma 1) and the last equality follows from the fact that in a binary tree, every node is either of degree 0 or 2. Statement (ii) can be shown in a similar way:

$$H_k(t) = \sum_{z \in \mathcal{L}_k} \sum_{a \in \Sigma} \sum_{i \in \{0,2\}} m_{z,(a,i)}^t \log_2\left(\frac{m_z^t}{m_{z,(a,i)}^t}\right)$$

$$= \sum_{z \in \mathcal{L}_k} \sum_{i \in \{0,2\}} \left(\sum_{a \in \Sigma} m^t_{z,(a,i)} \right) \log_2 \left(\frac{m^t_z}{\sum_{a \in \Sigma} m^t_{z,(a,i)}} \right)$$

$$+ \sum_{z \in \mathcal{L}_k} \sum_{a \in \Sigma} \sum_{i \in \{0,2\}} m^t_{z,(a,i)} \log_2 \left(\frac{\sum_{a \in \Sigma} m^t_{z,(a,i)}}{m^t_{z,(a,i)}} \right)$$

$$\leq \sum_{i \in \{0,2\}} n^t_i \log_2 \left(\frac{|t|}{n^t_i} \right) + \sum_{z \in \Sigma^k} \sum_{a \in \Sigma} \sum_{i \in \{0,2\}} n^t_{z,i,a} \log_2 \left(\frac{n^t_{z,i}}{n^t_{z,i,a}} \right)$$

$$= H^{\deg}(t) + H^{\deg,\ell}_k(t),$$

where the inequality follows again from the log-sum inequality. $\qquad\square$

4.2   Unlabeled Unranked Trees

In this subsection, we consider unranked trees $t \in \mathcal{T}(\Sigma)$ over the unary alphabet $\Sigma = \{a\}$. As $\Sigma = \{a\}$, the fixed dummy symbol used to pad k-histories and k-label-histories is $\square = a$. Moreover, note that in order to compute $H_k(t)$ for an unranked tree $t \in \mathcal{T}(\Sigma)$, we have to consider $\mathrm{fcns}(t)$, which is an unlabeled binary tree (we must take $\square = a$ by our conventions for the dummy symbol; hence the fresh $\square$-labeled leaves in $\mathrm{fcns}(t)$ are labeled with a, too). As in the case of unlabeled binary trees, we observe that some entropy measures, in particular those that involve labels, only attain trivial values for unranked unlabeled trees. More precisely, for every tree $t \in \mathcal{T}(\{a\})$ we have $H^\ell_k(t) = H^{\deg,\ell}_k(t) = 0$, as every node has the same label a, and $H^{\deg}(t) = H^{\ell,\deg}_k(t)$, as every node has the same k-label-history and the same label. Moreover, we get $H^\ell_k(t) + H^{\ell,\deg}_k(t) = H^{\deg}(t) + H^{\deg,\ell}_k(t) = H^{\deg}(t) + H^\ell_k(t) = H^{\deg}(t)$. By this observation, we only compare $H_k(t)$ with $H^{\deg}(t)$ for $t \in \mathcal{T}(\{a\})$ in this subsection. By Lemmas 4 and 5, there exists a family of unlabeled trees $(t_n)_{n \geq 1}$ such that $|t_n| = \Theta(n)$ and for which $H_k(t_n)$ is exponentially smaller than $H^{\deg}(t_n)$. For general unranked unlabeled trees, we have the following result; see [16] for the proof.

Theorem 4. *For every unlabeled unranked tree t with $|t| \geq 2$ and integer $k \geq 1$, we have $H_k(t) \leq 2H^{\deg}(t) + 2 \log_2(|t|) + 4$.*

As $H^{\deg}(t) = H^\ell_k(t) + H^{\ell,\deg}_k(t) = H^{\deg}(t) + H^{\deg,\ell}_k(t)$ for every tree $t \in \mathcal{T}(\{a\})$ and $k \geq 0$, we obtain the following corollary from Theorem 4:

Corollary 1. *For every unlabeled unranked tree $t \in \mathcal{T}(\{a\})$ with $|t| \geq 2$ and integer $k \geq 1$, we have $H_k(t) \leq 2(H^{\deg}(t) + H^{\deg,\ell}_k(t)) + 2 \log_2(|t|) + 4$, and $H_k(t) \leq 2(H^\ell_k(t) + H^{\ell,\deg}_k(t)) + 2 \log_2(|t|) + 4$.*

We note that there exist families of unranked trees over a non-unary alphabet, for which the degree entropy is exponentially smaller than the kth-order label-shape tree entropy. This is not very surprising as the label-shape entropy incorporates the node labels, while the degree entropy does not.

4.3 Labeled Unranked Trees

In this section, we consider general unranked labeled trees $t \in \mathcal{T}(\Sigma)$ over arbitrary alphabets Σ. The entropies to be compared are $H_k(t)$, $H^{\deg}(t) + H_k^{\deg,\ell}(t)$, $H_k^\ell(t) + H_k^{\ell,\deg}(t)$ and $H^{\deg}(t) + H_k^\ell(t)$. Somewhat surprisingly it turns out that $H_k^\ell(t) + H_k^{\ell,\deg}(t)$ is at most $H^{\deg}(t) + H_k^{\deg,\ell}(t)$ for every tree t:

Theorem 5. Let $t \in \mathcal{T}(\Sigma)$. Then $H_k^\ell(t) + H_k^{\ell,\deg}(t) \le H^{\deg}(t) + H_k^{\deg,\ell}(t)$.

Proof. We have

$$H_k^\ell(t) + H_k^{\ell,\deg}(t)$$

$$= \sum_{z \in \Sigma^k} \sum_{a \in \Sigma} n_{z,a}^t \log_2\left(\frac{n_z^t}{n_{z,a}^t}\right) + \sum_{z \in \Sigma^k} \sum_{a \in \Sigma} \sum_{i=0}^{|t|} n_{z,i,a}^t \log_2\left(\frac{n_{z,a}^t}{n_{z,i,a}^t}\right)$$

$$= \sum_{z \in \Sigma^k} \sum_{a \in \Sigma} \sum_{i=0}^{|t|} n_{z,i,a}^t \log_2\left(\frac{n_z^t}{n_{z,a}^t}\right) + \sum_{z \in \Sigma^k} \sum_{a \in \Sigma} \sum_{i=0}^{|t|} n_{z,i,a}^t \log_2\left(\frac{n_{z,a}^t}{n_{z,i,a}^t}\right)$$

$$= \sum_{z \in \Sigma^k} \sum_{a \in \Sigma} \sum_{i=0}^{|t|} n_{z,i,a}^t \log_2\left(\frac{n_z^t}{n_{z,i,a}^t}\right)$$

$$= \sum_{z \in \Sigma^k} \sum_{a \in \Sigma} \sum_{i=0}^{|t|} n_{z,i,a}^t \log_2\left(\frac{n_z^t}{n_{z,i}^t}\right) + \sum_{z \in \Sigma^k} \sum_{a \in \Sigma} \sum_{i=0}^{|t|} n_{z,i,a}^t \log_2\left(\frac{n_{z,i}^t}{n_{z,i,a}^t}\right)$$

$$= \sum_{z \in \Sigma^k} \sum_{i=0}^{|t|} n_{z,i}^t \log_2\left(\frac{n_z^t}{n_{z,i}^t}\right) + \sum_{z \in \Sigma^k} \sum_{a \in \Sigma} \sum_{i=0}^{|t|} n_{z,i,a}^t \log_2\left(\frac{n_{z,i}^t}{n_{z,i,a}^t}\right)$$

$$\le H^{\deg}(t) + H_k^{\deg,\ell}(t),$$

where the final inequality follows from the log-sum inequality (Lemma 1).   □

As a corollary of Lemma 2 and Theorem 5 it turns out that $H^{\deg}(t) + H_k^{\deg,\ell}(t)$ and $H_k^\ell(t) + H^{\deg}(t)$ are equivalent up to constant factors.

Corollary 2. Let $t \in \mathcal{T}(\Sigma)$. Then

$$H^{\deg}(t) + H_k^{\deg,\ell}(t) \le H^{\deg}(t) + H_k^\ell(t) \le 2H^{\deg}(t) + H_k^{\deg,\ell}(t).$$

In the rest of the section we present three examples showing that in all cases that are not covered by Theorem 5 we can achieve a non-constant (in most cases even exponential) separation between the corresponding entropies.

Lemma 7. (i) $|t_n| = 2n + 1$,
(ii) $H_k(t_n) \le \log_2(e) + \log_2\left(n - \lfloor\frac{k-1}{2}\rfloor\right) + 2$,
(iii) $H_k^{\deg,\ell}(t_n) = 2n$ and hence $H^{\deg}(t_n) + H_k^{\deg,\ell}(t_n) \ge 2n$, and
(iv) $H_k^\ell(t_n) \ge 2n$ and hence $H_k^\ell(t_n) + H_k^{\ell,\deg}(t_n) \ge 2n$.

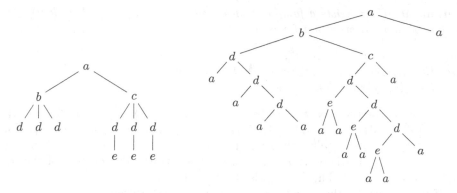

Fig. 2. The binary tree t_3 from Lemma 8 (left) and its first-child next-sibling encoding fcns(t_3) (right).

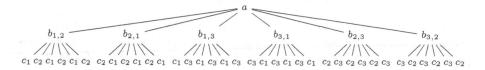

Fig. 3. The tree $t_{3,2}$ from Lemma 9.

For the tree t_n in Lemma 7 one can take $t_n = a(bcbc \cdots bc)$ with n occurrences of b (respectively, c). Lemma 7 shows that there are not only families of binary trees, but also families of unranked (non-binary) trees $(t_n)_{n \geq 1}$ (for which we have to compute $H_k(t_n)$ via the fcns-endcoding) such that $|t_n| = \Theta(n)$ and $H_k(t_n)$ is exponentially smaller than $H^{\deg}(t_n) + H_k^{\deg,\ell}(t_n)$ and $H_k^{\ell}(t_n) + H_k^{\ell,\deg}(t_n)$.

Lemma 8. *There exists a family of unranked trees $(t_n)_{n \geq 1}$ such that for all $n \geq 1$ and $1 \leq k \leq n$:*

(i) $|t_n| = 3n + 3$,
(ii) $H_k(t_n) \geq 2(n - k + 1)$,
(iii) $H^{\deg}(t_n) + H_k^{\deg,\ell}(t_n) \geq 2n$ and
(iv) $H_1^{\ell}(t_n) + H_1^{\ell,\deg}(t_n) = 3 \log_2(3)$.

For the tree t_n in Lemma 7 one can take $t_n = a(b(dd \cdots d) c(d(e)d(e) \cdots d(e)))$ with $2n$ occurrences of d. The tree t_3 is shown in Fig. 2. Note that we clearly need $\Omega(\log n)$ bits to represent this tree (since we have to represent its size). This does not contradict Theorem 2 and the $\mathcal{O}(1)$-bound for $H_1^{\ell}(t_n) + H_1^{\ell,\deg}(t_n)$ in Lemma 8, since we have the additional additive term of order $o(|t|)$ in Theorem 2.

In the following lemma, $n^{\underline{k}} = n(n-1) \cdots (n-k+1)$ is the falling factorial.

Lemma 9. *There exists a family of unranked trees* $(t_{n,k})_{n \geq 1}$, *where* $k(n) \leq n$ *may depend on* n, *such that for all* $n \geq 1$:

(i) $|t_{n,k}| = 1 + n^{\underline{k}} + k \cdot n \cdot n^{\underline{k}}$,
(ii) $H^{\deg}(t_{n,k}) + H_1^{\ell}(t_{n,k}) \leq \mathcal{O}(n \cdot n^{\underline{k}} \cdot k \cdot \log k)$ *and*
(iii) $H_{k-1}(t_{n,k}) \geq \Omega(n \cdot n^{\underline{k}} \cdot k \cdot \log(n - k + 1))$.

The label set of the tree $t_{n,k}$ is $\{a\} \cup \{b_u \mid u \in [n]^{\underline{k}}\} \cup \{c_i \mid 1 \leq i \leq n\}$, where $[n]^{\underline{k}} = \{(i_1, i_2 \ldots, i_k) \mid 1 \leq i_1, \ldots, i_k \leq n, i_j \neq i_l \text{ for } j \neq l\}$. For $u = (i_1, i_2, \ldots, i_k) \in [n]^{\underline{k}}$ define the tree $t_u = b_u((c_{i_1} c_{i_2} \cdots c_{i_k})^n)$; then $t_{n,k}$ is $a(t_{u_1} t_{u_2} \cdots t_{u_m})$, where $u_1, u_2, \ldots, u_m$ is an arbitrary enumeration of the set $[n]^{\underline{k}}$ (hence, $m = n^{\underline{k}}$). The tree $t_{3,2}$ is shown in Fig. 3.

If $k \in (\log n)^{\mathcal{O}(1)}$ then the trees $t_{n,k}$ from Lemma 9 satisfy

$$\frac{H^{\deg}(t_{n,k}) + H_1^{\ell}(t_{n,k})}{H_{k-1}(t_{n,k})} \leq \mathcal{O}\left(\frac{\log k}{\log(n - k + 1)}\right) = o(1).$$

This yields a relatively weak separation between $H^{\deg}(t) + H_1^{\ell}(t)$ and $H_k(t)$. In contrast, in Lemmas 7 and 8 we achieved an exponential separation. It remains open, whether such an exponential separation is also possible for $H_1^{\ell} + H^{\deg}$ and H_k. In other words, does there exist a family of trees t_n such that $H_k(t_n) \in \Omega(n)$ and $H^{\deg}(t_n) + H_1^{\ell}(t_n) \in \mathcal{O}(\log n)$?

5   Experiments

We finally complement our theoretical results with experimental data. We computed the entropies $H^{\deg}$, H_k, H_k^{ℓ}, $H_k^{\ell,\deg}$ and $H_k^{\deg,\ell}$ (for $k \in \{0, 1, 2, 4\}$) for 13 XML files from XMLCompBench (http://xmlcompbench.sourceforge.net). Table 1 shows the values for H_k, $H^{\deg} + H_k^{\ell}$, $H_k^{\ell} + H_k^{\ell,\deg}$ and $H^{\deg} + H_k^{\deg,\ell}$ (which can be achieved up to lower order terms by compressors). It turns out that for all XML trees used in this comparison the kth-order label-shape entropy (for $k > 0$) from [14] is significantly smaller than the entropies from [12]. In the full version [16, Table 2] the reader finds also the values for H_k^{ℓ}, $H_k^{\ell,\deg}$ and $H_k^{\deg,\ell}$ (divided by the tree size so that the table fits on the page). Additionally, we computed in [16] the label-shape entropy H_k for a modified version of each XML tree where all labels are replaced by a single dummy symbol, i.e., we considered the underlying, unlabeled tree as well (in [16, Table 2] this value is denoted by H_k'). Note again that the label-shape entropy H_k is the only measure for which this modification is interesting. In the setting of unlabeled trees, our experimental data indicate that neither the label-shape entropy nor the degree entropy (which is the upper bound on the number of bits needed by the data structure in [17] ignoring lower order terms; see also Theorem 1) is favorable.

Table 1. Values of the four entropies compared in this paper for various XML trees.

XML	k	H_k	$H^{\deg} + H_k^{\ell}$	$H_k^{\ell} + H_k^{\ell,\deg}$	$H^{\deg} + H_k^{\deg,\ell}$
BaseBall	0	202 568.08	153 814.94	146 066.64	146 066.64
	1	6 348.08	145 705.73	137 957.42	145 323.26
	2	2 671.95	145 705.73	137 957.42	145 323.26
	4	1 435.11	145 705.73	137 957.42	145 323.26
DBLP	0	18 727 523.44	14 576 781.00	12 967 501.16	12 967 501.16
	1	2 607 784.68	12 137 042.56	10 527 690.38	12 076 935.39
	2	2 076 410.50	12 136 974.71	10 527 595.96	12 076 845.69
	4	1 951 141.63	12 136 966.29	10 527 586.31	12 076 836.82
EXI-Array	0	1 098 274.54	962 858.05	649 410.59	649 410.59
	1	4 286.39	387 329.51	73 882.05	387 304.76
	2	4 270.18	387 329.51	73 882.05	387 304.76
	4	4 263.82	387 329.51	73 882.05	387 304.76
EXI-factbook	0	530 170.92	481 410.05	423 012.12	423 012.12
	1	11 772.65	239 499.01	181 101.08	204 649.84
	2	5 049.98	239 499.01	181 101.08	204 649.84
	4	4 345.42	239 499.01	181 101.08	204 649.84
EnWikiNew	0	2 118 359.59	1 877 639.22	1 384 034.65	1 384 034.65
	1	243 835.84	1 326 743.94	833 139.36	1 095 837.20
	2	78 689.86	1 326 743.94	833 139.36	1 095 837.20
	4	78 687.51	1 326 743.94	833 139.36	1 095 837.20
EnWikiQuote	0	1 372 201.38	1 229 530.04	894 768.55	894 768.55
	1	156 710.30	871 127.39	536 365.91	717 721.09
	2	51 557.50	871 127.39	536 365.91	717 721.09
	4	51 557.31	871 127.39	536 365.91	717 721.09
EnWikiVersity	0	2 568 158.43	2 264 856.93	1 644 997.36	1 644 997.36
	1	278 832.56	1 594 969.93	975 110.35	1 311 929.24
	2	74 456.55	1 594 969.93	975 110.35	1 311 929.24
	4	74 456.41	1 594 969.93	975 110.35	1 311 929.24
Nasa	0	3 022 100.11	2 872 172.41	2 214 641.55	2 214 641.55
	1	292 671.36	1 368 899.76	701 433.91	1 226 592.72
	2	168 551.10	1 363 699.16	696 194.53	1 221 474.16
	4	147 041.08	1 363 699.16	696 194.53	1 221 474.16
Shakespeare	0	655 517.90	521 889.47	395 890.85	395 890.85
	1	138 283.88	370 231.89	244 047.64	347 212.36
	2	125 837.77	370 061.20	243 843.87	347 041.31
	4	123 460.80	370 057.77	243 838.09	347 037.86
SwissProt	0	18 845 126.39	16 063 648.44	13 755 427.39	13 755 427.39
	1	3 051 570.48	11 065 924.67	8 757 703.61	10 238 734.83
	2	2 634 911.88	11 065 924.67	8 757 703.61	10 238 734.83
	4	2 314 609.48	11 065 924.67	8 757 703.61	10 238 734.83
Treebank	0	16 127 202.92	15 669 672.80	12 938 625.09	12 938 625.09
	1	7 504 481.18	12 301 414.61	9 482 695.67	9 925 567.44
	2	5 607 499.40	11 909 330.06	9 051 186.33	9 559 968.40
	4	4 675 093.61	11 626 935.89	8 736 301.14	9 285 544.85
USHouse	0	36 266.08	34 369.06	28 381.43	28 381.43
	1	10 490.44	24 249.78	17 968.41	19 438.19
	2	9 079.97	24 037.34	17 569.59	19 216.99
	4	6 308.98	23 634.87	16 830.00	18 783.36
XMark1	0	1 250 525.41	1 186 214.34	988 678.93	988 678.93
	1	167 586.81	592 634.17	394 639.43	523 996.29
	2	131 057.35	592 625.76	394 565.79	523 969.97
	4	127 157.34	592 037.39	393 770.73	523 432.87

References

1. Bille, P., Gawrychowski, P., Gørtz, I.L., Landau, G.M., Weimann, O.: Top tree compression of tries. In: Proceedings of the ISAAC 2019, LIPIcs, vol. 149, pp. 4:1–4:18. Schloss Dagstuhl - Leibniz-Zentrum für Informatik (2019)
2. Bille, P., Gørtz, I.L., Landau, G.M., Weimann, O.: Tree compression with top trees. Inf. Comput. **243**, 166–177 (2015)
3. Bousquet-Mélou, M., Lohrey, M., Maneth, S., Noeth, E.: XML compression via DAGs. Theory Comput. Syst. **57**(4), 1322–1371 (2015)
4. Cover, T.M., Thomas, J.A.: Elements of Information Theory, 2nd edn. Wiley, Hoboken (2006)
5. Ferragina, P., Luccio, F., Manzini, G., Muthukrishnan, S.: Structuring labeled trees for optimal succinctness, and beyond. In: Proceeding of the FOCS 2005, pp. 184–196. IEEE Computer Society (2005)
6. Ferragina, P., Luccio, F., Manzini, G., Muthukrishnan, S.: Compressing and indexing labeled trees, with applications. J. ACM **57**(1), 4:1–4:33 (2009)
7. Flajolet, P., Sipala, P., Steyaert, J.-M.: Analytic variations on the common subexpression problem. In: Paterson, M.S. (ed.) ICALP 1990. LNCS, vol. 443, pp. 220–234. Springer, Heidelberg (1990). https://doi.org/10.1007/BFb0032034
8. Gagie, T.: Large alphabets and incompressibility. Inf. Process. Lett. **99**(6), 246–251 (2006)
9. Ganardi, M., Hucke, D., Lohrey, M., Benkner, L.S.: Universal tree source coding using grammar-based compression. IEEE Trans. Inf. Theory **65**(10), 6399–6413 (2019)
10. Ganardi, M., Jez, A., Lohrey, M.: Balancing straight-line programs. In: Proceedings of the FOCS 2019, pp. 1169–1183. IEEE Computer Society (2019)
11. Ganczorz, M.: Entropy bounds for grammar compression. CoRR, abs/1804.08547 (2018)
12. Ganczorz, M.: Using statistical encoding to achieve tree succinctness never seen before. In: Proceedings of the STACS 2020, LIPIcs, vol. 154, pp. 22:1–22:29. Schloss Dagstuhl - Leibniz-Zentrum für Informatik (2020)
13. Gascón, A., Lohrey, M., Maneth, S., Reh, C.P., Sieber, K.: Grammar-based compression of unranked trees. Theory Comput. Syst. **64**(1), 141–176 (2020)
14. Hucke, D., Lohrey, M., Benkner, L.S.: Entropy bounds for grammar-based tree compressors. In: Proceedings of the ISIT 2019, pp. 1687–1691. IEEE (2019)
15. Hucke, D., Lohrey, M., Benkner, L.S.: Entropy bounds for grammar-based tree compressors. CoRR, abs/1901.03155 (2019)
16. Hucke, D., Lohrey, M., Benkner, L.S.: A comparison of empirical tree entropies. CoRR, abs/2006.01695 (2020)
17. Jansson, J., Sadakane, K., Sung, W.-K.: Ultra-succinct representation of ordered trees with applications. J. Comput. Syst. Sci. **78**(2), 619–631 (2012)
18. Lohrey, M., Maneth, S., Mennicke, R.: XML tree structure compression using RePair. Inf. Syst. **38**(8), 1150–1167 (2013)
19. Manzini, G.: An analysis of the Burrows-Wheeler transform. J. ACM **48**(3), 407–430 (2001)
20. Ochoa, C., Navarro, G.: RePair and all irreducible grammars are upper bounded by high-order empirical entropy. IEEE Trans. Inf. Theory **65**(5), 3160–3164 (2019)
21. Prezza, N.: On locating paths in compressed cardinal trees. CoRR, abs/2004.01120 (2020)

Efficient Enumeration of Distinct Factors Using Package Representations

Panagiotis Charalampopoulos[1,2], Tomasz Kociumaka[3],
Jakub Radoszewski[2], Wojciech Rytter[2], Tomasz Waleń[2],
and Wiktor Zuba[2(✉)]

[1] Department of Informatics, King's College London, London, UK
panagiotis.charalampopoulos@kcl.ac.uk
[2] Institute of Informatics, University of Warsaw, Warsaw, Poland
{jrad,rytter,walen,w.zuba}@mimuw.edu.pl
[3] Department of Computer Science, Bar-Ilan University, Ramat Gan, Israel
kociumaka@mimuw.edu.pl

Abstract. We investigate properties and applications of a new compact representation of string factors: families of *packages*. In a string T, each package (i, ℓ, k) represents the factors of T of length ℓ that start in the interval $[i, i+k]$. A family $\mathcal{F}$ of packages represents the set $\mathsf{Factors}(\mathcal{F})$ defined as the union of the sets of factors represented by individual packages in $\mathcal{F}$. We show how to efficiently enumerate $\mathsf{Factors}(\mathcal{F})$ and showcase that this is a generic tool for enumerating important classes of factors of T, such as powers and antipowers. Our approach is conceptually simpler than problem-specific methods and provides a unifying framework for such problems, which we hope can be further exploited.

We also consider a special case of the problem in which every occurrence of every factor represented by $\mathcal{F}$ is captured by some package in $\mathcal{F}$. For both applications mentioned above, we construct an efficient package representation that satisfies this property.

We develop efficient algorithms that, given a family $\mathcal{F}$ of m packages in a string of length n, report all distinct factors represented by these packages in $\mathcal{O}(n \log^2 n + m \log n + |\mathsf{Factors}(\mathcal{F})|)$ time for the general case and in the optimal $\mathcal{O}(n + m + |\mathsf{Factors}(\mathcal{F})|)$ time for the special case. We can also compute $|\mathsf{Factors}(\mathcal{F})|$ in $\mathcal{O}(n \log^2 n + m \log n)$ time in the general case and in $\mathcal{O}(n + m)$ time in the special case.

In particular, we improve over the state-of-the-art $\mathcal{O}(nk^4 \log k \log n)$-time algorithm for computing the number of distinct k-antipower factors, by providing an algorithm that runs in $\mathcal{O}(nk^2)$ time, and we obtain an alternative linear-time algorithm to enumerate distinct squares.

P. Charalampopoulos—Partially supported by ERC grant TOTAL under the EU's Horizon 2020 Research and Innovation Programme (agreement no. 677651).
T. Kociumaka—Supported by ISF grants no. 1278/16 and 1926/19, by a BSF grant no. 2018364, and by an ERC grant MPM under the EU's Horizon 2020 Research and Innovation Programme (grant no. 683064).
J. Radoszewski, T. Waleń and W. Zuba—Supported by the Polish National Science Center, grant no. 2018/31/D/ST6/03991.

© Springer Nature Switzerland AG 2020
C. Boucher and S. V. Thankachan (Eds.): SPIRE 2020, LNCS 12303, pp. 247–261, 2020.
https://doi.org/10.1007/978-3-030-59212-7_18

Keywords: Square in a string · Antipower · Longest previous factor array · String synchronising set

1   Introduction

There are many interesting subsets of factors of a given string T of length n which can be described very concisely (sometimes in $\mathcal{O}(n)$ space, even for subsets of quadratic size). In this paper, we consider compact descriptions, called *package representations*, defined in terms of weighted intervals: each interval $[i, i + k]$ gives starting positions of factors and the weight ℓ gives the common length of these factors. Formally, $\mathcal{F}$ is a set of triples (i, ℓ, k).

By $\mathsf{Factors}(\mathcal{F})$ we denote the set of factors in a given text T of length n that are represented by packages from $\mathcal{F}$. More formally,

$$\mathsf{Factors}(\mathcal{F}) = \{T[j \mathinner{.\,.} j + \ell) : j \in [i, i + k] \text{ and } (i, \ell, k) \in \mathcal{F}\}.$$

A package representation $\mathcal{F}$ is called *special* if it represents all occurrences of $\mathsf{Factors}(\mathcal{F})$. Formally, $\mathcal{F}$ is special if for every factor $F \in \mathsf{Factors}(\mathcal{F})$ and for every occurrence $T[j \mathinner{.\,.} j + \ell) = F$, there is a triple $(i, \ell, k) \in \mathcal{F}$ such that $j \in [i, i + k]$. Special representations describe *all* occurrences of factors with a given property.

We consider the following subsets of factors.

Powers. A square is a string of the form X^2. In general, for an integer $k > 1$, a k-power is a string of the form X^k. This notion can be generalized to rational exponents $\gamma > 1$, setting $X^\gamma = X^k X[1 \mathinner{.\,.} r]$ for $\gamma = k + r/|X|$, where k and $r < |X|$ are non-negative integers.

Antipowers. A k-antipower (for an integer $k \geq 2$) is a concatenation of k pairwise distinct strings of the same length. Antipowers were introduced in [15] and have already attracted considerable attention [1,2,4,14,25].

Example 1. Consider a string $T = \mathsf{abababababa}$. The squares in T can be represented by a set of packages $\mathcal{F} = \{(1, 4, 7), (1, 8, 3)\}$. The package $(1, 4, 7)$ represents all the squares of length 4 and the package $(1, 8, 3)$—those of length 8.

Our problem can be related to computing the *subword complexity* of the string T; see, e.g., [31]. Let us recall that the subword complexity is a function which gives, for every $\ell \in [1, n]$, the number of different factors of T of length ℓ. The subword complexity of a given string can be computed using the suffix tree in linear time. Our algorithm can be easily augmented to determine, for each length ℓ, the number of length-ℓ factors in $\mathsf{Factors}(\mathcal{F})$.

Our results. We compute $|\mathsf{Factors}(\mathcal{F})|$ in $\mathcal{O}(n \log^2 n + m \log n)$ time in the general case and in $\mathcal{O}(n + m)$ time in the special case, both for any length-n string T over an integer alphabet. The solution to the general case uses string synchronising sets and runs, whereas the solution to the special case is based on the longest previous factor array. Our algorithms for special package representations yield new simple algorithms for reporting and counting powers and antipowers. In particular, we present the first linear-time algorithms to count and enumerate distinct γ-powers for a given rational constant $\gamma > 1$; Crochemore et al. [11] showed how to do this for integer γ only. For k-antipowers, we improve the previously known best time complexity.

2   Algorithms for Special Package Representations

Let $T = T[1] \cdots T[n]$. The longest previous factor array $LPF[1 .. n]$ is defined as

$$LPF[i] = \max\{\ell \geq 0 : T[i .. i + \ell) = T[j .. j + \ell) \text{ for some } j \in [1, i - 1]\}.$$

This array can be computed in $\mathcal{O}(n)$ time [9, 10]. Let

$$U_\ell = \{j \in [1, n] : LPF[j] \geq \ell\}$$
$$\mathsf{Pairs}(\mathcal{F}) = \bigcup_{(i, \ell, k) \in \mathcal{F}} \{(j, \ell) : j \in [i, i + k] \setminus U_\ell\}.$$

The algorithms are based on the following crucial observation that links the solution to the special case with the LPF table.

Observation 2. *If $\mathcal{F}$ is a special package representation, then* $\mathsf{Factors}(\mathcal{F}) = \{T[j .. j + \ell) : (j, \ell) \in \mathsf{Pairs}(\mathcal{F})\}$ *and* $|\mathsf{Factors}(\mathcal{F})| = |\mathsf{Pairs}(\mathcal{F})|$.

2.1   Reporting Distinct Factors

Due to Observation 2, reporting all distinct factors reduces to computing the set $\mathsf{Pairs}(\mathcal{F})$. We can assume that packages representing factors of the same length are disjoint; this can be achieved by merging overlapping packages in a preprocessing step that can be executed in $\mathcal{O}(n)$ time using radix sort.

The definition of $\mathsf{Pairs}(\mathcal{F})$ yields the following (inefficient) algorithm. It constructs the sets U_ℓ for all $\ell = n, \ldots, 1$ and, for each of them, generates all elements of the set $\mathsf{Pairs}(\mathcal{F})$ with the second component equal to ℓ.

Algorithm 1: High-level structure of the algorithm.

$U := \emptyset; \mathcal{P} := \emptyset$

for $\ell := n$ **down to** 1 **do**
  $U := U \cup \{j : LPF[j] = \ell\}$ $// U = U_\ell$
  foreach $(i, \ell, k) \in \mathcal{F}$ **do**
    foreach $j \in [i, i + k] \setminus U$ **do**
      $\mathcal{P} := \mathcal{P} \cup \{(j, \ell)\}$ $//$ Ultimately, $\mathcal{P} = \mathsf{Pairs}(\mathcal{F})$

Next, we describe an efficient implementation of Algorithm 1 based on the union-find data structure. In our algorithm, the elements of the data structure are $[1, n+1]$ and the sets stored in the data structure always form intervals. The operation $\mathtt{Find}(i)$ returns the rightmost element of the interval containing i, and the operation $\mathtt{Union}(i)$ joins the intervals containing elements i and $i-1$.

Algorithm 2: Implementation of Algorithm 1.

$\mathcal{P} := \emptyset$
for $i := 0$ **to** $n+1$ **do** Create set $\{i\}$
for $\ell := n$ **down to** 1 **do**
    foreach j *such that* $LPF[j] = \ell$ **do** $\mathtt{Union}(j)$
    foreach $(i, \ell, k) \in \mathcal{F}$ **do**
        $j := \mathtt{Find}(i-1)+1$
        while $j \leq i+k$ **do**
            $\mathcal{P} := \mathcal{P} \cup \{(j, \ell)\}$
            $j := \mathtt{Find}(j)+1$

Theorem 3. *In the case of special package representations, all elements of* $\mathsf{Factors}(\mathcal{F})$ *can be reported (without duplicates) in* $\mathcal{O}(n+m+|\mathsf{Factors}(\mathcal{F})|)$ *time.*

Proof. We use Algorithm 2. The set U_ℓ is stored in the union-find data structure so that for each interval $[i,j]$ in the data structure, $i \notin U_\ell$ and $[i+1,j] \subseteq U_\ell$.

The elements of $\mathcal{F}$ are sorted by the second component using radix sort. The union-find data structure admits at most n union operations and $m+|\mathsf{Factors}(\mathcal{F})|$ find operations. We use a data structure for a special case of the union-find problem, where the sets of the partition have to form integer intervals at all times, so that each operation takes $\mathcal{O}(1)$ amortized time [18]. □

2.2 Counting Distinct Factors

Let us start with a warm-up algorithm. Recall that, in a preprocessing, we made sure that packages representing factors of the same length are disjoint.

By Observation 2, for each $(i, \ell, k) \in \mathcal{F}$, it suffices to count the number of elements in $LPF[i..i+k]$ that are smaller than ℓ. This can be done using range queries in time $O((n+m)\sqrt{\log n})$.

Let us proceed to a linear-time algorithm. We start with a simple fact.

Fact 4. *For every length-n text T, we have $\sum_{i=1}^{n-1} |LPF[i+1] - LPF[i]| = \mathcal{O}(n)$.*

Proof. The claim follows from the fact that $LPF[i+1] \geq LPF[i]-1$ for $i \in [1, n-1]$. To prove this inequality, let $\ell = LPF[i]$. We have $T[i..i+\ell) = T[j..j+\ell)$ for some $j < i$. Hence, $T[i+1..i+\ell) = T[j+1..j+\ell)$, so $LPF[i+1] \geq \ell-1$. □

We reduce the counting problem to answering off-line a linear number of certain queries. The off-line structure of the computation is crucial for efficiency.

Theorem 5. *In the case of special package representations, $|\mathsf{Factors}(\mathcal{F})|$ can be computed in $\mathcal{O}(n+m)$ time.*

Proof. Consider the following queries:

$$Q(i, \ell) = |[1, i] \setminus U_\ell| = |\{j \in [1, i] : LPF[j] < \ell\}|.$$

Then, the counting version of our problem reduces to efficiently answering such queries. Indeed, by Observation 2, we have

$$|\mathsf{Factors}(\mathcal{F})| = \sum_{(i,\ell,k)\in\mathcal{F}} Q(i + k, \ell) - Q(i - 1, \ell).$$

Thus, we have to answer $\mathcal{O}(m)$ queries of the form $Q(i, \ell)$. An off-line algorithm answering q queries in $\mathcal{O}(n + q)$ time would be sufficient for our purposes.

We maintain an array $A[1 .. n]$ such that during the ith phase of the algorithm:

$$A[\ell] = \begin{cases} i - Q(i, \ell) & \text{if } \ell > LPF[i], \\ Q(i, \ell) & \text{otherwise.} \end{cases}$$

Since $LPF[1] = 0$, the array needs to be filled with 1's for the first phase. Next, we observe that $i + 1 - Q(i + 1, \ell) = i - Q(i, \ell)$ if $\ell > \max(LPF[i + 1], LPF[i])$ and $Q(i+1, \ell) = Q(i, \ell)$ if $\ell \leq \min(LPF[i + 1], LPF[i])$. Hence, in the transition from the ith phase to the $(i + 1)$th phase, we only need to update $\mathcal{O}(|LPF[i + 1] - LPF[i]|)$ entries of A. By Fact 4, the cost of maintaining the array A for $i = 1$ to n is $\mathcal{O}(n)$ in total. Each query $Q(i, \ell)$ can be answered in $\mathcal{O}(1)$ time during the ith phase.

Consequently, we can answer off-line q queries $Q(i, \ell)$ in $\mathcal{O}(n+q)$ time, assuming that the queries are sorted by the first component. Sorting can be performed in $\mathcal{O}(n + q)$ time using radix sort. □

3 Applications

In this section, we show three applications of special package representations.

3.1 Squares

It is known that a string of length n contains at most $\frac{11}{6}n$ distinct squares [12, 17], and the same bound hold for γ-powers with $\gamma \geq 2$. Moreover, all the distinct square factors in a string over an integer alphabet can be reported in $\mathcal{O}(n)$ time [6, 11, 20]. The algorithm from [11] can report distinct string powers of a given integer exponent using a run-based approach via Lyndon roots. Hence, it can report distinct squares and cubes in particular. We show that our generic approach—which is also much simpler—applies to this problem.

A *generalised run* in a string T is a triple (i, j, p) such that:

- $T[i .. j]$ has a period p (not necessarily the shortest) with $2p \leq j - i + 1$,
- $T[i - 1] \neq T[i - 1 + p]$ if $i > 1$, and $T[j + 1] \neq T[j + 1 - p]$ if $j < n$.

A *run* is a generalised run for which p is the shortest period of $T[i \mathinner{.\,.} j]$. The number of runs and generalised runs is $\mathcal{O}(n)$ and they can all be computed in $\mathcal{O}(n)$ time; see [5,30].

Proposition 6. *All distinct squares in a string of length n can be computed in $\mathcal{O}(n)$ time.*

Proof. A generalised run (i, j, p) induces squares $T[k \mathinner{.\,.} k + 2p)$ for all $k \in [i, j - 2p + 1]$. Moreover, each occurrence of a square is induced by exactly one generalised run. For every generalised run (i, j, p), we add package $(i, 2p, j - 2p - i + 1)$ to $\mathcal{F}$; see Fig. 1. Then, we solve the factors problem using Theorem 3. $\square$

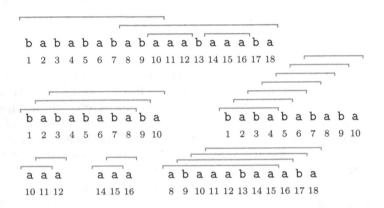

Fig. 1. Four runs (presented at the top) generate a package representation (below) of all squares as a set of five packages: $\{(1, 8, 2), (1, 4, 6), (10, 2, 1), (14, 2, 1), (8, 8, 3)\}$. One of the runs induces two generalised runs: with periods 2 and 4.

3.2   Powers with Rational Exponents

Proposition 6 can be easily generalised to powers of arbitrary exponent $\gamma \geq 2$. For exponents $\gamma < 2$, however, we need α-gapped repeats apart from the generalised runs. An α-*gapped repeat* (for $\alpha \geq 1$) in a string T is a quadruple (i_1, j_1, i_2, j_2) such that $i_1 \leq j_1 < i_2 \leq j_2$, and the factors $T[i_1 \mathinner{.\,.} j_1] = T[i_2 \mathinner{.\,.} j_2] = U$ and $T[j_1 + 1 \mathinner{.\,.} i_2 - 1] = V$ satisfy $|UV| \leq \alpha|U|$. The two occurrences of U are called the *arms* of the α-gapped repeat and $|UV|$ is called the *period* of the α-gapped repeat. In other words, a gapped repeat is a string $S = T[i_1 \mathinner{.\,.} j_2]$ associated with one of its periods larger than $\frac{1}{2}|S|$. Consequently, the same factor $T[i_1 \mathinner{.\,.} j_2]$ can induce many α-gapped repeats.

An α-gapped repeat is called *maximal* if its arms cannot be extended simultaneously with the same character to either direction. The number of maximal α-gapped repeats in a string of length n is $\mathcal{O}(n\alpha)$ and they can all be computed in $\mathcal{O}(n\alpha)$ time assuming an integer alphabet [19].

Theorem 7. *For a given rational number $\gamma > 1$, all distinct γ-powers in a length-n string can be counted in $\mathcal{O}(\frac{\gamma}{\gamma-1}n)$ time and enumerated in $\mathcal{O}(\frac{\gamma}{\gamma-1}n + $ output) time.*

Proof. Each γ-power X^γ with $\gamma < 2$ is a $\frac{1}{\gamma-1}$-gapped repeat with period $|X|$, and therefore it is contained in a maximal $\frac{1}{\gamma-1}$-gapped repeat or in a generalised run with the same period; see [29]. Moreover, each γ-power X^γ with $\gamma \geq 2$ is contained in a generalised run with period $|X|$.

In other words, to generate all γ-powers, for each generalised run and $\frac{1}{\gamma-1}$-gapped repeat (if $\gamma < 2$) with period p, we need to consider all factors contained in it of length γp, provided that γp is an integer; see Fig. 2.

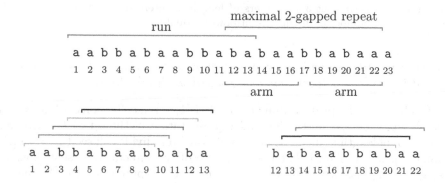

Fig. 2. A string with a (generalised) run with period 6 and a maximal $\frac{6}{5}$-gapped repeat (hence, also a maximal 1.5-gapped repeat) with period 6. The run and the gapped repeat generate 1.5-powers of length 9. Equal 1.5-powers are drawn with the same color; in total, the string contains 6 distinct 1.5-powers of length 9. (Color figure online)

We proceed as follows. For each generalised run (i, j, p), if γp is an integer and $j - i + 1 \geq \gamma p$, then we insert $(i, \gamma p, j - i + 1 - \gamma p)$ to $\mathcal{F}$. Moreover, if $\gamma < 2$, then for each maximal $\frac{1}{\gamma-1}$-gapped repeat (i_1, j_1, i_2, j_2) with period $p = i_2 - i_1$, if γp is an integer, then we insert $(i_1, \gamma p, j_2 - i_1 + 1 - \gamma p)$ to $\mathcal{F}$. By the above discussion, the constructed family $\mathcal{F}$ is a special package representation of all γ-powers. The claim follows by Theorems 3 and 5.   □

Remark 8. For every fixed rational number $\gamma < 2$, strings of length n may contain $\Omega(n^2)$ distinct γ-powers. Specifically, if $\gamma = 2 - \frac{x}{y}$, where x and y are coprime positive integers, then the number of γ-powers in $\mathsf{a}^m\mathsf{ba}^m$ is $\Theta(\frac{m^2x}{y^2})$ [28].

3.3   Antipowers

In [25], it was shown how to report all occurrences of k-antipowers in $\mathcal{O}(nk \log k + $ output) time and count them in $\mathcal{O}(nk \log k)$ time. In [26], it was shown that the number of distinct k-antipower factors in a string of length n can be computed in $\mathcal{O}(nk^4 \log k \log n)$ time. Below, we show how to improve the latter result.

```
                        b a|a b|a b
                        b b|a a|b a
                        a b|b a|a b
                        b a|b b|a a
                        b b|a b|b a
                        a b|b a|b b
                        b a|b b|a b
                        b b|a b|b a
            A₆    ━━━━━━━━━━  ━━
                   b b a b b a b b a a b a b
```

Fig. 3. Here, we consider 3-antipowers of length $\ell = 6$. The set of their starting positions is $A_6 = [1, 5] \cup [7, 7]$. Note that the first and the fourth antipower are the same, so we have only 5 distinct 3-antipowers of length 6. Interestingly $A_\ell = \emptyset$ for $\ell \neq 6$. Hence, the total number of distinct 3-antipowers equals 5.

Theorem 9. *All distinct k-antipower factors of a string of length n can be reported in $\mathcal{O}(nk^2 + \mathsf{output})$ time and counted in $\mathcal{O}(nk^2)$ time.*

Proof. The interval representation of a set $A \subseteq [1, n]$ is a collection of all maximal intervals in A. Let A_ℓ be the interval representation of the set of k-antipower factors of T of length ℓ (it can be non-empty only if k divides ℓ); see Fig. 3. In [25, Lemma 13], it was shown that the total size of the interval representations of sets $A_1, \ldots, A_n$ is $\mathcal{O}(nk^2)$. Moreover, they can be computed in $\mathcal{O}(nk^2)$ time.

For each ℓ and each interval $[i, j] \in A_\ell$, we insert $(i, \ell, j - i)$ to $\mathcal{F}$. The conclusion follows by Theorems 3 and 5.   □

4   Enumerating General Package Representations

For most of this section, we will focus on computing $|\mathsf{Factors}(\mathcal{F})|$. In the end, we will briefly explain how our solution can be adapted to enumerate $\mathsf{Factors}(\mathcal{F})$.

We consider highly periodic and non-highly-periodic factors separately (a precise definition follows). In both cases, we will employ the solution of Kociumaka et al. [26] for the so-called PATH PAIRS PROBLEM, which we define below.

We say that $\mathcal{T}$ is a *compact tree* if it is a rooted tree with positive integer weights on edges. If an edge weight is $e > 1$, this edge contains $e - 1$ implicit nodes. A *path* in a compact tree is an upwards or downwards path that connects two explicit nodes.

PATH PAIRS PROBLEM

Input: Two compact trees $\mathcal{T}$ and $\mathcal{T}'$ containing up to N explicit nodes each, and a set Π of M pairs (π, π') of equal-length paths, where π is a path going downwards in $\mathcal{T}$ and π' is a path going upwards in $\mathcal{T}'$.

Output: $|\bigcup_{(\pi, \pi') \in \Pi} \mathsf{Induced}(\pi, \pi')|$, where by $\mathsf{Induced}(\pi, \pi')$ we denote the set of pairs of (explicit or implicit) nodes (u, u') such that, for some i, the ith node on π is u and the ith node on π' is u'.

Lemma 10 ([26]). *The* PATH PAIRS PROBLEM *can be solved in time* $\mathcal{O}(N + M \log N)$ *assuming that the weighted heights of the input trees do not exceed* N.

4.1   Non-Highly-Periodic Factors

Our solution uses the string synchronising sets recently introduced by Kempa and Kociumaka [22].

Informally, in the simpler case that T is cube-free, a τ-synchronising set of T consists in a small set of positions of T, called here *synchronisers*, such that each length-τ fragment of T contains at least one synchroniser, and the synchronisers within two long enough matching fragments of T are consistent.

Formally, for a string T and a positive integer $\tau \leq \frac{1}{2}n$, a set $S \subseteq [1, n - 2\tau + 1]$ is a τ-*synchronising set* of T if it satisfies the following two conditions:

1. If $T[i \mathbin{..} i + 2\tau) = T[j \mathbin{..} j + 2\tau)$, then $i \in S$ if and only if $j \in S$.

2. For $i \in [1, n - 3\tau + 2]$, $S \cap [i \mathbin{..} i + \tau) = \emptyset$ if and only if $\mathsf{per}(T[i \mathbin{..} i + 3\tau - 2]) \leq \frac{1}{3}\tau$.

Theorem 11 ([22]). *Given a string T of length n over an integer alphabet and a positive integer $\tau \leq \frac{1}{2}n$, one can construct in $\mathcal{O}(n)$ time a τ-synchronising set of T of size $\mathcal{O}(\frac{n}{\tau})$.*

As in [22], for a τ-synchronising set S, let $\mathsf{succ}_S(i) := \min\{j \in S \cup \{n - 2\tau + 2\} : j \geq i\}$ and $\mathsf{pred}_S(i) := \max\{j \in S \cup \{0\} : j \leq i\}$.

Lemma 12 ([22]). *If a factor U of T with $|U| \geq 3\tau - 1$ and $\mathsf{per}(U) > \frac{1}{3}\tau$ occurs at positions i and j in T, then $\mathsf{succ}_S(i) - i = \mathsf{succ}_S(j) - j \leq |U| - 2\tau$.*

By U^R we denote the reversal of a string U. We show the following result.

Lemma 13 (Aperiodic Lemma). *Assume that we are given a text T of length n, a positive integer $x \leq \frac{1}{3}n$, and a family $\mathcal{F}$ of m packages that represent factors of lengths in $[3x, 9x)$ and shortest periods greater than $\frac{1}{3}x$. Then, $|\mathsf{Factors}(\mathcal{F})|$ can be computed in $\mathcal{O}((n + m) \log n)$ time.*

Proof. We compute an x-synchronising set S of T in $\mathcal{O}(n)$ time using Theorem 11 and build the suffix trees $\mathcal{T}$ and $\mathcal{T}'$ of T and T^R, respectively, in $\mathcal{O}(n)$ time [13].

Let us now focus on all packages representing factors of a fixed length ℓ. By relying on Lemma 12, we will intuitively assign each factor to its first synchroniser.

Let us denote $\mathcal{A}_\ell = \bigcup\{[i, i + k] : (i, \ell, k) \in \mathcal{F}\}$. For each $j \in \mathcal{A}_\ell$, let $s = \mathsf{succ}_S(j)$ and consider $P_j = T[j \mathbin{..} s]$ and $Q_j = [s + 1 \mathbin{..} j + \ell)$; see Fig. 4.

Note that, by Lemma 12, $s - j \leq x$ and, as $j \leq n - \ell + 1$, we have $s \leq n - 2x + 1$. Thus, $s \in S$. Hence, Lemma 12 implies that, for any $j, j' \in \mathcal{A}_\ell$ such that $T[j \mathbin{..} j + \ell) = T[j' \mathbin{..} j' + \ell)$, we have $P_j = P_{j'}$ and $Q_j = Q_{j'}$. Consequently, our problem reduces to computing the size of the set $\mathcal{P}_\ell = \{(P_j, Q_j) : j \in \mathcal{A}_\ell\}$. In turn, in our instance of the PATH PAIRS PROBLEM, we want to count the pairs of nodes $u \in \mathcal{T}'$, $v \in \mathcal{T}$ such that $(\mathcal{L}(u)^R, \mathcal{L}(v)) \in \mathcal{P}_\ell$, where $\mathcal{L}(u)$ is the label

Fig. 4. The elements of an x-synchronising set S of string T are denoted by asterisks. The position j is an element of $[i, i + k]$ for some package (i, ℓ, k). The red asterisk denotes the synchroniser $s = \text{succ}_S(j)$. (Color figure online)

of the path from the root to the node u. It remains to show how to compute path pairs that induce exactly these pairs of nodes. To this end, we design a line-sweeping algorithm.

We initialize an empty set Π that will eventually store the desired pairs of paths. We will scan the text T in a left-to-right manner with two fingers: f_p for packages and f_s for synchronisers, both initially set to 0. We maintain an invariant that $f_p \leq f_s$. Whenever $f_p - 1 = f_s \leq n - 2x$, we set $f_s = \text{succ}_S(f_s + 1)$.

The finger f_p is repeatedly incremented until it reaches f_s. For each maximal interval $[i, j] \subseteq \mathcal{A}_\ell$ that f_p encounters, we do the following: If $j > f_s$, we split the interval into $[i, i + f_s]$ and $[f_s + 1, j]$ and consider the first of them as $[i, j]$. Let

$$X_1 = T[i .. f_s], \ X_2 = T[j .. f_s], \ Y_1 = T[f_s + 1 .. i + \ell), \ Y_2 = T[f_s + 1 .. j + \ell).$$

For $k = 1, 2$, let u_k be the locus of X_k^R in $\mathcal{T}'$ and v_k be the locus of Y_k in $\mathcal{T}$. If either of these loci is an implicit node, we make it explicit. Finally, we add to Π the pair of paths u_1-to-u_2 in $\mathcal{T}'$ and v_1-to-v_2 in $\mathcal{T}$.

Let us denote the number of packages representing factors of length ℓ by m_ℓ. As there are $\mathcal{O}(\frac{n}{x})$ synchronisers, the line-sweeping algorithm can be performed in $\mathcal{O}(\frac{n}{x} + m_\ell)$ time. Thus, the number of paths (and extra explicit nodes) that we introduce in the two suffix trees is also $\mathcal{O}(\frac{n}{x} + m_\ell)$.

Over all $\ell \in [3x, 9x)$, we have

$$\mathcal{O}\left(x \cdot \frac{n}{x} + \sum_{\ell=3x}^{9x-1} m_\ell \right) = \mathcal{O}(n + m)$$

pairs of paths. The only operations that we need to explain how to perform efficiently are (a) computing the loci of strings in $\mathcal{T}$ and $\mathcal{T}'$ and (b) making all of them explicit. Part (a) can be implemented using an efficient algorithm for answering a batch of weighted ancestor queries from [24]. In part (b), we process the weighted ancestors in an order of non-decreasing weights, after globally sorting them using radix sort. The whole construction works in $\mathcal{O}(n + m)$ time. We obtain an instance of the PATH PAIRS PROBLEM with $N, M = \mathcal{O}(n + m)$. The suffix trees are of weighted height $\mathcal{O}(n)$, so Lemma 10 completes the proof. $\square$

4.2   Highly Periodic Factors

A string U is called *periodic* if $2 \cdot \mathsf{per}(U) \le |U|$ and *highly periodic* if $3 \cdot \mathsf{per}(U) \le |U|$.

The *Lyndon root* of a periodic string U is the lexicographically smallest rotation of its length-$\mathsf{per}(U)$ prefix. If L is the Lyndon root of a periodic string U, then U can be uniquely represented as (L, y, a, b) for $0 \le a, b < |L|$ such that $U = L[|L| - a + 1 .. |L|]L^y L[1 .. b]$. We call this the *Lyndon representation* of U.

In $\mathcal{O}(n)$ time, one can compute the Lyndon representations of all runs [11]. The unique run that extends a periodic factor of T can be computed in $\mathcal{O}(1)$ time after $\mathcal{O}(n)$-time preprocessing [5,27]. This allows computing its Lyndon representation in $\mathcal{O}(1)$ time.

For highly periodic factors, we will use *Lyndon roots* instead of synchronisers. The rest of this subsection is devoted to proving the following lemma.

Lemma 14 (Periodic Lemma). *Given a text T of length n and a set $\mathcal{F}$ of m packages of highly periodic factors, $|\mathsf{Factors}(\mathcal{F})|$ can be computed in $\mathcal{O}((n + m) \log n)$ time.*

Proof. For each $(i, \ell, k) \in \mathcal{F}$ and $j \in [i, i + k]$, the fragment $T[j .. j + \ell)$ has a (unique) Lyndon representation (L, y, a, b) for some Lyndon root L. Let

$$P_{j,\ell} = L[|L| - a + 1 .. |L|] \text{ and } Q_{j,\ell} = L^y L[1 .. b]$$

Our problem consists in computing the size of the set

$$\mathcal{P} = \{(P_{j,\ell}, Q_{j,\ell}) : j \in [i, i + k], (i, \ell, k) \in \mathcal{F}\}$$

Let $\mathcal{T}$ and $\mathcal{T}'$ be the suffix trees of T and T^R, respectively. Then, we want to compute the number of pairs of nodes $u \in \mathcal{T}'$ and $v \in \mathcal{T}$ with $(\mathcal{L}(u)^R, \mathcal{L}(v)) \in \mathcal{P}$. We will show how this reduces to an instance of the PATH PAIRS PROBLEM.

We have to appropriately define pairs of paths over $\mathcal{T}$ and $\mathcal{T}'$. Let us note that all the factors that each package of $\mathcal{F}$ represents have the same Lyndon root, since two strings with different periods at most $\frac{1}{3}\ell$ cannot overlap on $\ell - 1$ positions by the Fine and Wilf's periodicity lemma [16].

We initialize an empty set Π that will store pairs of paths. Let us consider a package $(i, \ell, k) \in \mathcal{F}$ such that $T[i .. i + k + \ell)$ is represented by (L, y, a, b). By periodicity, we may focus on the factors starting in the first (at most) $|L|$ positions of $T[i .. i + k + \ell)$.

To this end, let $t = \min\{|L|, k + 1\}$. We will insert at most two paths to Π, specified below:

- Let X_1 be the suffix of L of length a, X_2 be the suffix of L of length $a' = \max\{a - t, 0\}$, $Y_1 = L^\infty[1 .. \ell - a]$ and $Y_2 = L^\infty[1 .. \ell - a']$.

- If $t > a$, let X_1' be the suffix of L of length $|L| - 1$ and X_2' be the suffix of L of length $d = |L| + a - t$, $Y_1' = L^\infty[1 .. \ell - |L| + 1]$ and $Y_2' = L^\infty[1 .. \ell - d]$.

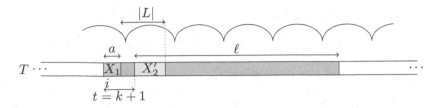

Fig. 5. The shaded part of the text denotes $T[i..i+k+\ell)$, which can be represented as (L, y, a, b), for some package $(i, \ell, k) \in \mathcal{F}_\ell$. X_1 is shaded in red, while X_2 is the empty string. We are in the case that $t = k + 1 > a$; X_2' is shaded in green. (Color figure online)

See Fig. 5 for an illustration. It can be readily verified that these pairs of paths induce exactly the required pairs of nodes.

For $k = 1, 2$, let u_k be the locus of X_k^R in $\mathcal{T}'$ and v_k be the locus of Y_k in $\mathcal{T}$. If either of these loci is an implicit node, we make it explicit. Finally, we add to Π the pair of paths u_1-to-u_2 in $\mathcal{T}'$ and v_1-to-v_2 in $\mathcal{T}$. Similarly for X_k's and Y_k's.

Let us consider the time complexity of the algorithm. The suffix trees $\mathcal{T}$ and $\mathcal{T}'$ can be computed in $\mathcal{O}(n)$ time [13]. Computing the loci of strings and making them explicit can be performed in $\mathcal{O}(n + m)$ time as in the proof of Lemma 13. We obtain an instance of the PATH PAIRS PROBLEM with $N = \mathcal{O}(n + m)$ and $M \leq 2m$. Lemma 10 completes the proof.   □

4.3   Wrap-Up

Let $\mathcal{F}_\ell$ be the set of triples from $\mathcal{F}$ with the second component ℓ. Note that one can easily compute the contributions of all packages representing factors whose length is bounded by 2 in $\mathcal{O}(n)$ time using radix sort; we can thus assume that all packages represent factors of length at least 3.

We will iterate over $x = 3^j$ for all integers $j \in [1, \lfloor \log_3 n \rfloor - 1]$. For each $\ell \in [3x, 9x)$, we want to replace $\mathcal{F}_\ell$ by two—not too large—sets of packages:

- $\mathcal{F}_\ell^p$ representing factors with shortest period at most $x/3$, and
- $\mathcal{F}_\ell^a$ representing factors with shortest period greater than $x/3$,

such that $\mathsf{Factors}(\mathcal{F}_\ell) = \mathsf{Factors}(\mathcal{F}_\ell^p) \cup \mathsf{Factors}(\mathcal{F}_\ell^a)$.

Our aim is to decompose each package in $\mathcal{F}_\ell$ in pieces (i.e., decompose $[i, i+k]$ into subintervals), such that all factors represented by each piece either have shortest period at most $x/3$ or none of them does. We then want to group the resulting pieces into the two sets.

Let $\mathcal{R}_x$ denote the set of runs of T with length at least $3x$ and period at most $x/3$. As shown in [23, Section 4.4], $|\mathcal{R}_x| = \mathcal{O}(n/x)$. Further, $\mathcal{R}_x$ can be computed in $\mathcal{O}(n)$ time, by filtering out the runs that do not satisfy the criteria.

Lemma 15. *Given $\mathcal{R}_x$, we can compute sets $\mathcal{F}_\ell^p$ and $\mathcal{F}_\ell^a$ in $\mathcal{O}(n/x + |\mathcal{F}_\ell|)$ time.*

Proof. Initially, let $I = \emptyset$. For each run $R = T[a .. b]$ with $\mathsf{per}(R) \leq x/3$ and $|R| \geq \ell$, we set $I := I \cup [a, b - \ell + 1]$. There are $\mathcal{O}(n/x)$ such runs and hence our representation of I consists of $\mathcal{O}(n/x)$ intervals.

Recall that packages are pairwise disjoint. We decompose a package $(i, \ell, k) \in \mathcal{F}_\ell$ as follows.

For each maximal interval $[r, t]$ in $[i, i + k] \cap I$ we insert $(r, \ell, t - r)$ to $\mathcal{F}_\ell^p$, while for each maximal interval $[r', t']$ in $[i, i+k] \setminus I$ we insert $(r', \ell, t' - r')$ to $\mathcal{F}_\ell^a$. This can be done in $\mathcal{O}(n/x + |\mathcal{F}_\ell|)$ time with a standard line-sweeping algorithm. $\qquad\square$

Now, let us put everything together. First of all, we compute $\mathcal{R}_x$ for each $x \in [1, \lfloor \log_3 n \rfloor - 1]$ in $\mathcal{O}(n \log n)$ total time. Then, for each ℓ, we replace $\mathcal{F}_\ell$ by $\mathcal{F}_\ell^p$ and $\mathcal{F}_\ell^a$ in $\mathcal{O}(n/\ell + |\mathcal{F}_\ell|)$ time, employing Lemma 15.

We process all $\mathcal{F}_\ell^p$'s together, as each factor U represented by them must be highly periodic; since $\mathsf{per}(U) \leq x/3$ and $|U| \geq 3x$ for some x, we surely have $3 \cdot \mathsf{per}(U) \leq |U|$. The total size of these sets is $\sum_{\ell=3}^{n} \mathcal{O}(n/\ell + |\mathcal{F}_\ell|) = \mathcal{O}(n \log n + m)$, and hence a call to Lemma 14 requires $\mathcal{O}(n \log^2 n + m \log n)$ time.

Then, we make a call to Lemma 13 for each $x \in [1, \lfloor \log_3 n \rfloor - 1]$, and the union of sets $\mathcal{F}_\ell^a$ for $\ell \in [3x, 9x)$. Again by Lemma 15, for each such ℓ, we have $|\mathcal{F}_\ell^a| = \mathcal{O}(n/x + |\mathcal{F}_\ell|)$.

The total time complexity required by the calls to Lemma 13 is:

$$\sum_{x=1}^{\lfloor \log_3 n \rfloor - 1} \mathcal{O}\left(n \log n + \sum_{\ell=3x}^{9x-1} |\mathcal{F}_\ell| \log n \right) = \mathcal{O}(n \log^2 n) + \sum_{\ell=3}^{n} \mathcal{O}(|\mathcal{F}_\ell| \log n)$$

$$= \mathcal{O}(n \log^2 n + m \log n).$$

We have thus proved the main result of this section.

Theorem 16. $|\mathsf{Factors}(\mathcal{F})|$ *can be computed in* $\mathcal{O}(n \log^2 n + m \log n)$ *time.*

4.4   Reporting Factors

The reporting version of the PATH PAIRS PROBLEM, where one is to output $\bigcup_{(\pi, \pi') \in \Pi} \mathsf{Induced}(\pi, \pi')$, can be solved in $\mathcal{O}(N + M \log N + \mathsf{output})$ time by a straightforward modification of the proof of Lemma 10.[1] We can also retrieve a pair of paths inducing each pair of nodes within the same time complexity (in order to be able to represent the relevant string as a factor of T).

Theorem 17. *All elements of* $\mathsf{Factors}(\mathcal{F})$ *can be reported (without duplicates) in* $\mathcal{O}(n \log^2 n + m \log n + \mathsf{output})$ *time.*

---

[1] The workhorse of Lemma 10 is computing the size of the union of certain 1D-intervals. For the reporting version, we simply have to report all elements of this union.

5   Final Remarks

Another natural representation of factors consists in a set of intervals $\mathcal{I}$, such that each $[i, j] \in \mathcal{I}$ represents all factors of $T[i \mathinner{.\,.} j]$. This problem is very closely related to the problem of property indexing [3,7,8,21]. Employing either of the optimal property indexes that were presented in [7,8], one can retrieve the (number of) represented factors in optimal time.

References

1. Alamro, H., Badkobeh, G., Belazzougui, D., Iliopoulos, C.S., Puglisi, S.J.: Computing the antiperiod(s) of a string. In: 30th Annual Symposium on Combinatorial Pattern Matching (CPM 2019). LIPIcs, vol. 128, pp. 32:1–32:11. Schloss Dagstuhl - Leibniz-Zentrum für Informatik (2019). https://doi.org/10.4230/LIPIcs. CPM.2019.32
2. Alzamel, M., et al.: Online algorithms on antipowers and antiperiods. In: Brisaboa, N.R., Puglisi, S.J. (eds.) SPIRE 2019. LNCS, vol. 11811, pp. 175–188. Springer, Cham (2019). https://doi.org/10.1007/978-3-030-32686-9_13
3. Amir, A., Chencinski, E., Iliopoulos, C.S., Kopelowitz, T., Zhang, H.: Property matching and weighted matching. Theor. Comput. Sci. **395**(2–3), 298–310 (2008). https://doi.org/10.1016/j.tcs.2008.01.006
4. Badkobeh, G., Fici, G., Puglisi, S.J.: Algorithms for anti-powers in strings. Inf. Process. Lett. **137**, 57–60 (2018). https://doi.org/10.1016/j.ipl.2018.05.003
5. Bannai, H., Tomohiro, I., Inenaga, S., Nakashima, Y., Takeda, M., Tsuruta, K.: The "runs" theorem. SIAM J. Comput. **46**(5), 1501–1514 (2017). https://doi.org/ 10.1137/15M1011032
6. Bannai, H., Inenaga, S., Köppl, D.: Computing all distinct squares in linear time for integer alphabets. In: 28th Annual Symposium on Combinatorial Pattern Matching (CPM 2017). LIPIcs, vol. 78, pp. 22:1–22:18. Schloss Dagstuhl - Leibniz-Zentrum für Informatik (2017). https://doi.org/10.4230/LIPIcs.CPM.2017.22
7. Barton, C., Kociumaka, T., Liu, C., Pissis, S.P., Radoszewski, J.: Indexing weighted sequences: neat and efficient. Inf. Comput. **270** (2020). https://doi.org/10.1016/j. ic.2019.104462
8. Charalampopoulos, P., Iliopoulos, C.S., Liu, C., Pissis, S.P.: Property suffix array with applications in indexing weighted sequences. ACM J. Exp. Algorithm. **25**(1) (2020). https://doi.org/10.1145/3385898
9. Crochemore, M., Hancart, C., Lecroq, T.: Algorithms on Strings. Cambridge University Press (2007)
10. Crochemore, M., Ilie, L.: Computing longest previous factor in linear time and applications. Inf. Process. Lett. **106**(2), 75–80 (2008). https://doi.org/10.1016/j. ipl.2007.10.006
11. Crochemore, M., Iliopoulos, C.S., Kubica, M., Radoszewski, J., Rytter, W., Waleń, T.: Extracting powers and periods in a word from its runs structure. Theor. Comput. Sci. **521**, 29–41 (2014). https://doi.org/10.1016/j.tcs.2013.11.018
12. Deza, A., Franek, F., Thierry, A.: How many double squares can a string contain? Discrete Appl. Math. **180**, 52–69 (2015). https://doi.org/10.1016/j.dam.2014.08. 016
13. Farach, M.: Optimal suffix tree construction with large alphabets. In: 38th Annual Symposium on Foundations of Computer Science (FOCS 1997), pp. 137–143. IEEE Computer Society (1997). https://doi.org/10.1109/SFCS.1997.646102

14. Fici, G., Postic, M., Silva, M.: Abelian antipowers in infinite words. Adv. Appl. Math. **108**, 67–78 (2019). https://doi.org/10.1016/j.aam.2019.04.001
15. Fici, G., Restivo, A., Silva, M., Zamboni, L.Q.: Anti-powers in infinite words. J. Comb. Theory Ser. A **157**, 109–119 (2018). https://doi.org/10.1016/j.jcta.2018.02.009
16. Fine, N.J., Wilf, H.S.: Uniqueness theorems for periodic functions. Proc. Am. Math. Soc. **16**(1), 109–114 (1965). https://doi.org/10.2307/2034009
17. Fraenkel, A.S., Simpson, J.: How many squares can a string contain? J. Comb. Theory Ser. A **82**(1), 112–120 (1998). https://doi.org/10.1006/jcta.1997.2843
18. Gabow, H.N., Tarjan, R.E.: A linear-time algorithm for a special case of disjoint set union. J. Comput. Syst. Sci. **30**(2), 209–221 (1985). https://doi.org/10.1016/0022-0000(85)90014-5
19. Gawrychowski, P., I, T., Inenaga, S., Köppl, D., Manea, F.: Tighter bounds and optimal algorithms for all maximal α-gapped repeats and palindromes. Theory Comput. Syst. **62**(1), 162–191 (2017). https://doi.org/10.1007/s00224-017-9794-5
20. Gusfield, D., Stoye, J.: Linear time algorithms for finding and representing all the tandem repeats in a string. J. Comput. Syst. Sci. **69**(4), 525–546 (2004). https://doi.org/10.1016/j.jcss.2004.03.004
21. Hon, W., Patil, M., Shah, R., Thankachan, S.V.: Compressed property suffix trees. Inf. Comput. **232**, 10–18 (2013). https://doi.org/10.1016/j.ic.2013.09.001
22. Kempa, D., Kociumaka, T.: String synchronizing sets: sublinear-time BWT construction and optimal LCE data structure. In: 51st Annual ACM SIGACT Symposium on Theory of Computing (STOC 2019), pp. 756–767. ACM (2019). https://doi.org/10.1145/3313276.3316368
23. Kociumaka, T.: Efficient data structures for internal queries in texts. Ph.D. thesis, University of Warsaw (2018). https://mimuw.edu.pl/~kociumaka/files/phd.pdf
24. Kociumaka, T., Kubica, M., Radoszewski, J., Rytter, W., Waleń, T.: A linear-time algorithm for seeds computation. ACM Trans. Algorithms **16**(2) (2020). https://doi.org/10.1145/3386369
25. Kociumaka, T., Radoszewski, J., Rytter, W., Straszyński, J., Waleń, T., Zuba, W.: Efficient representation and counting of antipower factors in words. In: Martín-Vide, C., Okhotin, A., Shapira, D. (eds.) LATA 2019. LNCS, vol. 11417, pp. 421–433. Springer, Cham (2019). https://doi.org/10.1007/978-3-030-13435-8_31
26. Kociumaka, T., Radoszewski, J., Rytter, W., Straszyński, J., Waleń, T., Zuba, W.: Efficient representation and counting of antipower factors in words (2020). https://arxiv.org/abs/1812.08101v3
27. Kociumaka, T., Radoszewski, J., Rytter, W., Waleń, T.: Internal pattern matching queries in a text and applications. In: 26th Annual ACM-SIAM Symposium on Discrete Algorithms (SODA 2015), pp. 532–551. SIAM (2015). https://doi.org/10.1137/1.9781611973730.36
28. Kociumaka, T., Radoszewski, J., Rytter, W., Waleń, T.: String powers in trees. Algorithmica **79**(3), 814–834 (2017). https://doi.org/10.1007/s00453-016-0271-3
29. Kolpakov, R.: Some results on the number of periodic factors in words. Inf. Comput. **270** (2020). https://doi.org/10.1016/j.ic.2019.104459
30. Kolpakov, R.M., Kucherov, G.: Finding maximal repetitions in a word in linear time. In: 40th Annual Symposium on Foundations of Computer Science (FOCS 1999), pp. 596–604. IEEE Computer Society (1999). https://doi.org/10.1109/SFFCS.1999.814634
31. Lothaire, M.: Algebraic Combinatorics on Words. Cambridge University Press (2002). https://doi.org/10.1017/cbo9781107326019

Combinatorics on Words

Lyndon Words, the Three Squares Lemma, and Primitive Squares

Hideo Bannai[1]([✉])[ID], Takuya Mieno[2,3][ID], and Yuto Nakashima[2][ID]

[1] M&D Data Science Center, Tokyo Medical and Dental University, Tokyo, Japan
`hdbn.dsc@tmd.ac.jp`
[2] Department of Informatics, Kyushu University, Fukuoka, Japan
`{takuya.mieno,yuto.nakashima}@inf.kyushu-u.ac.jp`
[3] Japan Society for the Promotion of Science, Tokyo, Japan

Abstract. We revisit the so-called "Three Squares Lemma" by Crochemore and Rytter [Algorithmica 1995] and, using arguments based on Lyndon words, derive a more general variant which considers three overlapping squares which do not necessarily share a common prefix. We also give an improved upper bound of $n \log_2 n$ on the maximum number of (occurrences of) primitively rooted squares in a string of length n, also using arguments based on Lyndon words. To the best of our knowledge, the only known upper bound was $n \log_\phi n \approx 1.441 n \log_2 n$, where ϕ is the golden ratio, reported by Fraenkel and Simpson [TCS 1999] obtained via the Three Squares Lemma.

1 Introduction

Periodic structures of strings have been and still are one of the most important and fundamental objects of study in the field of combinatorics on words [4], and the analysis and exploitation of their combinatorial properties are a key ingredient in the development of efficient string processing algorithms [17,18].

In this paper, we focus on *squares*, which are strings of the form u^2 ($= uu$) for some string u, which is called the *root* of the square. A well known open problem concerning squares is on the maximum number of distinct squares that can be contained in a string. Fraenkel and Simpson [13] showed that the maximum number of distinct square substrings of a string of length n is at most $2n$. Although slightly better upper bounds of $2n - \Theta(\log n)$ [16] and $\frac{11}{6}n$ [10] have been shown, it is conjectured that it is at most n [13], with a best known lower bound of $n - o(n)$ [13].

The "Three Squares Lemma" by Crochemore and Rytter [9] was the key lemma used by Fraenkel and Simpson to obtain the upper bound of $2n$.

© Springer Nature Switzerland AG 2020
C. Boucher and S. V. Thankachan (Eds.): SPIRE 2020, LNCS 12303, pp. 265–273, 2020.
https://doi.org/10.1007/978-3-030-59212-7_19

Lemma 1 (Three Squares Lemma (Lemma 10 of [9][1])). *Let u^2, v^2, w^2 be three prefixes of some string such that w is primitive and $|u| > |v| > |w|$. Then, $|u| \geq |v| + |w|$.*

Crochemore and Rytter further showed that the lemma implies that the number of primitively rooted squares that can start at any given position of a string is bounded by $\log_\phi |x|$, where $\phi = (1+\sqrt{5})/2$ is the golden ratio (Theorem 11 of [9]). Thus, it follows that the maximum number $psq(n)$ of occurrences of primitively rooted squares in a string of length n is less than $n \log_\phi n \approx 1.441 n \log_2 n$.

The original proof of the Three Squares Lemma by Crochemore and Rytter was based on the well known "Periodicity Lemma" by Fine and Wilf [11]. Concerning a similar problem on the maximum number of "runs" (maximally periodic substring occurrences such that the smallest period is at most half its length) that can be contained in a string, the Periodicity Lemma was also the tool of choice in its analysis [6,19,23,24]. However, this changed when Bannai et al. [1,2] applied arguments based on *Lyndon words* [22] to solve, by a very simple proof, a longstanding conjecture that the maximum number of runs in a string of length n is at most n. Using the same technique, the upper bound on the number of runs was further improved to $0.957n$ for binary strings [12]. Bannai et al. also showed a new algorithm for computing all runs in a string, which paved the way for algorithms with improved time complexity for general ordered alphabets to be developed [7,15,20].

In this paper, we take the first steps of investigating to what extent Lyndon words can be applied in the analysis of squares. We first give an alternate proof of the Three Squares Lemma by arguments based on Lyndon words, and extend it to show a more general variant which considers three overlapping squares which do not necessarily share a common prefix. Furthermore, we show a significantly improved upper bound of $n \log_2 n$ on the maximum number of occurrences of primitively rooted squares.

2   Preliminaries

Let Σ be an alphabet. An element of Σ is called a symbol. An element of Σ^* is called a string. The length of a string w is denoted by $|w|$. The empty string ε is the string of length 0. For any possibly empty strings x, y, z, if $w = xyz$, then x, y, z are respectively called a prefix, substring, suffix of w. They are a *proper* prefix, substring, or suffix if they are not equal to w. For any $1 \leq i \leq j \leq |w|$, $w[i..j]$ denotes the substring of w starting at position i and ending at position j. We assume that $w[0], w[|w| + 1] \neq w[i]$ for any $1 \leq i \leq |w|$. For any string x, let $x^1 = x$, and for any integer $k \geq 2$, let $x^k = x^{k-1}x$. If there exists no string x and integer $k \geq 2$ such that $w = x^k$, w is said to be *primitive*.

---

[1] In [9], u, v, w are all assumed to be primitive and $|u| > |v| + |w|$ was claimed, but it was noted in [13] that only primitivity of w is required, and that $|u| \geq |v| + |w|$ is the correct relation, giving $u = 01001010$, $v = 01001$, and $w = 010$ as an example when $|u| = |v| + |w|$.

A non-empty string w is said to be a *Lyndon word* [22] if w is lexicographically smaller than any of its non-empty proper suffixes. An important property of Lyndon words is that they cannot have a *border*, i.e., a non-empty substring that is both a proper suffix and prefix. Also, notice that whether a string is a Lyndon word or not depends on the choice of the lexicographic order. Unless otherwise stated, our results hold for any lexicographic order. However, we will sometimes require a pair of lexicographic orders $<_0$ and $<_1$, the former induced by an arbitrary total order on Σ, and the other induced by the opposite total order, i.e., for any $a, b \in \Sigma$, $a <_0 b$ if and only if $b <_1 a$.

An integer $1 \leq p \leq |w|$ is a *period* of string w if $w[i] = w[i + p]$ for all $i = 1, \ldots, |w| - p$. A string is a *repetition* if its smallest period p is at most half of its length. An occurrence $w[i..j] = v$ of a repetition v with smallest period p is a *maximal repetition* (or a *run*) in w, if the smallest periods of both $w[i-1..j]$ and $w[i..j+1]$ are not p.

For any repetition v, an *L-root* [8] λ_v is a substring of v that is a Lyndon word whose length is equal to the smallest period of v. It is easy to see that an L-root of a repetition always exists and is unique. We also define the *L-root interval* r_v in v as the substring corresponding to the maximal integer power in v of λ_v. Any repetition v can be written as $v = xr_v y$ where x (resp. y) is a possibly empty proper suffix (resp. prefix) of the L-root λ_v. Notice that for any square u^2, $|r_{u^2}| \geq |u|$. Also, for any square u^2, it can be shown that the smallest period p_u of u^2 is a divisor of $|u|$ and is equal to $|\lambda_{u^2}|$, which implies that it is also the smallest period of r_{u^2} and a divisor of $|r_{u^2}|$.

The next lemma shows that a Lyndon word can only occur in a run as a substring of the L-root of the run.

Lemma 2. *For any Lyndon word v, there is no Lyndon word $w = xyz$ for strings x, y, z such that x (resp. z) is a non-empty suffix (resp. prefix) of v.*

Proof. If such w exists, $v \leq x < xyz = w < z \leq v$, a contradiction. $\square$

3   Squares and L-Roots

We first prove a lemma concerning two squares.

Lemma 3. *Let u^2 and v^2 be squares where v^2 is a proper prefix of u^2. Then, the L-root interval r_{u^2} of u^2 is not a substring of v^2, and either r_{v^2} is a prefix of r_{u^2}, or r_{v^2} ends before r_{u^2} starts.*

Proof. Let p_u and p_v respectively be the smallest periods of u^2 and v^2. If r_{u^2} is a substring of v^2, then, $v^2 = xr_{u^2}y = wr_{v^2}z$ for some suffix x of λ_{u^2}, some prefix y of λ_{u^2}, some suffix w of λ_{v^2}, and some prefix z of λ_{v^2}. If $p_u \neq p_v$, then $r_{u^2} \neq r_{v^2}$ must hold since p_u and p_v are respectively their smallest periods. This implies either $|x| \neq |w|$ or $|y| \neq |z|$. However, that would contradict Lemma 2. If $p_u = p_v$, then it must be that $r_{u^2} = r_{v^2}$ due to their maximality. Since u is longer than v, and $p_u = p_v$ must also be a divisor of their lengths, u^2 must be at least $2p_u$ longer than v^2. However that would contradict the maximality of r_{u^2}, since at least one more copy of λ_{u^2} would fit inside u^2.

Next, suppose that r_{v^2} overlaps with r_{u^2}, and is not a prefix of r_{u^2}. Since r_{u^2} cannot be a substring of v^2 in which r_{v^2} is a substring, r_{u^2} starts in v^2, and ends after the end of v^2. There are two cases: (1) r_{v^2} starts after the beginning of r_{u^2} and ends in r_{u^2} (Fig. 1) or (2) r_{v^2} starts before r_{u^2}, and ends in r_{u^2} (Fig. 2).

Fig. 1. Case (1) of Lemma 3.   **Fig. 2.** Case (2) of Lemma 3.

Case (1) implies that $r_{u^2} = x r_{v^2} y$ for some non-empty proper suffix x of λ_{v^2} and some suffix y of r_{u^2}. Let $r_{v^2} = x'z$ where $|x'| = |x|$. Since $|x'| < p_v$, we have $x > r_{v^2} > x'$, and thus, $r_{u^2} = x r_{v^2} y > x' z y$. This can hold only if $|x|$ is a multiple of p_u, but this also implies $x = x'$ which is a contradiction.

Case (2) implies that a suffix of r_{v^2} overlaps with a prefix of r_{u^2}. Let $r_{v^2} = xy$, $r_{u^2} = yz$ where y is the overlap, and observe that $|x| < p_u$ due to the maximality of r_{u^2}. Notice that since u^2 has period p_u which is a divisor of $|r_{u^2}|$, x must also be a suffix of r_{u^2}, so we can write $r_{u^2} = wx$ for some w. From Lemma 2, x must be an integer power of λ_{v^2}, since otherwise, there would be an occurrence of λ_{v^2} crossing the boundary of x and y. Thus, r_{u^2} contains the Lyndon word λ_{v^2} of length p_v as a prefix and suffix, which can only hold if $p_u = p_v$. However, this contradicts the maximality of r_{u^2}.   □

To prove Lemma 1, we use the previous lemma, together with the following lemma used in the proof of the "runs" theorem [2] which connects L-roots of runs and longest Lyndon words starting at each position.

Lemma 4 (Lemma 3.3 of [2]). *For any run $w[i..j]$ with period p, consider the lexicographic order $< \in \{<_0, <_1\}$ such that $w[j+1] < w[j+1-p]$. Then, any occurrence of the L-root of the run $w[i..j]$ is the longest Lyndon word starting at that position.*

It is easy to see that for any repetition, there is a unique run with the same smallest period and L-root in which the repetition is contained. For any occurrence of a repetition in a string, we will refer to the lexicographic order considered in Lemma 4 as *the* lexicographic order of the repetition.

Proof (of Lemma 1). Consider the lexicographic order of w^2, i.e., L-root λ_{w^2} is a longest Lyndon word starting at the first position of r_{w^2}. From Lemma 3, the starting positions $b_{w^2}, b_{v^2}, b_{u^2}$ respectively of $r_{w^2}, r_{v^2}, r_{u^2}$ are non-decreasing. There are four cases: (1) $b_{w^2} < b_{v^2} < b_{u^2}$, (2) $b_{w^2} < b_{v^2} = b_{u^2}$, (3) $b_{w^2} = b_{v^2} <$

b_{u^2}, and (4) $b_{w^2} = b_{v^2} = b_{u^2}$, where inequality of the starting positions implies the disjointness of the L-root intervals.

Case (1): It follows that $r_{w^2}, r_{v^2}, r_{u^2}$ occur disjointly in u^2. Therefore, $2|u| \geq |r_{w^2}| + |r_{v^2}| + |r_{u^2}|$. Since $|r_{w^2}| \geq |w|, |r_{v^2}| \geq |v|, |r_{u^2}| \geq |u|$, we have $|u| \geq |w| + |v|$.

Case (2): It follows that r_{w^2} occurs disjointly before r_{u^2}, and r_{v^2} is a prefix of r_{u^2}. Since $r_{v^2} \geq |v|$, r_{w^2} is a substring of v and thus also of u. Due to u^2 and v^2, there are two other occurrences of r_{w^2} respectively $|u|$ and $|v|$ positions to the right. Since w is primitive, the smallest period of r_{w^2} is $|\lambda_{w^2}| = |w|$, and thus the two occurrences of r_{w^2} must be at least $|w|$ apart. Therefore, $|w| \leq |u| - |v|$, which implies $|u| \geq |v| + |w|$.

Case (3): By the assumption of the lexicographic order, λ_{w^2} is the longest Lyndon word starting at b_{w^2} and thus $|\lambda_{w^2}| \geq |\lambda_{v^2}|$. Since r_{w^2} is a prefix of r_{v^2}, it must hold that $\lambda_{w^2} = \lambda_{v^2}$ due to Lemma 2. Since $|v| > |w| = |\lambda_{w^2}| = |\lambda_{v^2}|$ and $|v|$ is a multiple of $|\lambda_{v^2}|$, we have $|v| \geq 2|\lambda_{v^2}|$. This implies $|r_{v^2}| \geq |v| + |\lambda_{v^2}|$. Also, since r_{u^2} occurs disjointly with r_{v^2} in u^2, we have $2|u| \geq |r_{v^2}| + |r_{u^2}|$, which implies $|u| \geq |r_{v^2}|$ since $|r_{u^2}| \geq |u|$. Then, $|u| \geq |r_{v^2}| \geq |v| + |\lambda_{v^2}| = |v| + |w|$.

Case (4): Analogously to the previous case, we have $\lambda_{w^2} = \lambda_{v^2} = \lambda_{u^2}$. This implies that $|u|, |v|$ are multiples of $|\lambda_{w^2}|$ and since $|u| > |v|$, we have $|u| \geq |v| + |\lambda_{v^2}| = |v| + |w|$. □

We note that actually, the proof of Lemma 3 does not require v^2 to be a prefix of u^2, but only that v^2 is a substring of u^2 that starts before r_{u^2}, so slightly stronger statements hold.

Corollary 1. *Let u^2 and v^2 be squares such that v^2 is a proper substring of u^2 that starts before the L-root interval r_{u^2} of u^2. Then, r_{u^2} is not a substring of v^2, and either the L-root interval r_{v^2} of v^2 is a prefix of r_{u^2}, or r_{v^2} ends before r_{u^2} starts.*

Corollary 2. *Let u^2, v^2, and w^2 be squares such that v^2 is a proper substring of u^2 that starts before r_{u^2}, and w^2 is a proper substring of v^2 that starts before r_{u^2} and r_{v^2}, where r_{u^2}, r_{v^2} are respectively the L-root intervals of u^2, v^2 with respect to the lexicographic order of w. If w is primitive, then $|u| \geq |v| + |w|$.*

4 Tighter Upper Bound for $psq(n)$

There can be $\Theta(n^2)$ occurrences of non-primitively rooted squares in a string of length n (e.g. a unary string). However, as mentioned in the introduction, Lemma 1 implies an upper bound of $n \log_\phi n \simeq 1.441 n \log_2 n$ for $psq(n)$, i.e., the maximum number of occurrences of primitively rooted squares in a string of length n. On the other hand, the best known lower bound is given by Fibonacci words, which contain $\frac{2(3-\phi)}{5\log_2 \phi} F_n \log_2 F_n + O(F_n)$ occurrences of primitive squares [14], where F_n is the length of the n-th Fibonacci word, ϕ is the golden ratio, and $\frac{2(3-\phi)}{5\log_2 \phi} \approx 0.7962$. Below, we prove a significantly improved upper bound for $psq(n)$.

Theorem 1. $psq(n) \leq n \log_2 n$.

Each primitively rooted square of w is a substring of a run of w. Let $runs(w)$ denote the set of runs in w. Conversely, each run $\rho \in runs(w)$ with length ℓ_ρ and period p_ρ contains exactly $\ell_\rho - 2p_\rho + 1$ primitively rooted squares as substrings. Let λ_ρ be an L-root of a run ρ with respect to the lexicographical order of ρ. If we consider the rightmost occurrence of λ_ρ in ρ, there exist strings x_ρ, y_ρ such that $\rho = x_\rho \lambda_\rho y_\rho$ and y_ρ is a possibly empty proper prefix of λ_ρ. Since $|\lambda_\rho| \geq |y_\rho| + 1$, the number of primitively rooted squares in ρ is $\ell_\rho - 2p_\rho + 1 = |x_\rho| + |\lambda_\rho| + |y_\rho| - 2|\lambda_\rho| + 1 \leq |x_\rho|$. Thus, the total sum of $|x_\rho|$ for all runs in w gives an upper bound on the number of occurrences of primitively rooted squares in w. We will show that this total sum is bounded by $n \log_2 n$ for any string w of length n, which will yield Theorem 1.

To this end, we use the notion of Lyndon trees [2,3]. The *Lyndon tree* of a Lyndon word w is an ordered full binary tree defined recursively as follows[2]: If $|w| = 1$, then the Lyndon tree of w is a single node labeled w, and if $|w| \geq 2$, then the root is labeled w, and the left and right children of w are respectively the Lyndon trees of u and v, where $w = uv$ and v is the lexicographically smallest proper suffix of w. Note that this is known as the standard factorization of w [5,21], and u, v are guaranteed to be Lyndon words.

From Lemma 4 and Lemma 5 below, we have that for any string w, the right nodes of the two Lyndon trees of w with respect to $<_0$ and $<_1$ contain all L-roots of all runs in w.

Lemma 5 (Lemma 5.4 of [2]). *Let w be a Lyndon word. For any interval $[i..j]$ except for $[1..|w|]$, $[i..j]$ corresponds to a right node of the Lyndon tree if and only if $w[i..j]$ is the longest Lyndon word that starts at i.*

Thus, as before, we have that $\rho = x_\rho \lambda_\rho y_\rho = x'_\rho \lambda_\rho^k y_\rho$, where $\lambda_\rho^k = r_\rho$ is the L-root interval of ρ, x'_ρ is a possibly empty proper suffix of λ_ρ, and that each occurrence of λ_ρ corresponds to a right node in one the Lyndon trees. Now, $|x_\rho| = |x'_\rho| + (k-1)|\lambda_\rho|$, and we distribute this sum among each of the k occurrences of the L-root as follows: $|x'_\rho|$ for the leftmost occurrence (i.e., the periodicity only extends $|x'_\rho|$ symbols to the left of the occurrence), or $|\lambda_\rho|$ otherwise (i.e., the periodicity extends at least $|\lambda_\rho|$ symbols to the left of the occurrence).

Next, consider how long the periodicity can extend to the left of each occurrence of λ_ρ by looking at the Lyndon tree. Since λ_ρ corresponds to a right node, $w_\rho = z_\rho \lambda_\rho$ for some Lyndon words w_ρ and z_ρ. When $|z_\rho| \leq |\lambda_\rho|$, z_ρ cannot be a suffix of λ_ρ, since that would imply that $w_\rho = z_\rho \lambda_\rho < \lambda_\rho < z_\rho$, a contradiction. Thus, for the occurrence of L-root λ_ρ in w_ρ, the periodicity can extend at most $|z_\rho|$ symbols (more precisely, $|z_\rho| - 1$ symbols).

Let $\mathcal{S}(n)$ denote the maximum of the total sum of all $|x_\rho|$ for all potential L-roots λ_ρ that correspond to a right node in a (single) Lyndon tree for any string of length n. From the above arguments, we have $\mathcal{S}(n) = 0$ if $n = 1$, and otherwise,

---

[2] If w is not a Lyndon word, we simply consider the Lyndon word obtained by prepending to w a symbol smaller than any symbol in w.

$\mathcal{S}(n) \leq \max\{\mathcal{S}(n_1) + \mathcal{S}(n_2) + \min\{n_1, n_2\} \mid n_1, n_2 > 0 \text{ and } n_1 + n_2 = n\}$. We can show by induction that $\mathcal{S}(n)$ can be bounded by $\frac{n}{2}\log_2 n$.

Lemma 6. $\mathcal{S}(n) \leq \frac{n}{2}\log_2 n$.

Proof. Clearly, when $|n| = 1$, $0 = \mathcal{S}(n) \leq \frac{1}{2}\log_2 1 = 0$. For $n \geq 2$, assume that the lemma holds for any value less than n. Then,

$$\mathcal{S}(n) \leq \max\left\{\mathcal{S}(n_1) + \mathcal{S}(n_2) + \min\{n_1, n_2\} \mid n_1, n_2 \neq 0 \text{ and } n_1 + n_2 = n\right\}$$

$$\leq \max\left\{\frac{n - k_n}{2}\log_2(n - k_n) + \frac{k_n}{2}\log_2 k_n + k_n \;\middle|\; 1 \leq k_n \leq \frac{n}{2}\right\}$$

$$= \frac{1}{2}\max\left\{((n - k_n)\log_2(n - k_n) + k_n\log_2 k_n + 2k_n) \;\middle|\; 1 \leq k_n \leq \frac{n}{2}\right\}$$

$$\leq \frac{1}{2}\left(\left(n - \frac{n}{2}\right)\log_2\left(n - \frac{n}{2}\right) + \frac{n}{2}\log_2\frac{n}{2} + n\right)$$

$$= \frac{1}{2}\left(n\log_2\frac{n}{2} + n\right) = \frac{n}{2}\log_2 n.$$

The third inequality follows since the second derivative of the above function is positive and thus the function is maximized when $k_n = n/2$. □

Now, since any occurrence of an L-root corresponds to a right node in one of the two Lyndon trees, we have

$$psq(n) \leq \max_{w \in \Sigma^n} \sum_{\rho \in runs(w)} |x_\rho| \leq 2 \cdot \mathcal{S}(n) \leq n\log_2 n.$$

Acknowledgments. We would like to thank the anonymous reviewers for pointing out and correcting errors in the submitted version of the paper.

This work was supported by JSPS KAKENHI Grant Numbers JP20H04141 (HB), JP20J11983 (TM), and JP18K18002 (YN).

References

1. Bannai, H., Tomohiro, I., Inenaga, S., Nakashima, Y., Takeda, M., Tsuruta, K.: A new characterization of maximal repetitions by Lyndon trees. In: Indyk, P. (ed.) Proceedings of the Twenty-Sixth Annual ACM-SIAM Symposium on Discrete Algorithms (SODA 2015), San Diego, CA, USA, 4–6 January 2015, pp. 562–571. SIAM (2015). https://doi.org/10.1137/1.9781611973730.38
2. Bannai, H., Tomohiro, I., Inenaga, S., Nakashima, Y., Takeda, M., Tsuruta, K.: The "runs" theorem. SIAM J. Comput. **46**(5), 1501–1514 (2017). https://doi.org/10.1137/15M1011032
3. Barcelo, H.: On the action of the symmetric group on the free Lie algebra and the partition lattice. J. Comb. Theory Ser. A **55**(1), 93–129 (1990). https://doi.org/10.1016/0097-3165(90)90050-7
4. Berstel, J., Perrin, D.: The origins of combinatorics on words. Eur. J. Comb. **28**(3), 996–1022 (2007). https://doi.org/10.1016/j.ejc.2005.07.019

5. Chen, K.T., Fox, R.H., Lyndon, R.C.: Free differential calculus, IV. The quotient groups of the lower central series. Ann. Math. **68**(1), 81–95 (1958). https://doi.org/10.2307/1970044

6. Crochemore, M., Ilie, L.: Maximal repetitions in strings. J. Comput. Syst. Sci. **74**(5), 796–807 (2008). https://doi.org/10.1016/j.jcss.2007.09.003

7. Crochemore, M., et al.: Near-optimal computation of runs over general alphabet via non-crossing LCE queries. In: Inenaga, S., Sadakane, K., Sakai, T. (eds.) SPIRE 2016. LNCS, vol. 9954, pp. 22–34. Springer, Cham (2016). https://doi.org/10.1007/978-3-319-46049-9_3

8. Crochemore, M., Iliopoulos, C.S., Kubica, M., Radoszewski, J., Rytter, W., Walen, T.: Extracting powers and periods in a word from its runs structure. Theor. Comput. Sci. **521**, 29–41 (2014). https://doi.org/10.1016/j.tcs.2013.11.018

9. Crochemore, M., Rytter, W.: Squares, cubes, and time-space efficient string searching. Algorithmica **13**(5), 405–425 (1995). https://doi.org/10.1007/BF01190846

10. Deza, A., Franek, F., Thierry, A.: How many double squares can a string contain? Discrete Appl. Math. **180**, 52–69 (2015). https://doi.org/10.1016/j.dam.2014.08.016

11. Fine, N.J., Wilf, H.S.: Uniqueness theorems for periodic functions. Proc. Am. Math. Soc. **16**(1), 109–114 (1965). https://doi.org/10.1090/S0002-9939-1965-0174934-9

12. Fischer, J., Holub, Š., I, T., Lewenstein, M.: Beyond the runs theorem. In: Iliopoulos, C., Puglisi, S., Yilmaz, E. (eds.) SPIRE 2015. LNCS, vol. 9309, pp. 277–286. Springer, Cham (2015). https://doi.org/10.1007/978-3-319-23826-5_27

13. Fraenkel, A.S., Simpson, J.: How many squares can a string contain? J. Comb. Theory Ser. A **82**(1), 112–120 (1998). https://doi.org/10.1006/jcta.1997.2843

14. Fraenkel, A.S., Simpson, J.: The exact number of squares in Fibonacci words. Theor. Comput. Sci. **218**(1), 95–106 (1999). https://doi.org/10.1016/S0304-3975(98)00252-7

15. Gawrychowski, P., Kociumaka, T., Rytter, W., Walen, T.: Faster longest common extension queries in strings over general alphabets. In: Grossi, R., Lewenstein, M. (eds.) 27th Annual Symposium on Combinatorial Pattern Matching (CPM 2016), 27–29 June 2016, Tel Aviv, Israel. LIPIcs, vol. 54, pp. 5:1–5:13. Schloss Dagstuhl - Leibniz-Zentrum für Informatik (2016). https://doi.org/10.4230/LIPIcs.CPM.2016.5

16. Ilie, L.: A note on the number of squares in a word. Theor. Comput. Sci. **380**(3), 373–376 (2007). https://doi.org/10.1016/j.tcs.2007.03.025. (Combinatorics on Words)

17. Kempa, D., Kociumaka, T.: String synchronizing sets: sublinear-time BWT construction and optimal LCE data structure. In: Charikar, M., Cohen, E. (eds.) Proceedings of the 51st Annual ACM SIGACT Symposium on Theory of Computing (STOC 2019), Phoenix, AZ, USA, 23–26 June 2019, pp. 756–767. ACM (2019). https://doi.org/10.1145/3313276.3316368

18. Knuth, D.E., Morris Jr., J.H., Pratt, V.R.: Fast pattern matching in strings. SIAM J. Comput. **6**(2), 323–350 (1977). https://doi.org/10.1137/0206024

19. Kolpakov, R.M., Kucherov, G.: Finding maximal repetitions in a word in linear time. In: 40th Annual Symposium on Foundations of Computer Science (FOCS 1099), 17–18 October 1999, New York, NY, USA, pp. 596–604. IEEE Computer Society (1999). https://doi.org/10.1109/SFFCS.1999.814634

20. Kosolobov, D.: Computing runs on a general alphabet. Inf. Proc. Lett. **116**(3), 241–244 (2016). https://doi.org/10.1016/j.ipl.2015.11.016

21. Carpi, A., D'Alessandro, F.: On the commutative equivalence of bounded semi-linear codes. In: Mercaş, R., Reidenbach, D. (eds.) WORDS 2019. LNCS, vol. 11682, pp. 119–132. Springer, Cham (2019). https://doi.org/10.1007/978-3-030-28796-2_9
22. Lyndon, R.C.: On Burnside's problem. Trans. Am. Math. Soc. **77**(2), 202–202 (1954). https://doi.org/10.2307/1990868
23. Puglisi, S.J., Simpson, J., Smyth, W.: How many runs can a string contain? Theor. Comput. Sci. **401**(1), 165–171 (2008). https://doi.org/10.1016/j.tcs.2008.04.020
24. Rytter, W.: The number of runs in a string: improved analysis of the linear upper bound. In: Durand, B., Thomas, W. (eds.) STACS 2006. LNCS, vol. 3884, pp. 184–195. Springer, Heidelberg (2006). https://doi.org/10.1007/11672142_14

Computational Biology

Efficient Construction of Hierarchical Overlap Graphs

Sung Gwan Park[1], Bastien Cazaux[2,3], Kunsoo Park[1](✉),
and Eric Rivals[3]

[1] Seoul National University, Seoul, Korea
{sgpark,kpark}@theory.snu.ac.kr
[2] University of Helsinki, Helsinki, Finland
bastien.cazaux@lirmm.fr
[3] LIRMM, Univ Montpellier, CNRS, Montpellier, France
rivals@lirmm.fr

Abstract. The hierarchical overlap graph (HOG for short) is an overlap encoding graph that efficiently represents overlaps from a given set
P of n strings. A previously known algorithm constructs the HOG in
$O(||P|| + n^2)$ time and $O(||P|| + n \times \min(n, \max\{|s| : s \in P\}))$ space,
where $||P||$ is the sum of lengths of the n strings in P. We present a
new algorithm of $O(||P|| \log n)$ time and $O(||P||)$ space to compute the
HOG, which exploits the segment tree data structure. We also propose
an alternative algorithm using $O(||P|| \frac{\log n}{\log \log n})$ time and $O(||P||)$ space
in the word RAM model of computation.

Keywords: Hierarchical overlap graph · Segment tree · Word RAM
model

1 Introduction

Genome sequencing is limited by sequencing technologies that yield sequencing reads which are orders of magnitude shorter than the entire genome. Hence,
obtaining a whole genome sequence from sequencing reads resorts to *DNA assembly*. This problem consists in recovering the target sequence from the overlaps
of reads by inferring their order and relative positions in the target sequence. It
translates into seeking a maximal path in a graph that encodes suffix-prefix overlaps between pairs of reads [7,22,25,26]. The development of DNA sequencing
goes along with several proposals of overlap encoding graphs, usually classified
into two categories of digraphs:

S. G. Park and K. Park—Supported by Institute for Information & communications
Technology Promotion(IITP) grant funded by the Korea government (MSIT) (No.
2018-0-00551, Framework of Practical Algorithms for NP-hard Graph Problems).
E. Rivals—ER thanks funding Labex NUMEV, GEM project (ANR 2011-LABX-076).

C. Boucher and S. V. Thankachan (Eds.): SPIRE 2020, LNCS 12303, pp. 277–290, 2020.
https://doi.org/10.1007/978-3-030-59212-7_20

- *Overlap Graph* [25] and its variants (like String Graph [22]), in which each input read is a node and an arc connecting a pair of reads represents the longest overlap between them, and
- "assembly" *de Bruijn Graph* [26], in which for a length k, each node represents a k-long substring (termed k-mer) and an arc connects two k-mers whenever the suffix of one matches the prefix of the other over length $k-1$.

The overlap relation is not symmetrical, which explains why directed, rather than undirected, graphs should be used in DNA assembly. Moreover, a pair of reads can have several overlaps (in the same direction), in which case a shorter overlap is necessarily nested into a longer one.

Recently, Cazaux and Rivals [8,9] proposed an alternative graph in which the input reads and substrings corresponding to suffix-prefix overlaps are nodes in the graph. This digraph, called *Extended Hierarchical Overlap Graph* (EHOG), encodes both the longest suffix relationship and the longest prefix relationship between nodes by using two kinds of arcs. To compact the EHOG even more, the *Hierarchical Overlap Graph* (HOG) which includes only maximal overlaps between reads was defined. A maximal overlap is a longest overlap for at least one pair of reads. By definition, therefore, the HOG is a subgraph of the EHOG. See Fig. 1 for examples of EHOG and HOG. Even if the EHOG and the HOG can be identical for some instances, the ratio of the EHOG size over the HOG size (in the number of nodes) can tend to infinity for some families of instances [9]. Thus, efficient algorithms to build the HOG are important from both practical and theoretical viewpoints. The advantages of the HOG/EHOG for storing overlaps compared to other graphs are discussed in [9].

Given a set of strings, the *shortest superstring* problem is the problem of finding a shortest superstring of the given strings. The shortest superstring problem has applications in DNA assembly and data compression [6,29]. Since the problem is MAX SNP-hard, there has been extensive research to get better approximation ratios, e.g., 3 in [6], $2\frac{2}{3}$ in [3], $2\frac{1}{2}$ in [29], and more recently $2\frac{11}{23}$ [21] and $2\frac{11}{30}$ [23]. These approximation algorithms are based on the overlap graph (or equivalent *distance graph*). In the overlap graph (or the distance graph), many distinct arcs may encode the same overlap, but this fact is not specified in the graph. In the HOG, all identical overlaps are encoded into a unique node, i.e., this fact is specified. Hence, the HOG has structurally more information than the overlap graph, and thus it has a great potential in studying DNA assembly and the shortest superstring problem.

Suppose that an input instance P consists of n strings, where no string is a substring of another. The norm of P, denoted by $||P||$, is defined as the sum of lengths of the strings in P. Computing an overlap graph from P is equivalent to solving the all-pair suffix-prefix problem, which is studied extensively [12,14,20,27]. The best asymptotic bound for this problem is $O(||P||+n^2)$ [14], which is optimal. Computing the EHOG from P takes linear time in the norm of P [9]. However, further limiting the set of overlap nodes to maximal overlaps, which enables us to build the HOG, is more challenging. A previously known algorithm achieves $O(||P||+n^2)$ time with $O(||P||+n\times\min(n,\max\{|s|:s\in P\}))$

space [9], which has the same time complexity as the all-pair suffix-prefix problem. The question of an optimal algorithm for computing the HOG remains open. In this paper we present an algorithm taking $O(||P|| \log n)$ time with $O(||P||)$ space in the standard RAM model, which exploits the segment tree data structure (Sect. 3). We also propose an alternative algorithm using $O(||P|| \frac{\log n}{\log \log n})$ time and $O(||P||)$ space in the word RAM model of computation [15] (Sect. 4). Throughout the paper, we assume that the size of the alphabet is constant.

2   Preliminaries

In this paper we consider strings over a finite alphabet Σ. Given a string s, the length of s is denoted by $|s|$. For any two integers $1 \leq i \leq j \leq |s|$, the substring of s which starts from i and ends at j is denoted by $s[i..j]$. Substring $s[i..j]$ is a prefix of s if $i = 1$, and a suffix of s if $j = |s|$. A prefix (suffix) of s is a proper prefix (suffix) of s if it is different from s. Given two strings s and t, string u is an overlap from s to t if u is a proper suffix of s and also a proper prefix of t. The longest overlap from s to t is denoted by $ov(s, t)$. Given a set $P = \{s_1, s_2, ..., s_n\}$ of strings, the sum of $|s_i|$'s is denoted by $||P||$.

2.1   Hierarchical Overlap Graph

We use definitions of *extended hierarchical overlap graph* and *hierarchical overlap graph* in [9].

Definition 1. Given a set $P = \{s_1, s_2, \ldots, s_n\}$ of strings, let $Ov^+(P)$ be the set of all overlaps from s_i to s_j for $1 \leq i, j \leq n$. The *Extended Hierarchical Overlap Graph* of P, denoted by $\mathrm{EHOG}(P)$, is a directed graph (V^+, E^+) where $V^+ = P \cup Ov^+(P) \cup \{\epsilon\}$ and $E^+ = E_1^+ \cup E_2^+$, where $E_1^+ = \{(x, y) \in V^+ \times V^+ \mid x$ is the longest proper prefix of $y\}$ and $E_2^+ = \{(x, y) \in V^+ \times V^+ \mid y$ is the longest proper suffix of $x\}$.

Definition 2. Given a set $P = \{s_1, s_2, \ldots, s_n\}$ of strings, let $Ov(P)$ be the set of the *longest* overlap from s_i to s_j for $1 \leq i, j \leq n$. The *Hierarchical Overlap Graph* of P, denoted by $\mathrm{HOG}(P)$, is a directed graph (V, E) where $V = P \cup Ov(P) \cup \{\epsilon\}$ and $E = E_1 \cup E_2$, where $E_1 = \{(x, y) \in V \times V \mid x$ is the longest proper prefix of $y\}$ and $E_2 = \{(x, y) \in V \times V \mid y$ is the longest proper suffix of $x\}$.

For example, Fig. 1 from [9] shows an Aho-Corasick trie [1], EHOG, and HOG built with $P = \{aabaa, aacd, cdb\}$. Note that EHOG is a contracted form of the Aho-corasick trie and HOG is a contracted form of EHOG, as described in [9]. Consequently, both EHOG and HOG, without failure links, are trees.

By definitions of EHOG and HOG, each node u in a graph represents a string, which is the concatenation of labels on the path from the root to u. If (u, v) is a tree arc (an edge in E_1^+ or E_1, solid line in Fig. 1) in an EHOG (resp. HOG), the string represented by u is the longest proper prefix of the string represented by v in the EHOG (resp. HOG). If (u, v) is a failure link (an edge in E_2^+ or E_2,

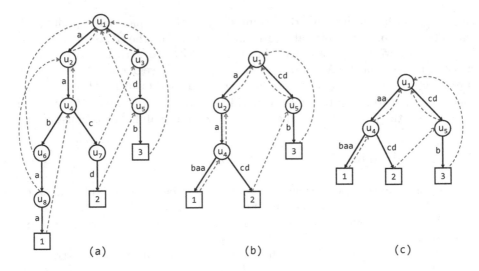

Fig. 1. Data structures built with $P = \{aabaa, aacd, cdb\}$. Dotted lines represent failure links of the nodes. (a) Aho-Corasick tri.e. (b) Extended hierarchical overlap graph. (c) Hierarchical overlap graph.

dotted line in Fig. 1) in an EHOG (resp. HOG), the string represented by v is the longest proper suffix of the string represented by u in the EHOG (resp. HOG). In this paper we use term 'node' to mean a node in EHOG or HOG, or a string represented by the node.

We can build an EHOG of $P = \{s_1, s_2, ..., s_n\}$ in $O(\|P\|)$ time and space [9]. Furthermore, if we know EHOG(P) and $Ov(P)$, we can compute HOG(P) in $O(\|P\|)$ time and space [9]. Therefore, the bottleneck of computing HOG(P) is to compute $Ov(P)$, which costs $O(\|P\| + n^2)$ time and $O(\|P\| + n \times \min(n, \max\{|s_i|\}))$ space in [9].

3   Main Algorithm

In this section we describe an algorithm to compute HOG from the given set $P = \{s_1, s_2, \ldots, s_n\}$ of strings in $O(\|P\| \log n)$ time.

3.1   New Approach to Compute HOG

First, we build an Aho-Corasick trie of P and renumber the strings (i.e., leaves) in lexicographic order. This can be done in $O(\|P\|)$ time, assuming that the size of the alphabet is constant. Next, we build EHOG(P) in $O(\|P\|)$ time [9]. Furthermore, for each node u in EHOG(P), we define an interval I(u) that contains every leaf node that is in the subtree of u (i.e. I(u) = $\{i \in [1..n] \mid u$ is a prefix of $s_i\}$). Since P is renumbered in lexicographic order, we can see that I(u) forms one interval.

Algorithm 1. Computing HOG using interval encoding

1: **procedure** BUILD-HOG-INTERVAL-ENCODING(EHOG(P))
2:     **for** $i \leftarrow 1$ **to** n **do**
3:         Initialize $B[1..n]$ to `false`
4:         u $\leftarrow$ leaf corresponding to s_i in EHOG(P)
5:         Mark u as included in HOG(P)
6:         **while** u $\neq$ **root do**
7:             u $\leftarrow$ failure link of u in EHOG(P)
8:             **if** $\exists\, j \in$ I(u) such that $B[j]$ is `false` **then**
9:                 Mark u as included in HOG(P)
10:            **for** $j \in$ I(u) **do**
11:                $B[j] \leftarrow$ `true`
12:        Build HOG(P) with marked nodes

Given EHOG(P), we compute $Ov(P)$ by discarding nodes that are not longest overlaps. If a string s is included in $Ov(P)$, s is a proper suffix of s_i and a proper prefix of s_j for some i and j by definition of $Ov(P)$. To compute all longest overlaps from s_i, we start from the i-th leaf s_i, follow the failure links repeatedly up to the root, and check whether the node we are looking at is the longest prefix of s_j for some j. (Note that every overlap between two strings in P is represented as a node in EHOG(P), and thus we can iterate through all overlaps from s_i by following the failure links starting from s_i.) While traversing the nodes through failure links (namely $v_0 = i$-th leaf $\rightarrow v_1 \rightarrow \cdots \rightarrow v_k =$ root), v_x ($1 \leq x \leq k$) is $ov(i,j)$ if and only if v_x is the first node that is a prefix of s_j during the traversal. More specifically, v_x should be the prefix of s_j and v_y's ($1 \leq y < x$) should not be the prefixes of s_j. To check whether there exists such j efficiently, we maintain a bit vector B of length n defined as follows. At the end of the iteration with v_x ($1 \leq x \leq k$), $B[j] = $ `true` if and only if there exists $1 \leq y \leq x$ such that v_y is a prefix of s_j. We can maintain B as defined by marking $B[j]$ for every $j \in$ I(v_x) as `true` during the iteration with v_x. Note that v_0 is always included in HOG(P) by definition and is not considered.

We can check whether v_x should be included in HOG(P) by using B. Suppose that there exists j such that $B[j] = $ `false` and $j \in$ I(v_x) at the beginning of the iteration with v_x. By the definition of $B[j]$ and I(v_x), $v_x = ov(i,j)$ and it should be included in HOG(P). On the other hand, if $B[j] = $ `true` at the beginning of the iteration with v_x, there exists a longer overlap from s_i to s_j than v_x and it should not be included in HOG(P). If we do this process for every leaf node, we can get the list of nodes that we should include in HOG(P). Algorithm 1 describes an algorithm to compute HOG(P).

For example, let's consider the example in Fig. 1(b). First, we consider the case with $i = 1$ in line 2. After we mark leaf 1 to be included in HOG(P) in line 5, we begin the loop with u $= u_4$, which is the failure link of leaf 1. We consider I(u_4) $= \{1, 2\}$ in array B. Since $B[1]$ and $B[2]$ are `false`, we mark u_4 to be included in HOG(P) and set $B[1]$ and $B[2]$ as `true`. We continue the loop with u $= u_2$ by following the failure link. Since there is no $j \in$ I(u_2) $= \{1, 2\}$ such

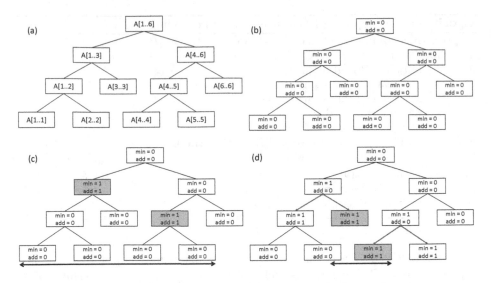

Fig. 2. Segment tree structure with $n = 6$. (a) The intervals that each node represents. (b) The values min and add that each node initially stores. (c) The values that each node stores after query 2 on $A[1..5]$. (d) The values that each node stores after query 1 on $A[3..4]$. Red arrows show that add values of the nodes are propagated to min and add values of their children. (Color figure online)

that $B[j]$ is false, we don't include u_2 in $HOG(P)$. We continue the loop with $u = u_1$. We consider $I(u_1) = \{1, 2, 3\}$ in array B. Since $B[3]$ is false, we mark u_1 to be included in $HOG(P)$ and set $B[3]$ as true. Since $u = u_1$ is the root, we finish the loop.

3.2 Improvement Using Segment Tree

To speed up Algorithm 1, we have to process these two types of queries efficiently.

i) Given an interval $[a..b]$, check whether there is any index $j \in [a..b]$ such that $B[j] = $ false (Lines 8–9).
ii) Given an interval $[a..b]$, set $B[j]$ as true for every $j \in [a..b]$ (Lines 10–11).

In order to process these queries, let's consider the following two types of queries on an integer array A. For an index j, $A[j] > 0$ means that $B[j] = $ true, while $A[j] = 0$ means that $B[j] = $ false.

1. Given an interval $[a..b]$, compute the minimum value among $A[a..b]$ (and check whether it is zero or not).
2. Given an interval $[a..b]$, add 1 to each element of $A[a..b]$.

We can see that one could use queries 1 and 2 to solve queries i and ii, respectively.

Algorithm 2. Computing minimum of an interval using segment tree

---

1: **procedure** SEGTREE-MIN(cnode, cinterval)
2:     **if** cnode.int ⊆ cinterval **then**
3:         **return** cnode.min
4:     **if** cnode.int ∩ cinterval = ∅ **then**
5:         **return** ∞
6:     left, right ← two children of cnode
7:     left.min += cnode.add, left.add += cnode.add
8:     right.min += cnode.add, right.add += cnode.add
9:     cnode.add = 0
10:     **return** min(SEGTREE-MIN(left, cinterval), SEGTREE-MIN(right, cinterval))

Algorithm 3. Add 1 to an interval using segment tree

1: **procedure** SEGTREE-UPDATE(cnode, cinterval)
2:     **if** cnode.int ⊆ cinterval **then**
3:         cnode.min += 1, cnode.add += 1
4:         **return**
5:     **if** cnode.int ∩ cinterval = ∅ **then**
6:         **return**
7:     left, right ← two children of cnode
8:     left.min += cnode.add, left.add += cnode.add
9:     right.min += cnode.add, right.add += cnode.add
10:     cnode.add = 0
11:     SEGTREE-UPDATE(left, cinterval)
12:     SEGTREE-UPDATE(right, cinterval)
13:     cnode.min = min(left.min, right.min)
14:     **return**

Let A be an integer array of length n. We use the segment tree data structure [5] to process queries 1 and 2 on A. The segment tree is a binary tree, which has n leaf nodes (they are 1, 2, ..., n) and has $O(\log n)$ height. Each leaf node represents one element, and each internal node represents an interval of elements. Figure 2(a) shows a segment tree for $n = 6$. For each node u in the segment tree, we define u.int as the interval that u represents. In Fig. 2(a), for instance, u.int for the root node is [1..6].

While processing the queries, each node u stores both the minimum value among the elements in u.int (denoted by u.min) and an added value to u.int (denoted by u.add). Since A should be initialized to zero, every value in the segment tree is also initialized to zero. Figure 2(b) shows an initial state of the segment tree.

Algorithms 2 and 3 show the algorithms to perform queries 1 and 2, respectively, in the segment tree, which use the *lazy propagation* technique in [19], though in [19] one computes the sum, while here we compute the minimum. If query 1 occurs, we follow the nodes recursively from top to down, starting from the root. Consider a node u during the recursion. If u.int is included in the query interval, we return u.min. If u.int is disjoint with the query interval, we return ∞ to indicate that there are no values to be considered in u.int. Otherwise, we propagate an added value to the child nodes, continue the process with the child nodes and return the minimum among them. Query 2 can be done in a similar way, but in this case we have to recompute the minimum value of a node after updating its child nodes, as shown in line 13 of Algorithm 3.

Figures 2(c) and 2(d) show an example of processing two queries, query 2 on $A[1..5]$ and query 1 on $A[3..4]$. In Fig. 2(c), we can see that two nodes representing $A[1..3]$ and $A[4..5]$ are updated in the segment tree. Note that min and add values of the descendant nodes are not updated yet. In Fig. 2(d), we access the two nodes representing $A[3..3]$ and $A[4..4]$ to compute the minimum value among $A[3..4]$. Note that add values in $A[1..3]$ and $A[4..5]$ are propagated to their children to ensure that appropriate min values are stored in $A[3..3]$ and $A[4..4]$.

We now prove the correctness of Algorithms 2 and 3. To the best of our knowledge, this is the first correctness proof for the folklore lazy propagation technique in [19]. The proof is non-trivial because Algorithms 2 and 3 work together, but their recursive structures differ. First, we need an invariant that holds for both algorithms, i.e., Invariant 1 below. Moreover, since Algorithm 2 makes recursive calls at the end, we need a top-down sub-invariant for Algorithm 2. In contrast, Algorithm 3 makes recursive calls in the middle, and thus we have to come up with a bottom-up sub-invariant for Algorithm 3.

Each node u in the segment tree maintains the following invariant while processing queries 1 and 2.

$$\min_{i \in \text{u.int}} A[i] = \text{u.min} + \sum_{\text{v}} \{\text{v.add} : \text{v is an ancestor of u}\}, \qquad (1)$$

where A is the conceptual array in the definitions of queries 1 and 2, and u is not an ancestor of itself.

Lemma 1. Invariant 1 holds after Algorithm 2 or 3 is called with cnode = root and cinterval = $[a..b]$ for query 1 or 2, respectively.

Proof. We prove the lemma by induction. Initially, Invariant 1 holds because $A[i] = 0$ for every index i, and u.min = 0 and u.add = 0 for every node u in the segment tree.

First we show that Invariant 1 holds after Algorithm 2 is called for query 1. The left-hand side (LHS) of Invariant 1 is unchanged since Algorithm 2 performs a query on A, but does not change it. However, the propagation of the add values in the segment tree may update the min and add values of other nodes in it. So we must prove that the right-hand side (RHS) of Invariant 1 remains the same too. When Algorithm 2 is called with cnode = root, it recurses through nodes in the segment tree (i.e., it goes down) until it reaches the base cases of recursion (which are handled in lines 2 and 4), and then it goes up by computing minima (in line 10). When Algorithm 2 goes down, we will show inductively that the RHS of Invariant 1 remains the same for every node in the segment tree after each execution of lines 6–9 (i.e., top-down sub-invariant for Algorithm 2). Consider one execution of lines 6–9. Since `left`, `right`, and `cnode` have their `min` and `add` changed, we show that the RHS of Invariant 1 remains the same for every node u in the subtree rooted at `cnode`.

- If u = `cnode`, `cnode.min` is not changed, and so the RHS of Invariant 1 remains the same.
- If u = `left` (similarly for u = `right`), `left.min` is increased as much as `cnode.add` is decreased, so the RHS of Invariant 1 remains the same.
- If u is a descendant of `left` (similarly for a descendant of `right`), `left.add` is increased as much as `cnode.add` is decreased. Since both `left` and `cnode` are u's ancestors, the RHS of Invariant 1 remains the same.

Therefore, the RHS of Invariant 1 remains the same for every node u in the segment tree when Algorithm 2 goes down.

When Algorithm 2 goes up (including the base cases of recursion), the RHS of Invariant 1 does not change for any node in the segment tree. Therefore, Invariant 1 holds after Algorithm 2 is called for query 1.

Now we show that Invariant 1 holds after Algorithm 3 is called for query 2. When Algorithm 3 is called with `cnode` = `root`, it goes down by recursion and then it goes up, like Algorithm 2. When Algorithm 3 goes down, one can show inductively that the RHS of Invariant 1 does not change after each execution of lines 7–10, in a way similar to Algorithm 2.

When Algorithm 3 goes up, we will show inductively that Invariant 1 holds for every node in the subtree rooted at `cnode` at the moment when SEGTREE-UPDATE (`cnode`, `cinterval`) returns (i.e., bottom-up sub-invariant for Algorithm 3). We first consider two base cases which are handled in lines 2 and 5.

- If `cnode.int` ⊆ `cinterval`, SEGTREE-UPDATE(`cnode`, `cinterval`) performs line 3 and returns in line 4. After line 3 is done, the RHS of Invariant 1 for `cnode` and its descendants increases by 1. Since every $A[i]$ for $i \in$ `cnode.int` increases by 1, the LHS of Invariant 1 for them also increases by 1 and Invariant 1 holds.
- If `cnode.int` ∩ `cinterval` = ∅, SEGTREE-UPDATE(`cnode`, `cinterval`) does nothing and returns in line 6, and thus the RHS of Invariant 1 remains the same for `cnode` and its descendants. Since every $A[i]$ for $i \in$ `cnode.int` remains the same, Invariant 1 holds.

Next, we consider the induction step, where we assume that Invariant 1 holds for `left`, `right` and their descendants by the bottom-up sub-invariant. Now we need to show that Invariant 1 holds for `cnode` when SEGTREE-UPDATE(`cnode`, `cinterval`) executes line 13 and returns. Suppose that `left.min` ≤ `right.min` (similarly for the case `left.min` > `right.min`). Consider Invariant 1 for `left` and `right`. Since `left` and `right` share the same ancestors, the summation parts of Invariant 1 for `left` and `right` are the same. So if `left.min` ≤ `right.min`,

---

Algorithm 4. Algorithm to compute HOG in $O(\|P\|\log n)$ time

---

1: **procedure** BUILD-HOG(EHOG(P))
2:     **for** $i \leftarrow 1$ **to** n **do**
3:         Initialize the segment tree
4:         u $\leftarrow$ leaf corresponding to s_i in EHOG(P)
5:         Mark u as included in HOG(P)
6:         **while** u $\neq$ root **do**
7:             u $\leftarrow$ failure link of u in EHOG(P)
8:             **if** SEGTREE-MIN(root, I(u)) = 0 **then**
9:                 Mark u as included in HOG(P)
10:            SEGTREE-UPDATE(root, I(u))
11:     Build HOG(P) with marked nodes

$\min\limits_{i \in \texttt{left.int}} A[i] \leq \min\limits_{i \in \texttt{right.int}} A[i]$ holds. Since $\texttt{cnode.int} = \texttt{left.int} \cup \texttt{right.int}$, the LHS of Invariant 1 for `cnode` is the same as that of `left`. The RHS of Invariant 1 for `cnode` is also the same as that of `left` because $\texttt{cnode.min} = \texttt{left.min}$ by line 13 and $\texttt{cnode.add} = 0$ by line 10.

Therefore, Invariant 1 holds for every node in the segment tree after Algorithm 3 is called with `cnode = root`.

Using Lemma 1, we can show the correctness of Algorithms 2 and 3 to solve queries 1 and 2.

Theorem 1. For any sequences of Algorithms 2 and 3 called with `cnode = root` and `cinterval = [a..b]`, Algorithm 2 (i.e., SEGTREE-MIN(root, cinterval)) returns a correct answer for query 1 with the given interval $[a..b]$.

Proof. By Lemma 1 Invariant 1 holds after every call on Algorithm 2 or 3. Furthermore, if we access node u by recursion in Algorithm 2, $\texttt{v.add} = 0$ for every ancestor v of u due to line 9 in Algorithm 2. Therefore, at the moment we access u, $\min\limits_{i \in \texttt{u.int}} A[i] = \texttt{u.min}$ always holds from Invariant 1.

Since Algorithm 2 computes the minimum of `u.min` for every u whose interval is included in the given interval $[a..b]$, it is equal to the minimum value among $A[a..b]$. Therefore, Algorithm 2 returns a correct answer for query 1.

Given the EHOG, Algorithm 4 describes how to compute the HOG using queries on the segment tree data structure. Algorithm 4 is almost identical to Algorithm 1. First, the condition ($\exists j \in \texttt{I(u)}$ such that $B[j]$ is `false`) on line 8 of Algorithm 1 is now performed by (SEGTREE-MIN(root, I(u)) = 0) on line 8 of Algorithm 4. Second, the update **for** loop of lines 10–11 in Algorithm 1 is performed using a single query on line 10 of Algorithm 4: SEGTREE-UPDATE(root, I(u)).

Since any interval $[a..b]$ can be represented by $O(\log n)$ nodes with a segment tree [5], Algorithms 2 and 3 can be done in $O(\log n)$ time. By using them, we can get an $O(\|P\|\log n)$ time algorithm to compute HOG(P), as shown in

Algorithm 4. Since HOG(P) and the segment tree take $O(\|P\|)$ and $O(n)$ space, respectively, the space complexity of building the HOG is $O(\|P\|)$.

4   Improvement Using the word RAM model

By using the word RAM model of computation [15] with w-bit machine words, where $w \geq \log n$, we show that we can compute the HOG from the given set P of n strings in $O(\|P\| \frac{\log n}{\log \log n})$ time.

Indeed, by using bitwise operations, we can improve queries 1 and 2 from $O(\log n)$ to $O(\log_w n) = O(\log_{\log n} n) = O(\frac{\log n}{\log \log n})$. To do so, we introduce the w-segment tree, which is the w-ary version of the segment tree as in [2,11].

4.1   Algorithms with Bitwise Operations

Unlike the original segment tree which is a binary tree, we define the w-segment tree as a tree with n leaves, a height of $O(\log_w n)$, and each node having at most w children. As in the segment tree, each internal node represents an interval of elements of P (i.e., $1, 2, \ldots, n$), and each leaf contains a single element (the interval of a node u is denoted by u.int). But, instead of storing for a node u the minimum value u.min and the added value u.add, we store two bit vectors of length w (v.Vmin and v.Vadd) for every internal node v. If a node u is the j-th child of its parent p, the j-th value of p.Vmin is true if u.min ≥ 1; false if u.min $= 0$ (same for Vadd).

To compute query 1 for a node u and an interval $[a, b]$, we begin by comparing the interval $[L, R] =$ u.int with $[a, b]$:

- If $[L, R] \subseteq [a, b]$, we return the j-th bit of p.Vmin, where u is the j-th child of its parent p.
- If $[L, R] \cap [a, b] = \emptyset$, we return true.
- Otherwise, we compute the positions i_a and i_b corresponding to a and b in $[0, w - 1]$:

$$i_a = \lfloor \tfrac{(a-L)w}{R-L+1} \rfloor \text{ and } i_b = \lfloor \tfrac{(b-L)w}{R-L+1} \rfloor.$$

If the j-th position of p.Vadd is equal to 1, all the values of u.Vmin and u.Vadd become 1, and the j-th position of p.Vadd becomes 0.
At the end, we recursively call the function on Child_{i_a} and Child_{i_b}, and return the minimum of two recursive calls and the values of u.Vmin between positions $i_a + 1$ and $i_b - 1$, where the minimum of the corresponding values of u.Vmin is computed as the following Boolean value:

$$(\text{u.Vmin AND } (2^{i_b} - 2^{i_a+1})) = (2^{i_b} - 2^{i_a+1}).$$

In a similar way, we can compute query 2 by using bitwise operations.

4.2   Using a Table for a Compressed space

Instead of a tree structure, we can use two tables to simulate the segment tree. Let

$$h = \frac{w^{\lceil \log_w n \rceil - 1} - 1}{w - 1} + \left\lceil \frac{n}{w} \right\rceil$$

denote the size of these tables, and let $\text{Tmin}[0..h - 1]$ and $\text{Tadd}[0..h - 1]$ be two tables of w-bit words initialized to $[0, \ldots, 0]$. We store Vmin's and Vadd's of Section 4.1 into Tmin and Tadd, respectively, in the BFS order of the w-segment tree (i.e., top to bottom, left to right) and run the algorithm described in Sect. 4.1 (see Algorithm 5). In the same way, we can build the algorithm corresponding to query 2 with bitwise operations.

Algorithm 5. Computing minimum of an interval using w-segment tree

1: **procedure** SEGTREEMINRAM$(k, [a, b])$
2:     $d \leftarrow \lfloor \log_w((w - 1)k + 1) \rfloor$ ▷ Depth of node k
3:     $x \leftarrow k - \frac{w^d - 1}{w - 1}$ ▷ Node k is the x-th node with depth d
4:     $Y \leftarrow w^{\lceil \log_w n \rceil - d}$ ▷ Node k represents an interval of length Y
5:     $L \leftarrow xY + 1$
6:     $R \leftarrow (x + 1)Y$
7:     $p \leftarrow \lfloor \frac{k-1}{w} \rfloor$ ▷ p is parent of node k
8:     $j \leftarrow (k - 1) \bmod w$ ▷ Node k is the j-th child of p
9:     $i_a \leftarrow \max(\lfloor \frac{(a-L)w}{Y} \rfloor, 0)$
10:    $i_b \leftarrow \min(\lfloor \frac{(b-L)w}{Y} \rfloor, w - 1)$
11:    **if** $(a \le L) \wedge (R \le b)$ **then**
12:        **return** $(\text{Tmin}[p] \text{ AND } 2^j) = 2^j$
13:    **if** $(R < a) \vee (b < L)$ **then**
14:        **return true**
15:    **if** $(\text{Tadd}[p] \text{ AND } 2^j) = 2^j$ **then**
16:        $\text{Tmin}[k] \leftarrow 2^w - 1$
17:        $\text{Tadd}[k] \leftarrow 2^w - 1$
18:        $\text{Tadd}[p] \leftarrow \text{Tadd}[p] \text{ AND } (2^w - 1 - 2^j)$
19:    **return** SEGTREEMINRAM$(wk + 1 + i_a, [a, b])$
             $\wedge$ SEGTREEMINRAM$(wk + 1 + i_b, [a, b])$
             $\wedge$ $(\text{Tmin}[k] \text{ AND } (2^{i_b} - 2^{i_a + 1}) = (2^{i_b} - 2^{i_a + 1}))$

By using a table to simulate the tree, we do not need to store the interval of each node and we can store the segment tree by using $O(n)$ bits.

Indeed, the tables Tmin and Tadd are of size h. As $\left\lceil \frac{n}{w} \right\rceil \le \frac{2n}{w}$ and $\frac{w^{\lceil \log_w n \rceil - 1} - 1}{w - 1} \le 2 \times w^{\lceil \log_w n \rceil - 2} \le \frac{2n}{w}$, we need at most $w \times 4 \times \frac{n}{w} = 4n$ bits to store each table.

That is, the space for the segment tree is reduced to $O(n)$ bits (i.e., $O(\frac{n}{\log n})$ words) by using the two tables, but the space complexity of building the HOG remains $O(\|P\|)$ due to the size of the HOG itself.

5   Conclusion

We have presented a new algorithm to compute the HOG in $O(\|P\| \log n)$ time and linear space, which improves upon an earlier solution, and a version of our algorithm using bitwise operations in the word RAM model of computation.

Several interesting questions concerning the HOG and EHOG deserve future work. The *reverse engineering* of indexing data structures, also termed *inference* or *recognition* problem, has attracted a lot of interest. The question is, for instance, given a tree, can one decide whether it is the suffix tree of some string or not? The reverse engineering problem has been studied, e.g., for the suffix tree [16] or the longest-common-prefix array [18]. In 2014, Gevezes and Pitsoulis investigated the reverse engineering of overlap graphs [10]: given a weighted directed graph G, find an instance P such that the overlap graph of P equals G. Clearly this question can be applied to the EHOG and HOG, where the weight on an arc (which is the length of the label on the arc) may or may not be given.

The sizes of the EHOG and HOG (in the number of nodes) can be equal, but they may differ considerably [9]. An average case analysis of their sizes could help understand their differences, and predict the memory required for storing them. Some results connected to this question exist in the literature, e.g., [24] for tries. The notion of *clusters* of word occurrences [4,13,17,28] can be helpful in investigating the number of nodes of the EHOG and HOG for a random set of words.

References

1. Aho, A.V., Corasick, M.J.: Efficient string matching: an aid to bibliographic search. Commun. ACM **18**(6), 333–340 (1975). https://doi.org/10.1145/360825.360855
2. Arge, L., Brodal, G.S., Georgiadis, L.: Improved dynamic planar point location. In: 47th Proceedings of FOCS, pp. 305–314 (2006). https://doi.org/10.1109/FOCS.2006.40
3. Armen, C., Stein, C.: A $2\frac{2}{3}$-approximation algorithm for the shortest superstring problem. In: CPM, pp. 87–101 (1996). https://doi.org/10.1007/3-540-61258-0_8
4. Bassino, F., Clement, J., Nicodeme, P.: Counting occurrences for a finite set of words: combinatorial methods. ACM Trans. Algorithms **8**(3), 31:1–31:28 (2012). https://doi.org/10.1145/2229163.2229175
5. Berg, M., Kreveld, M., Overmars, M., Schwarzkopf, O.: Computational Geometry: Algorithms and Applications, 3rd edn. Springer, Berlin (2008). https://doi.org/10.1007/978-3-540-77974-2
6. Blum, A., Jiang, T., Li, M., Tromp, J., Yannakakis, M.: Linear approximation of shortest superstrings. J. ACM **41**(4), 630–647 (1994). https://doi.org/10.1145/179812.179818
7. Cazaux, B., Juhel, S., Rivals, E.: Practical lower and upper bounds for the shortest linear superstring. In: SEA, pp. 18:1–18:14 (2018). https://doi.org/10.4230/LIPIcs.SEA.2018.18
8. Cazaux, B., Rivals, E.: A linear time algorithm for shortest cyclic cover of strings. J. Discrete Algorithms **37**, 56–67 (2016). https://doi.org/10.1016/j.jda.2016.05.001
9. Cazaux, B., Rivals, E.: Hierarchical overlap graph. Inf. Process. Lett. **155**, 105862 (2020). https://doi.org/10.1016/j.ipl.2019.105862

10. Gevezes, T.P., Pitsoulis, L.S.: Recognition of overlap graphs. J. Comb. Optim. **28**(1), 25–37 (2013). https://doi.org/10.1007/s10878-013-9663-3

11. Giora, Y., Kaplan, H.: Optimal dynamic vertical ray shooting in rectilinear planar subdivisions. ACM Trans. Algorithms **5**(3) (2009). https://doi.org/10.1145/1541885.1541889

12. Gonnella, G., Kurtz, S.: Readjoiner: a fast and memory efficient string graph-based sequence assembler. BMC Bioinform. **13**(1), 82 (2012). https://doi.org/10.1186/1471-2105-13-82

13. Guibas, L.J., Odlyzko, A.M.: Periods in strings. J. Comb. Theory Ser. A **30**(1), 19–42 (1981). https://doi.org/10.1016/0097-3165(81)90038-8

14. Gusfield, D., Landau, G.M., Schieber, B.: An efficient algorithm for the all pairs suffix-prefix problem. Inf. Process. Lett. **41**(4), 181–185 (1992). https://doi.org/10.1016/0020-0190(92)90176-V

15. Hagerup, T.: Sorting and searching on the word RAM. In: Morvan, M., Meinel, C., Krob, D. (eds.) STACS 1998. LNCS, vol. 1373, pp. 366–398. Springer, Heidelberg (1998). https://doi.org/10.1007/BFb0028575

16. Tomohiro, I., Inenaga, S., Bannai, H., Takeda, M.: Inferring strings from suffix trees and links on a binary alphabet. Discret. Appl. Math. **163**, 316–325 (2014). https://doi.org/10.1016/j.dam.2013.02.033

17. Jacquet, P., Szpankowski, W.: Autocorrelation on words and its applications: analysis of suffix trees by string-ruler approach. J. Comb. Theory Ser. A **66**(2), 237–269 (1994). https://doi.org/10.1016/0097-3165(94)90065-5

18. Karkkainen, J., Piatkowski, M., Puglisi, S.J.: String inference from longest-common-prefix array. In: ICALP. LIPIcs, vol. 80, pp. 62:1–62:14 (2017). https://doi.org/10.4230/LIPIcs.ICALP.2017.62

19. Laaksonen, A.: Guide to Competitive Programming. UTCS. Springer, Cham (2017). https://doi.org/10.1007/978-3-319-72547-5

20. Lim, J., Park, K.: A fast algorithm for the all-pairs suffix-prefix problem. Theoret. Comput. Sci. **698**, 14–24 (2017). https://doi.org/10.1016/j.tcs.2017.07.013

21. Mucha, M.: Lyndon words and short superstrings. In: SODA, pp. 958–972. SIAM (2013). https://doi.org/10.1137/1.9781611973105.69

22. Myers, E.W.: The fragment assembly string graph. Bioinformatics **21**(Suppl. 2), ii79–ii85 (2005). https://doi.org/10.1093/bioinformatics/bti1114

23. Paluch, K.: Better approximation algorithms for maximum asymmetric traveling salesman and shortest superstring (2014). https://arxiv.org/abs/1401.3670

24. Park, G., Hwang, H., Nicodeme, P., Szpankowski, W.: Profiles of tries. SIAM J. Comput. **38**(5), 1821–1880 (2009). https://doi.org/10.1137/070685531

25. Peltola, H., Soderlund, H., Tarhio, J., Ukkonen, E.: Algorithms for some string matching problems arising in molecular genetics. In: IFIP Congress, pp. 53–64 (1983)

26. Pevzner, P.A., Tang, H., Waterman, M.S.: An eulerian path approach to DNA fragment assembly. Proc. Natl. Acad. Sci. **98**(17), 9748–9753 (2001). https://doi.org/10.1073/pnas.171285098

27. Rachid, M.H., Malluhi, Q.: A practical and scalable tool to find overlaps between sequences. BioMed Res. Int. **2015** (2015). https://doi.org/10.1155/2015/905261

28. Robin, S., Rodolphe, F., Schbath, S.: DNA, Words and Models. Cambridge University Press, Cambridge (2005)

29. Sweedyk, Z.: A $2\frac{1}{2}$-approximation algorithm for shortest superstring. SIAM J. Comput. **29**(3), 954–986 (2000). https://doi.org/10.1137/S0097539796324661

Tailoring r-index for Document Listing Towards Metagenomics Applications

Dustin Cobas[1], Veli Mäkinen[2], and Massimiliano Rossi[3]([⊠])

[1] CeBiB — Center for Biotechnology and Bioengineering, Department of Computer Science, University of Chile, Santiago, Chile
dcobas@dcc.uchile.cl
[2] Department of Computer Science, University of Helsinki, Helsinki, Finland
veli.makinen@helsinki.fi
[3] Department of Computer and Information Science and Engineering, University of Florida, Gainesville, USA
rossi.m@ufl.edu

Abstract. A basic problem in *metagenomics* is to assign a sequenced read to the correct species in the reference collection. In typical applications in genomic epidemiology and viral metagenomics the reference collection consists of a set of species with each species represented by its highly similar strains. It has been recently shown that accurate read assignment can be achieved with k-mer hashing-based *pseudoalignment*: a read is assigned to species A if each of its k-mer hits to a reference collection is located only on strains of A. We study the underlying primitives required in pseudoalignment and related tasks. We propose three space-efficient solutions building upon the *document listing with frequencies* problem. All the solutions use an r-*index* (Gagie *et al.*, SODA 2018) as an underlying index structure for the text obtained as concatenation of the set of species, as well as for each species. Given t species whose concatenation length is n, and whose Burrows-Wheeler transform contains r runs, our first solution, based on a grammar-compressed document array with precomputed queries at non terminal symbols, reports the frequencies for the **ndoc** distinct documents in which the pattern of length m occurs in $\mathcal{O}(m + \log(n)\mathbf{ndoc})$ time. Our second solution is also based on a grammar-compressed document array, but enhanced with bitvectors and reports the frequencies in $\mathcal{O}(m + ((t/w)\log n + \log(n/r))\mathbf{ndoc})$ time, over a machine with wordsize w. Our third solution, based on the interleaved LCP array, answers the same query in $\mathcal{O}(m + \log(n/r)\mathbf{ndoc})$ time. We implemented our solutions and tested them on real-world and synthetic datasets. The results show that all the solutions are fast on highly-repetitive data, and the size overhead introduced by the indexes are comparable with the size of the r-index.

Keywords: Metagenomics · r-index · Document listing.

© Springer Nature Switzerland AG 2020
C. Boucher and S. V. Thankachan (Eds.): SPIRE 2020, LNCS 12303, pp. 291–306, 2020.
https://doi.org/10.1007/978-3-030-59212-7_21

1   Introduction

Metagenomics is the study of genomic material recovered directly from environmental samples. Thus, conversely to genomic samples, metagenomic samples consist of genome sequences of a community of organisms sharing the same environment, highlighting the microbial diversity in the environmental samples. The samples of genome sequences are collected using shotgun sequencing. This creates a mixture of genome fragments from all organisms in the environment. One important step in metagenomics is to assign each fragment to its owner, allowing to identify and quantify species. This step is called read assignment [19], and it is the basic step in most metagenomic analysis workflows such as in genomic epidemiology [25], and viral epidemiology [6].

Read assigners were first implemented using computationally expensive read aligners [19,23,38]. In [37] the authors showed that similar results are achieved replacing the read aligners with the computationally less expensive k-mer hashing methods. Read assigners based on k-mer set indexing are referred to as *pseudoaligners*. Efficient indexing of k-mer sets, including *colored de Bruijn graphs* [20], has been deeply investigated and we refer the reader to the survey [27] for further reading. Pseudoaligners such as Kallisto [4], MetaKallisto [34], and Themisto [25] use *colored de Bruijn graphs* and are based on the following pseudoalignment criterion. Given a set of references $T_1, \ldots, T_t$ (representing t distinct species), and read P, the read P is pseudoaligned with T_i if there exists a k-mer of P that occurs in T_i and for all other k-mers u of P, either u occurs in T_i or u does not occur in $T_1, \ldots, T_t$. This approach is motivated by the fact that the species are usually quite dissimilar, but the strains inside the species are highly similar.

In this paper, we study some basic primitives that are required in different variations of the pseudoalignment criteria. We argue that the specific criterion given above is just one example of a family of criteria, and it is important to study the general framework rather than tailoring the methods to a very narrow setting. Towards this goal of obtaining general results, instead of studying directly k-mers of a pattern, we focus here on searching the complete pattern. We continue the discussion in Sect. 6 on how to integrate the results with k-mer based criteria.

We modelled this read assignment problem as a *document listing with frequencies* problem, where the set of species is a collection and each species is a document formed by the concatenation of its strains. Given a pattern P we want to report all documents where P occurs, and their frequencies. This problem was first introduced in [35] and further refined in [3] and [15] (details in Sect. 3). We propose three solutions. All solutions use an r-index [14] as text index for the concatenation of all documents. The first solution is an extension to frequencies of the solution proposed in [9] in which a grammar-compressed document array is used, and for each non terminal node, precomputed answers are stored. The second and the third solution are based on the *term frequency* approach presented in [33] which uses an additional index for all documents. The key idea is to find the leftmost and rightmost occurrence of the pattern P in the index of each document, by searching the pattern in the index of the concatenation of all documents. To do this, the second solution uses the grammar-compressed document array of [9] enhanced with bitvectors at non terminal nodes marking which descendant contains the leftmost and rightmost occurrence of the pattern in each document. The third solution relies on a modified version of the interleaved longest common prefix array [13]. We implemented our solutions and we tested them using real-world and synthetic datasets.

2 Basics

A string $S[1..n]$ is a sequence of n characters over an alphabet Σ of size $\sigma = |\Sigma|$. A document T is a string terminated by a special symbol \$ $\notin \Sigma$ that is lexicographically smaller than all characters in Σ. A collection $D = \{T_1, T_2, \ldots, T_t\}$ is a set of t documents, which is usually represented as the concatenation of its documents, i.e. $\mathcal{D}[1..n] = T_1 T_2 \cdots T_t$. When it is clear from the context, we will refer to T_i as document i. Given a string $S[1..n]$, let $\text{rank}_c(S, i)$ be the number of occurrences of symbol c in $S[1..i]$, and let $\text{select}_c(S, j)$ be the position of the j-th symbol c in $S[1..n]$. When string S is from alphabet $\{0, 1\}$, we call it a bitvector. For bitvector S it holds $\text{rank}_0(S, i) = i - \text{rank}_1(S, i)$.

Given a string S over an alphabet σ, the *suffix array* [26] $\text{SA}[1..n]$ of S is an array of integers providing the starting position of the suffixes of S sorted in lexicographic order. The *inverse suffix array* $\text{ISA}[1..n]$ of S is an array of integers that, for each suffix of S, provides the position of the suffix in the suffix array. In particular we have that for all $1 \leq i \leq n$, $\text{SA}[\text{ISA}[i]] = i$.

A *compressed suffix array* [31] $\text{CSA}[1..n]$ is a space-efficient representation of the suffix array whose size $|\text{CSA}|$ in bits is usually bounded by $\mathcal{O}(n \log \sigma)$. We denote by $t_{search}(m)$ the time to find the interval of the suffix array corresponding to all occurrences of $P[1..m]$, while by $t_{lookup}(n)$ the time necessary to access any value $\text{SA}[i]$.

The r-index [14] is a compressed text index whose main components are a run-length encoded *Burrows-Wheeler* transform (BWT) [5] and the sample of the suffix array at the beginning and at the end of each run of the BWT. We denote by r the number of equal character runs of the BWT. The r-index of the document $T[1..n]$ can be computed in $\mathcal{O}(n)$ time and occupies $\mathcal{O}(r \log(n/r))$ space. We can find all occurrences of a given pattern $P[1..m]$ in the document $T[1..n]$ in time $\mathcal{O}(m + occ)$ time. The r-index supports SA and ISA queries in $\mathcal{O}(\log(n/r))$ time and $\mathcal{O}(r \log(n/r))$ space[1].

Given a collection $D = \{T_1, \ldots, T_t\}$ of t documents and its concatenation $\mathcal{D} = T_1 T_2 \cdots T_t$ of length n, the *document array* [28] $\text{DA}[1..n]$ stores in each position i the index of the document which the suffix $\text{SA}[i]$ belongs to.

Given a document $T[1..n]$, the *longest common prefix* array $\text{LCP}_T[1..n]$ stores in each position $2 \leq i \leq n$ the length of the longest common prefix between the two strings $T[\text{SA}[i-1]..n]$ and $T[\text{SA}[i]..n]$.

---

[1] Throughout the paper, we report the space in words, where not otherwise specified.

Given a collection $D = \{T_1, \ldots, T_t\}$ whose concatenation is $\mathcal{D}[1..n]$, the interleaved longest-common-prefix array ILCP$[1..n]$ is defined in [13] as the interleaving of the LCP arrays of the documents $T_1, \ldots, T_t$ in the order they appear in the suffix array of $\mathcal{D}$, i.e., if SA$[i]$ is the lexicographically j-th suffix of the k-th document, ILCP$[i] = $ LCP$_k[j]$. Let the ILCP array be run-length encoded in ρ runs. Then, it can be represented using two arrays: LILCP$[1..\rho]$ contains the prefix sums of the lengths of the ρ runs; VILCP$[1..\rho]$ contains the values of these runs. Furthermore, the LILCP array can be replaced by a sparse bitvector $L[1..n]$ such that LILCP$[i] = \texttt{select}_1(L, i)$.

Given a string $S[1..n]$, a *straight line grammar* for S is a context-free grammar $\mathcal{G}$ that uniquely generates the string S. We denote by $\mathcal{T}$ the parse tree of S. Given a node $t \in \mathcal{T}$, t is a *terminal* node if t has no children, t is a *non terminal* node otherwise. Each node $t \in \mathcal{T}$ uniquely identifies an interval of S denoted by $S[\ell_t..r_t]$. For the ease of explanation we say that a character c occurs in t by meaning that the character c occurs in $S[\ell_t..r_t]$. The parse tree $\mathcal{T}$ is *binary* if its maximum arity is 2, and $\mathcal{T}$ is *balanced* if every substring is covered by $\mathcal{O}(\log n)$ *maximal nodes*, which are the highest nodes of the tree whose expansions form a partition of the substring. Computing the smallest grammar is an NP-hard problem [22], but various $\mathcal{O}(\log(n/\mathcal{G}^*))$-approximation exists. We consider those that are binary and balanced [7,21,32].

3   Related Work

In this section we define three problems and report solutions and techniques from the literature that are used in our approach. For a complete overview we refer the reader to the survey [29].

Problem 1 (Document listing). Given a collection $D = \{T_1, T_2, \ldots, T_t\}$, and a pattern P, return the set of documents $L \subseteq D$ where P occurs.

Muthukrishnan [28] proposed the first solution to Problem 1 in optimal time and linear space. He defined the *document array* DA and used a suffix tree [36] to find all occurrences of the pattern P represented as an interval $[s_p..e_p]$. Then, he proposed a recursive algorithm to find all distinct documents ndoc in DA$[s_p..e_p]$ in optimal time $\mathcal{O}(\text{ndoc})$.

Sadakane [33] replaced the suffix tree with a compressed suffix array CSA and the document array with a bitvector marking the starting position of each document in text order. He also replaced the data structures to find all distinct documents ndoc in DA$[s_p..e_p]$ with a succinct version using only $\mathcal{O}(n)$ bits. With this solution, Problem 1 can be solved in $\mathcal{O}(t_{search}(m) + \text{ndoc}\, t_{lookup}(n))$ using a data structures of $|\text{CSA}| + \mathcal{O}(n)$ bits.

Gagie *et al.* [13] introduced the ILCP array whose property stated in Lemma 1 allows to apply almost verbatim the technique used by Sadakane to find distinct elements in DA$[s_p..e_p]$. The solution uses a run-length compressed suffix array RLCSA [24] which allows to answer the queries of Problem 1 in $\mathcal{O}(t_{search}(m) + \text{ndoc}\, t_{lookup}(n))$ time.

Claude and Munro [8] proposed the first grammar-based document listing later improved by Navarro in [30]. Cobas and Navarro [9], later proposed a practical variant in which they store the *document array* as a binary balanced straight line grammar. Then, they precompute and store the answers for all non terminal nodes of the grammar. The queries are answered by using a CSA to find the interval DA$[s_p..e_p]$ and merging the precomputed answers for the $\mathcal{O}(\log n)$ non terminal symbols covering DA$[s_p..e_p]$. This leads to a solution that solves Problem 1 in $\mathcal{O}(t_{search}(m) + \text{ndoc} \log n)$ time.

Problem 2 (Term frequency). Given $D = \{T_1, T_2, \ldots, T_t\}$, and a pattern P, for each document $T \in D$ return the number of occurrences of P in T.

Sadakane [33], addressed also the *term frequency* problem. The solution to Problem 1 is enhanced building a compressed suffix array CSA for each document. Given the interval $[s_p..e_p]$ of all occurrences of the pattern P, he uses the data structure to find the distinct documents in DA$[s_p..e_p]$ and their leftmost and rightmost occurrences. Those positions are then mapped into an interval in the CSA of the document. The sizes of these intervals represent the frequencies of the documents. This approach solves Problem 2 in $\mathcal{O}(t_{search}(m) + \text{ndoc} t_{lookup}(n))$ time.

Problem 3 (Document listing with frequencies). Given $D = \{T_1, T_2, \ldots, T_t\}$, and a pattern P, return the set of documents where P occurs and their frequencies.

Välimäki and Mäkinen [35] first proposed Problem 3 and showed that the document listing problem can be solved using a rank and select data structure on the document array, to simulate Muthukrishnan's [28] solution. In addition, after locating the interval SA$[s_p..e_p]$ of all occurrences of P in $\mathcal{D}$, the frequencies for each distinct document in DA$[s_p..e_p]$ are computed using a rank array on the document array, i.e., the number of occurrences of P in document T_i are $\text{rank}_i(\text{DA}, s_e) - \text{rank}_i(\text{DA}, s_p - 1)$. Using a *wavelet tree* [18] to represent the document array, given a pattern $P[1..m]$, Problem 3 can be solved in $\mathcal{O}(t_{search}(m) + \text{ndoc} \log t)$ time.

Belazzougui *et al.* [3] built a *monotone minimum perfect hash function* [1] on the document array. Combining Muthukrishnan's [28] and Sadakane's [33] approaches, it is possible to find the leftmost and rightmost occurrence of the pattern P in the i-th document. Using the constant time rank on the document array, Problem 3 can be solved in $\mathcal{O}(t_{search}(m) + \text{ndoc})$ time.

Gagie *et al.* [15] proposed a solution based on *wavelet trees* [18], that does not rely on Muthukrishnan's [28] solution. The idea is to use the *range quantile* [16] problem to find the i-th smallest value in the range DA$[s_p..e_p]$. Then, retrieve its frequency as the length of the interval corresponding to $[s_p..e_p]$ in its leaf in the wavelet tree. With this approach Problem 3 can be solved in $\mathcal{O}(t_{search}(m) + \text{ndoc} \log t)$ time.

4   The Document Listing with Frequencies

We are now ready to describe our *document listing with frequencies* approaches. We propose three different solutions, which rearrange and adapt different concepts of previous work. The first solution is based on the solution for the *document listing* proposed in [9]. We grammar compress DA, and for all non terminal nodes, we precompute and store the results of *document listing with frequencies* queries. The second solution combines Sadakane's approach [33] for the *term frequency* problem, with the grammar compressed document array. We enhance the grammar compressed document array with bitvectors in each non terminal, to locate the leftmost and rightmost occurrences of each document in the corresponding interval in the document array. The third solution combines Sadakane's approach [33] for the *term frequency* problem with the ILCP array. In this case we use two copies of the ILCP array to locate the leftmost and rightmost occurrences of each document in the corresponding interval in the document array.

As a common step in all three approaches, given a collection $D = \{T_1[1..n_1], \ldots, T_t[1..n_t]\}$, we build one r-index for the concatenation of the documents D. Given the pattern $P[1..m]$, in order to find the frequencies of the occurrences of the pattern in each document, we first find all occurrences of the pattern P in the concatenation of all documents D using the r-index in $\mathcal{O}(m)$ time and $\mathcal{O}(r \log(n/r))$ space. All occurrences of the pattern P are identified as an interval in the suffix array of D, i.e. $SA[s_p..e_p]$.

For the second and the third approach we also build an r-index for each document T_i, for $1 \leq i \leq t$. The r-index for $T_1, \ldots, T_t$ can be built in $\mathcal{O}(\sum_{i=1}^{t} n_i) = \mathcal{O}(n)$ time and occupying $\mathcal{O}(\sum_{i=1}^{t} r_i \log(n_i/r_i)) = \mathcal{O}(Rt \log(n/r_k))$ space, where r_i is the number of runs in the BWT of T_i, $R = \sum_{i=1}^{t} r_i$, and $k = \mathrm{argmin}(r_1, \ldots, r_t)$.

4.1   Precomputed Document List with Frequencies

Following the ideas for the *document listing* problem proposed in [9], we grammar compress DA producing a binary and balanced grammar of ν non-terminals, that can be stored in $\mathcal{O}(r \log(n/r))$ space [14]. Let T be the parse tree of the document array $DA[1..n]$, given a non terminal node $nt \in T$ let $DA[s_{nt}..e_{nt}]$ be its expansion. For all non terminal nodes $nt \in T$, we precompute and store the list D_{nt} of the distinct documents in $DA[s_{nt}..e_{nt}]$ with their frequencies. The lists are stored in ascending order.

Query. Given the range $[s_p..e_p]$ of all occurrences of P, we find maximal nodes of the parse tree T that cover $DA[s_p..e_p]$. Since the grammar is binary and balanced, the number of maximal non terminal nodes covering $DA[s_p..e_p]$ is $\mathcal{O}(\log n)$. Those nodes can be found in $\mathcal{O}(\log n)$ time traversing the parse tree T from the root towards the interval $DA[s_p..e_p]$. We use an atomic heap [12] to merge the $\mathcal{O}(\log n)$ lists and compute the frequencies of the documents, by inserting the head of each list in the heap; extracting the minimum and inserting the next element from the same list. While extracting the document, we compute the frequencies for each document. The atomic heap allows to insert end extract the minimum in constant amortized time, thus the total time to compute the output is $\mathcal{O}(\text{ndoc}\log n)$ since each document can appear in each list.

Summarizing, we can answer to Problem 3 in $\mathcal{O}(m + \text{ndoc}\log n)$ time, using $\mathcal{O}(r\log(n/r) + t \times \nu)$ space.

4.2 Grammar-Compressed Document Array with Bitvectors

Let $\mathcal{T}$ be the parse tree of the document array $\mathsf{DA}[1..n]$ with ν non-terminals. For each non terminal node $nt \in \mathcal{T}$ we store if the i-th document occurs in the expansion of nt and, if so, whether the leftmost (resp. rightmost) occurrence is in the left child or in the right child of nt. Let ℓ and r be the left child and right child of nt, respectively. The above information can be stored into two bitvectors L_{nt} and R_{nt} of length t, such that for all documents $i = 1, \ldots, t$, $\mathsf{L}_{nt}[i] = 0$ if the leftmost occurrence of the i-th document is in ℓ, and 1 otherwise, and $\mathsf{R}_{nt}[i] = 1$ if the rightmost occurrence of the i-th document is in r, and 0 otherwise. Note that if $\mathsf{L}_{nt}[i] > \mathsf{R}_{nt}[i]$, then the i-th document does not occur in nt.

For the i-th document it holds that $\mathsf{L}_{nt}[i] = \mathsf{L}_{\ell}[i] \wedge \overline{\mathsf{R}_{\ell}[i]}$ and $\mathsf{R}_{nt}[i] = \overline{\mathsf{L}_r[i]} \vee \mathsf{R}_r[i]$ where $\overline{x}$ is $1 - x$. We compute L_{nt} and R_{nt} for each non terminal node in a bottom up fashion and we store them. Considering that non terminal nodes associated to the same non terminal symbol have the same subtree, we can compute the L_{nt} and R_{nt} bitvectors only once for each non terminal symbol. Thus, the whole running time of the algorithm is $\mathcal{O}((t/w) \times \nu)$ using bit parallelism on words of w bits.

Query. Let $t_1, \ldots, t_k$ be the $k = \mathcal{O}(\log n)$ maximal non terminals that cover the interval corresponding to $\mathsf{DA}[s_p..e_p]$. We build a binary tree $\mathcal{T}'$ having as leaves the nodes corresponding to $t_1, \ldots, t_k$. Each internal node stores a pair of bitvectors L and R, computed using the rules described above. The height of $\mathcal{T}'$ is $\mathcal{O}(\log\log n)$. To retrieve the leftmost and rightmost occurrences of each document, we start from the root of $\mathcal{T}'$, for each document present in the root, we descend the tree, using the information stored in the bitvectors, to find first the leftmost, and then the rightmost occurrence of the document.

We perform exactly two traversals of the tree for each document that occurs at least once in the interval, since the L and R bitvectors store the information that a document does not appear in the interval of the node. Using bit parallelism on words of size w, we can find the leftmost and rightmost occurrence of each document in $\mathcal{O}(\text{ndoc}(t/w)(\log n + \log\log n))$ time.

Once we have computed the leftmost and rightmost occurrences ℓ_i and r_i for each document i, we use random access to SA of the r-index to find their corresponding suffix values $\mathsf{SA}[\ell_i]$ and $\mathsf{SA}[r_i]$ in the concatenation of the documents. We, then, find the corresponding suffix values in the document T_i, and, using random access to ISA we find the leftmost and rightmost occurrence ℓ_i' and r_i' in the suffix array of the document T_i. The size of this interval is the number of occurrences of the pattern P in T_i, i.e. $r_i' - \ell_i' + 1$.

Keeping all together, we can answer queries to Problem 3 in $\mathcal{O}(m + ((t/w)\log n + \log(n/r))\mathbf{ndoc})$ time, using $\mathcal{O}(r\log(n/r) + Rt\log(n/r_k) + (t/w)\times\nu)$ space.

4.3   Double Run-Length Encoded ILCP

We first introduce a variation of the interleaved LCP array (ILCP) introduced in [13] called *double run-length encoded* ILCP, denoted by ILCP*. The ILCP* is composed by the array VILCP* storing the values of the runs, and the array LILCP* storing their lengths. Given the run-length encoded ILCP array for the collection $D = \{T_1, T_2, \ldots, T_t\}$ we merge together consecutive runs whose elements are from the same document, keeping the smallest value as the value of the run. Formally, let ρ be the number of runs of ILCP, let $\ell_1 = 1$ and $r_1 = \mathsf{LILCP}[1]$, and for all $i = 2, \ldots, \rho$ let $\ell_i = \sum_{j=1}^{i-1}\mathsf{LILCP}[j]$ and $r_i = \ell_i + \mathsf{LILCP}[i] - 1$. Moreover, for all $1 \le i \le j \le n$, let $|\mathsf{DA}[i..j]| = |\{\mathsf{DA}[k] \mid i \le k \le j\}|$.

Definition 1. *Let us assume that we have computed the run-length encoding up to position i of VILCP, the next run of* ILCP* *is defined as follows. Let $\ell = max\{k \mid |\mathsf{DA}[\ell_i..r_k]| = 1\}$ if $|\mathsf{DA}[\ell_i..r_i]| = 1$ and 0 otherwise. Then $\mathsf{VILCP}^\star[j] = min\{\mathsf{VILCP}[i..i+\ell]\}$, and $\mathsf{LILCP}^\star[j] = \sum_{k=i}^{i+\ell}\mathsf{LILCP}[k]$.*

The ILCP has a nice property described in [13] that we are going to recall.

Lemma 1 ([13, Lemma 1]). *Given a collection $D = \{T_1, \ldots, T_t\}$ whose concatenation is $\mathcal{D}[1..n]$, let SA be its suffix array, and let DA be its document array. Let $\mathsf{SA}[s_p..e_p]$ be the interval corresponding to the occurrences of the pattern $P[1..m]$ in $\mathcal{D}$. Then, the leftmost occurrences of the distinct document identifiers in $\mathsf{DA}[s_p..e_p]$ are in the same positions as the values strictly less than m in $\mathsf{ILCP}[s_p..e_p]$.*

Extending Lemma 1 to ILCP* we have that:

Lemma 2. *Given a collection $D = \{T_1, \ldots, T_t\}$ whose concatenation is $\mathcal{D}[1..n]$, let SA be its suffix array, and let DA be its document array. Let $\mathsf{SA}[s_p..e_p]$ be the interval corresponding to the occurrences of the pattern $P[1..m]$ in $\mathcal{D}$. Then, the leftmost occurrences of the distinct document identifiers in $\mathsf{DA}[s_p..e_p]$ are in the same positions as the values strictly less than m in $\mathsf{ILCP}^\star[s_p..e_p]$. If there are two values smaller than m for one document, we consider the leftmost one.*

Proof. For the runs of ILCP* that are also runs of ILCP, the property of Lemma 1 holds. We have to show that the same property holds also for runs of values from the same document.

Let $[s_p..e_p]$ be the interval of all occurrences of P in the text. If a *same-document* run has value greater than or equals to m, then all occurrences in the run have ILCP value larger than or equals to m, hence by Lemma 1 the property is satisfied. If the considered run has value strictly smaller than m we have to consider three cases. The first case to consider is if the run is entirely included in

ILCP$[s_p..e_p]$, than the head of the run is the value strictly less than m, otherwise the head of the run would not be in the interval ILCP$[s_p..e_p]$. The second case to consider is if the run is not entirely included in ILCP$[s_p..e_p]$, and the run is broken by the left boundary of the interval, then, the leftmost occurrence of the document is in s_p. The last case is if the run is broken by the right boundary of the interval, then, if there is another run containing a value smaller than m for document i, by Lemma 1 the leftmost occurrence is the head of the other run, otherwise the leftmost occurrence is the head of the run crossing the right boundary.

Thus, considering the last run in the interval as a special case, we can apply the same approach as in [13]. Then we consider the last run, checking if it is a *same-document* run or not, and if it is, we check if the same document has already been found by the algorithm.

We build the double run-length encoded LCP array on $\mathcal{D}$. We, then, build a *range minimum query* data structure [11] on VILCP* and a bitvector L$[1..n]$ such that LILCP$^\star[i]$ = $\mathtt{select}_1(\mathsf{L}, i)$. This allows, together with Lemma 2, to use Sadakane's approach to find distinct documents to VILCP*. This allows us to retrieve the leftmost occurrences of the distinct documents. To retrieve the rightmost occurrence, we build the ILCP array using the *right* LCP, i.e. the LCP array defined as follows. We store in each position $1 \leq i \leq n-1$ the length of the longest common prefix between the two strings $T[\mathsf{SA}[i]..n]$ and $T[\mathsf{SA}[i+1]..n]$. In this case, we have that the rightmost occurrences of the distinct documents in DA$[s_p..e_p]$ correspond to values of the ILCP strictly smaller than m. In particular, all properties that apply to the ILCP array also apply to the ILCP array defined array using the *right* LCP. We, then, also double run-length encode it.

Query. Given the interval $[s_p..e_p]$, as in [13], we apply Sadakane's technique to find distinct elements in DA, to find distinct values in both the *double run-length encoded* ILCP arrays. Provided the positions of the leftmost and rightmost occurrences of each document, we then use the r-index to find the corresponding value of the suffix array. We map those positions back in the original document, and, using random access to ISA of the document, we obtain the interval $[s_p'..e_p']$ in the suffix array of the document, whose size corresponds to the frequency of the document.

Keeping all together, we can answer queries to Problem 3 in $\mathcal{O}(m + \log(n/r)\mathtt{ndoc})$ time, using $\mathcal{O}(r\log(n/r) + Rt\log(n/r_k) + |\mathsf{ILCP}^\star s|)$ space, where $|\mathsf{ILCP}^\star s|$ is the size of both the ILCP* arrays.

5   Experimental Result

We implemented the data structures and measured their performance on real-world datasets. Experiments were performed on a server with Intel(R) Xeon(R) CPU E5-2407 processors @ 2.40 GHz and 250 GiB RAM running Debian Linux kernel 4.9.0-11-amd64. The compiler was g++ version 6.3.0 with -O3 -DNDEBUG

Table 1. Statistics for document collections (small, medium, and large variants): *Collection* name; *Size* in megabytes; *R-Index* bits per symbol (bps); *Docs*, number of documents; *Seqs*, average number of sequences (or versions) per each document; number of *Patterns*; For the synthetic collections (second group), we sum-up variants that use 10 or 100 base documents with the different mutation probabilities.

Collection	Size	R-Index	Docs	Seqs	Patterns
Species	105	11.79	3	10	7658
	631	3.15	3	60	20 536
Page	110	0.60	60	147	7658
	641	0.38	190	164	14 286
Concat	95		10	1000	7538–10 832
	95		100	100	10 614–13 165

options. Runtimes were recorded with Google Benchmark framework[2]. The source code is available online at: github.com/duscob/dret

Datasets. To evaluate our proposals, we experimented on different real and synthetic datasets. We used a variation of the dataset described by Mäklin *et al.* [25], and some of the datasets tested by Cobas and Navarro [9]. These are available at `zenodo.org` and `jltsiren.kapsi.fi/RLCSA`, respectively. Table 1 summarizes some statistics on the collections and patterns used in the queries.

Real Datasets. We used two repetitive datasets from real-life scenarios: `Species` and `Page`. `Species` collection is composed of sequences of *Enterococcus faecalis*[3], *Escherichia coli*[4] and *Staphylococcus aureus*[5] species. We created three documents, one per species, containing sequences of different strains of the corresponding species. We created two variants of `Species` dataset with 10 and 60 strains per document. `Page` is a collection composed of pages extracted from Finnish-language Wikipedia. Each document groups an article and all its previous revisions. We tested on two variants of `Page` collection of different sizes: the smaller composed of 60 pages and 8834 revisions, and the bigger with 190 pages and 31208 revision.

Synthetic Datasets. Synthetic collections allow us to explore the performance of our solutions on different repetitive scenarios. We experimented on the `Concat` datasets, very similar to `Page`. Each `Concat` collection contains $d = \{10, 100\}$ documents. Each document groups a base document and $10000/d$ versions of this. We generate the different versions of a base document with a mutation

---

[2] github.com/google/benchmark.
[3] DOI: 10.5281/zenodo.3724100.
[4] DOI: 10.5281/zenodo.3724112.
[5] DOI: 10.5281/zenodo.3724135.

probability R. Notice that we have a Concat dataset for each combination of $d = \{10, 100\}$ and $R = \{0.001, 0.003, 0.01, 0.03\}$. A mutation is a substitution by a different random symbol. The base documents sequences of 1000 symbols randomly extracted from English file of Pizza&Chili [10].

Queries. The query patterns for Species collections are substrings of lengths $m = \{8, 12, 16\}$ extracted from the dataset. In the case of Page datasets, the patterns are Finnish words of length $m \geq 5$ that appears in the collections. For Concat collections, the queries are terms selected from an MSN query log. See Gagie *et al.* [13] for more details.

Implementation Details. All our implementations use the r-index as text index. We use the implementation of [14] available at github.com/nicolaprezza/ r-index. Since the implementation does not support random access to the *suffix array* SA and to the *inverse suffix array* ISA, we used a grammar-compressed *differential suffix array* and *differential inverse suffix array*—the differential versions store the difference between two consecutive values of the array —. Mäkinen *et al.* [24] show that SA of repetitive collections contains large *self-repetitions* which are suitable to be compressed using a grammar compressor like balanced Re-Pair.

Since we use the random access to SA and ISA to retrieve the frequencies of the distinct documents, we implemented also a variant using a *wavelet tree* on the document array, as in [35], to support the rank functionalities over the *document array* DA. For our experiments, we use the sdsl-lite [17] implementation of the wavelet tree.

Algorithms. We plugged-in our proposal with two different approaches to calculate the frequencies from the occurrences. All implementations marked with -ISA uses the random access to SA and ISA to retrieve the frequencies, while the one marked with -WT uses the *wavelet tree*.

- GCDA-PDL: *Grammar-Compressed Document Array with Precomputed Document Lists.* Solution described in Sect. 4.1, using balanced Re-Pair[6] for DA and sampling the sparse tree as in [9].
- GCDA: *Grammar-Compressed Document Array.* Solution described in Sect. 4.2, using balanced Re-Pair for DA and bit-vectors stored in the non-terminals. We implemented the variants: GCDA-ISAs and GCDA-WT.
- ILCP: *Interleaved Longest Common Prefix.* Solution described in Sect. 4.3, using ILCP array (not double run-length encoded). We implemented the variants: ILCP-ISAs and ILCP-WT.
- ILCP⋆: *double run-length encoded Interleaved Longest Common Prefix.* Solution described in Sect. 4.3, using ILCP⋆ array. We implemented the variants: ILCP⋆-ISAs and ILCP⋆-WT.

---

[6] www.dcc.uchile.cl/gnavarro/software/repair.tgz.

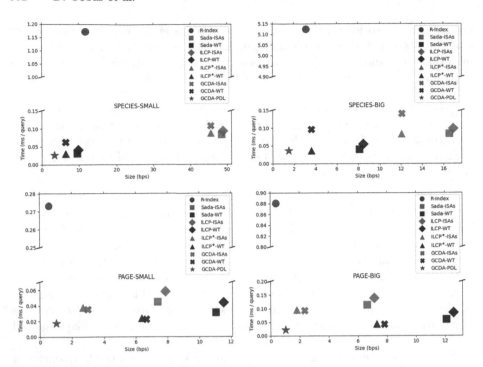

Fig. 1. Document listing with frequencies on Species and Page datasets. The x axis shows the total size of the index in bits per symbol (bps). The broken y axis shows the average time per query.

- Sada: *Sadakane*. The algorithm proposed in [33]. We provide the variants: Sada-ISAs and Sada-WT.
- R-Index: r-*index*. Bruteforce algorithm that scans all occurrences of the pattern, counting the frequencies.

Note that in all our algorithms we do not use the random access to SA and ISA of the r-index, thus we do not need to store the samples. The only exception is R-Index which needs the samples to compute the frequencies.

Results. Figure 1 contains our experimental results for document listing with frequencies on real datasets. We show the trade-off between time and space for all tested indexes on different variants of the collections Species and Page.

The two variants of Species collections are composed of few large documents (only three, one per species). In this scenario, GCDA-PDL proves to be the best solution, finding the document frequencies in 27–36 μs (μsec) per each pattern in average, and requiring only 1.5–3.5 bits per symbol (bps). GCDA-PDL is the fastest and smallest index, requiring even less space than R-Index, since GCDA-PDL does not store the samples. The large size of the sampling scheme for collections with low repetitiveness has also been observed in [14]. The best

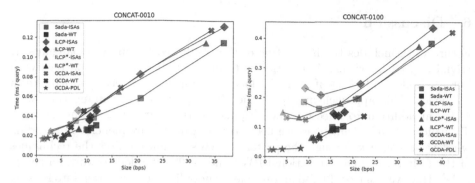

Fig. 2. Document listing with frequencies on synthetic collection `Concat`. The x axis shows the total size of the index in bits per symbols (bps). The y axis shows the average time per query. `R-index` is omitted from the plots due to its excessively high time.

competitor is ILCP*-WT, being almost as fast (30–36 μs per query) as GCDA-PDL, but requiring 1.85–2.4 times more space. In these collections, -WT indexes perform better than -ISAs solutions. They can answer the queries at least 1.45 times faster, while they are 2–7 times smaller. In terms of space, GCDA-WT represents a good option, improving even the space required by R-Index in some cases, but much slower than GCDA-PDL and ILCP*-WT.

`Page` collections that contain more documents than `Species` collections: 60 documents in its small version and 190 in the bigger one. Again GCDA-PDL turns up as the best index. It uses less than 1.05 bps and answers the queries in 17–22 μs. R-Index requires the least space among the solutions, 0.38–0.60 bps, but is 15.86–40.35 times slower. The second overall-best index is ILCP*-ISAs, with 1.80–2.69 bps and query times of 37–95 μs, closely followed by GCDA-ISAs. On the `Page` variants, -WT indexes are faster than its counterparts -ISAs, but 1.47–4.05 times bigger.

On real datasets GCDA-PDL outperforms the rest of the competitors, but the ILCP*-variants are also relevant solutions obtaining a good space/time tradeoff.

The comparison of the indexes on synthetic collections `Concat` are shown in Fig. 2. These kinds of collections allow us to observe the indexes' behavior as the repetitiveness varies. Each plot combines the results for the different mutation probabilities of a given collection and number of base documents. The plots show the increasing mutation rates using variations of the same color, from lighter to darker.

GCDA-PDL outperforms all the other indexes. For the collections composed of 10 base documents, our index obtains the best space/time tradeoff, requiring 1.22–3.84 bps with a query time of 16–19 μs. Only GCDA-WT and ILCP*-WT obtain competitive query times, but they are 2.20–4.20 times bigger. R-Index requires the least space for lower mutation rates, but it is 79–83 times slower than GCDA-PDL (note that the R-Index data for this collection is not shown in Fig. 2 due to its high query times). In the case of the collections composed of 100 base documents, GCDA-PDL dominates the space/time map.

6 Discussion

Future work includes the integration of the results with real pseudoaligners. A trivial approach for such integration is to query each k-mer of a pattern with our methods, and check if a single document (species) receives positive term frequency. This approach multiplies the $O(m)$ part of the running time with $O(k)$, in addition to affecting the output-sensitive part of the running time. To avoid the $O(k)$ multiplier, we need to maintain the frequencies in a sliding window of length k through the pattern. Such solution requires the techniques of the fully-functional bidirectional BWT index [2] extended to work on the r-index. However, one could also modify the pseudoalignment criterion into looking at maximal runs of k-mer hits, in the order of the (reverse) pattern. For this, our methods are readily applicable: just do backward search with the pattern P until obtaining an empty interval with suffix $P[i..m]$. Report term frequency of $P[i+1..m]$ if $m - i \geq k$. Continue analogous process backward searching $P[1..i]$. If all the maximal runs of k-mer hits report a single document (species) T_i, assign P to T_i. The $O(m)$ part of the running time remains unaffected, and the output-sensitive part remains smaller than with the sliding window approach.

Acknowledgments. We wish to thank Antti Honkela and Tommi Mäklin for introducing us the need for better solutions to the pseudo-alignment problem. Some initial solutions were discussed during summer 2019 with Jarno Alanko, Travis Gagie, and Gonzalo Navarro at the Dagstuhl seminar: 25 Years of the Burrows-Wheeler Transform. This led to the plan of tackling this problem during the visits (supported by the EU's Horizon 2020 research and innovation programme under Marie Skłodowska-Curie grant agreement No 690941 (BIRDS) and the Academy of Finland (grant 309048)) of DC and MR to Helsinki. MR is supported by the National Science Foundation (NSF) IIS (Grant No. 1618814). DC is supported by the National Agency for Research and Development (ANID)/Scholarship Program/DOCTORADO BECAS CHILE/2020-21200906 and by Google's Latin America Research Awards 2019. We also wish to thank the anonymous reviewers for their insightful comments and suggestions.

References

1. Belazzougui, D., Boldi, P., Pagh, R., Vigna, S.: Monotone minimal perfect hashing: searching a sorted table with O(1) accesses. In: Proceedings of the Twentieth Annual ACM-SIAM Symposium on Discrete Algorithms, SODA 2009, pp. 785–794. SIAM (2009)
2. Belazzougui, D., Cunial, F.: Fully-functional bidirectional Burrows-Wheeler indexes and infinite-order de Bruijn graphs. In: 30th Annual Symposium on Combinatorial Pattern Matching, CPM 2019. LIPIcs, vol. 128, pp. 10:1–10:15. Schloss Dagstuhl - Leibniz-Zentrum für Informatik (2019)
3. Belazzougui, D., Navarro, G., Valenzuela, D.: Improved compressed indexes for full-text document retrieval. J. Discrete Algorithms **18**, 3–13 (2013)
4. Bray, N.L., Pimentel, H., Melsted, P., Pachter, L.: Near-optimal probabilistic RNA-Seq quantification. Nat. Biotechnol. **34**(5), 525–527 (2016)
5. Burrows, M., Wheeler, D.: A block sorting lossless data compression algorithm. Technical report 124, Digital Equipment Corporation (1994)

6. Carroll, D., et al.: The global virome project. Science **359**(6378), 872–874 (2018)
7. Charikar, M., et al.: The smallest grammar problem. IEEE Trans. Inf. Theory **51**(7), 2554–2576 (2005)
8. Claude, F., Munro, J.I.: Document listing on versioned documents. In: Kurland, O., Lewenstein, M., Porat, E. (eds.) SPIRE 2013. LNCS, vol. 8214, pp. 72–83. Springer, Cham (2013). https://doi.org/10.1007/978-3-319-02432-5_12
9. Cobas, D., Navarro, G.: Fast, small, and simple document listing on repetitive text collections. In: Brisaboa, N.R., Puglisi, S.J. (eds.) SPIRE 2019. LNCS, vol. 11811, pp. 482–498. Springer, Cham (2019). https://doi.org/10.1007/978-3-030-32686-9_34
10. Pizza & Chili repetitive corpus: http://pizzachili.dcc.uchile.cl/repcorpus.html. Accessed 16 April 2020
11. Fischer, J., Heun, V.: Space-efficient preprocessing schemes for range minimum queries on static arrays. SIAM J. Comput. **40**(2), 465–492 (2011)
12. Fredman, M.L., Willard, D.E.: Trans-dichotomous algorithms for minimum spanning trees and shortest paths. J. Comput. Syst. Sci. **48**(3), 533–551 (1994)
13. Gagie, T., et al.: Document retrieval on repetitive string collections. Inform. Retrieval J. **20**(3), 253–291 (2017)
14. Gagie, T., Navarro, G., Prezza, N.: Fully functional suffix trees and optimal text searching in BWT-runs bounded space. J. ACM **67**(1), 2:1–2:54 (2020)
15. Gagie, T., Navarro, G., Puglisi, S.J.: New algorithms on wavelet trees and applications to information retrieval. Theor. Comput. Sci. **426**, 25–41 (2012)
16. Gagie, T., Puglisi, S.J., Turpin, A.: Range quantile queries: another virtue of wavelet trees. In: Karlgren, J., Tarhio, J., Hyyrö, H. (eds.) SPIRE 2009. LNCS, vol. 5721, pp. 1–6. Springer, Heidelberg (2009). https://doi.org/10.1007/978-3-642-03784-9_1
17. Gog, S., Beller, T., Moffat, A., Petri, M.: From theory to practice: plug and play with succinct data structures. In: Gudmundsson, J., Katajainen, J. (eds.) SEA 2014. LNCS, vol. 8504, pp. 326–337. Springer, Cham (2014). https://doi.org/10.1007/978-3-319-07959-2_28
18. Grossi, R., Gupta, A., Vitter, J.S.: High-order entropy-compressed text indexes. In: Proceedings of the Fourteenth Annual ACM-SIAM Symposium on Discrete Algorithms, 12–14 January 2003, Baltimore, Maryland, USA, pp. 841–850. ACM/SIAM (2003)
19. Huson, D.H., Auch, A.F., Qi, J., Schuster, S.C.: Megan analysis of metagenomic data. Genome Res. **17**(3), 377–386 (2007)
20. Iqbal, Z., Caccamo, M., Turner, I., Flicek, P., McVean, G.: De novo assembly and genotyping of variants using colored de Bruijn graphs. Nat. Genet. **44**(2), 226–232 (2012)
21. Jez, A.: A really simple approximation of smallest grammar. Theor. Comput. Sci. **616**, 141–150 (2016)
22. Lehman, E., Shelat, A.: Approximation algorithms for grammar-based compression. In: Proceedings of the Thirteenth Annual ACM-SIAM Symposium on Discrete Algorithms, pp. 205–212. Society for Industrial and Applied Mathematics (2002)
23. Lindner, M.S., Renard, B.Y.: Metagenomic abundance estimation and diagnostic testing on species level. Nucleic Acids Res. **41**(1), e10–e10 (2013)
24. Mäkinen, V., Navarro, G., Sirén, J., Välimäki, N.: Storage and retrieval of highly repetitive sequence collections. J. Comput. Biol. **17**(3), 281–308 (2010)

25. Mäklin, T., Kallonen, T., Alanko, J., Mäkinen, V., Corander, J., Honkela, A.: Genomic epidemiology with mixed samples. BioRxiv (2020). Supplement: Pseudoalignment in the mGEMS pipeline
26. Manber, U., Myers, E.W.: Suffix arrays: a new method for on-line string searches. SIAM J. Comput. **22**(5), 935–948 (1993)
27. Marchet, C., Boucher, C., Puglisi, S.J., Medvedev, P., Salson, M., Chikhi, R.: Data structures based on k-mers for querying large collections of sequencing datasets. bioRxiv p. 866756 (2019)
28. Muthukrishnan, S.: Efficient algorithms for document retrieval problems. In: Proceedings of the thirteenth annual ACM-SIAM symposium on Discrete algorithms (SODA), pp. 657–666. Society for Industrial and Applied Mathematics (2002)
29. Navarro, G.: Spaces, trees, and colors: the algorithmic landscape of document retrieval on sequences. ACM Comput. Surv. (CSUR) **46**(4), 52 (2014)
30. Navarro, G.: Document listing on repetitive collections with guaranteed performance. Theoret. Comput. Sci. **772**, 58–72 (2019)
31. Navarro, G., Mäkinen, V.: Compressed full-text indexes. ACM Comput. Surv. **39**(1), 2 (2007)
32. Rytter, W.: Application of Lempel-Ziv factorization to the approximation of grammar-based compression. Theor. Comput. Sci. **302**(1–3), 211–222 (2003)
33. Sadakane, K.: Succinct data structures for flexible text retrieval systems. J. Discrete Algorithms **5**(1), 12–22 (2007)
34. Schaeffer, L., Pimentel, H., Bray, N., Melsted, P., Pachter, L.: Pseudoalignment for metagenomic read assignment. Bioinform. **33**(14), 2082–2088 (2017)
35. Välimäki, N., Mäkinen, V.: Space-efficient algorithms for document retrieval. In: Ma, B., Zhang, K. (eds.) CPM 2007. LNCS, vol. 4580, pp. 205–215. Springer, Heidelberg (2007). https://doi.org/10.1007/978-3-540-73437-6_22
36. Weiner, P.: Linear pattern matching algorithms. In: 14th Annual Symposium on Switching and Automata Theory, Iowa City, Iowa, USA, 15–17 October 1973, pp. 1–11. IEEE Computer Society (1973)
37. Wood, D.E., Salzberg, S.L.: Kraken: ultrafast metagenomic sequence classification using exact alignments. Genome Biol. **15**(3), R46 (2014)
38. Xia, L.C., Cram, J.A., Chen, T., Fuhrman, J.A., Sun, F.: Accurate genome relative abundance estimation based on shotgun metagenomic reads. PloS one **6**(12), e27992 (2011)

Author Index

Printed in the United States
By Bookmasters